Human Factors in Safety-critical Systems

Human Factors in Safety-critical Systems

Edited by
Felix Redmill and Jane Rajan

Butterworth-Heinemann
Linacre House, Jordan Hill, Oxford OX2 8DP
A division of Reed Educational and Professional Publishing Ltd

ℛ A member of the Reed Elsevier plc group

OXFORD BOSTON JOHANNESBURG
MELBOURNE NEW DELHI SINGAPORE

First published 1997

© Reed Educational and Professional Publishing Ltd 1997

British Library Cataloguing in Publication Data
Human Factors in safety-critical systems
  1. Reliability (Engineering) 2. Human engineering
  I. Redmill, Felix, 1944– II. Rajan, Jane
  620.8'0285

ISBN 0 7506 2715 8

Library of Congress Cataloguing in Publication Data
Human factors in safety-critical systems/edited by Felix Redmill and
    Jane Rajan.
    p. cm
    Includes index.
    ISBN 0 7506 2715 8
    1. Human computer interaction. 2. Industrial safety.
    I. Redmill, Felix. II. Rajan, Jane.
    Qa76.9 H85H8675
    005.1–dc21

Printed and bound in Great Britain by Biddles, Guildford and Kings Lynn

Contents

Preface

Rationale

In all accidents, there is a human cause. Until recently, this was considered to extend only as far as the humans in direct contact with the equipment, but the inquiries into a number of recent disasters have emphasised the fact that it extends back along a causative chain in which even the highest management and policy makers are links. And a weak link, no matter how far back, can introduce a latent fault into a system which, much later and given a particular set of circumstances, can have a disastrous effect. The subject of human factors is therefore achieving increasing recognition as a primary component in the field of safety.

The 'safety-critical systems' referred to in the title of this book are computer systems in applications where failure (for whatever reason) would endanger human life, property, or the environment. Such systems already extend into almost every industry sector and are rapidly increasing in scope and number. Software-based systems are now almost certainly the first and most economical choice for providing control, wherever it is required. They are included in the control of railway signalling, civil aviation, road transport, chemical plant, off-shore oil exploration, and medical equipment, to name only some safety-critical applications.

Traditionally, computer systems, for whatever purpose, have been designed and built only by 'software engineers' and systems analysts. Now it is recognised that 'safety-critical systems' is in fact a new field which comprises a number of disciplines, the most important being software engineering, safety engineering, human factors, and management. This book brings these together. Flowing around these disciplines and binding them together into the composite 'safety-critical systems' are the less tangible and more fluid quality and safety cultures. They too are addressed.

Readership

That such a book as this is needed can hardly be in question. It is required by both managers and practitioners in all the disciplines mentioned above, as well as by researchers. The fact that the field of safety-critical systems comprises a number of diciplines necessarily means that practitioners in any one of them are not *au fait* with the principles and state of the art of the others. This book lays down the principles and state of the art of human factors, in the context of safety, not only for human factors students and practitioners, but also for managers and engineers in the software and safety disciplines. Each chapter is written by an acknowledged expert in the field and clearly sets out the principles of the subject in hand, in a structured manner and without using jargon, cliché or esoteric language. The content of the book is therefore accessible to all.

Content

The book is not merely a collection of chapters by experts. It was planned by the editors, and appropriate recognised experts were invited to prepare the defined chapters. There were cycles of editing and rewriting until the desired result was achieved.

The book is divided into three Parts. The first sets the scene, with Chapter 1 offering an extensive introduction to safety-critical systems, and Chapters 2 and 3 introducing the principles of human error and human reliability, and defining the state of the art of human reliability assessment. This part of the book provides the newcomer to both the safety and the human-factors disciplines with the wherewithall to understand the subsequent chapters.

Part 2 addresses human-computer interaction and the interface across which this occurs. Too often interfaces are specified and designed only by analysts and software engineers who have no training in human factors, with the frequent results that, in the first place tasks are ineffectively allocated between computers and humans and, in the second, human-computer interaction is inefficient, incorrect, boring, and, in safety-critical situations, sometimes hazardous. The chapters in this part are written to raise awareness generally, offer sound practical advice to practitioners, and indicate the state of the art to researchers. Not only are technical aspects covered, but so also is the important and often neglected subject of training.

Part 3 presents the managerial and socio-technical aspects of safety-critical systems, and explains their impact on safety. The issues described are frequently neglected, have in the past been played down by engineers, but are now recognised as providing a crucial perspective in the quest for safety. The penultimate chapter covers the interesting subject of 'violations', which are intentional breaches of rules, often for apparently good reasons, but which can introduce hazards. The final chapter deals with the human influence on the safety case.

The coverage is broad, and it is hoped that the book will be of value to practitioners, researchers and students in the fields of safety engineering, software engineering, and management, as well as human factors.

In the book, the terms 'man' and 'he' are used to denote male and female persons. No discrimination or offence is intended by this. It should also be noted that the views expressed are those of the authors of the chapters and not necessarily those of their employers or past employers.

Felix Redmill and Jane Rajan
October 1996.

1

Safety-critical systems and human reliability

1

Introducing safety-critical systems

1.1 INTRODUCTION

The first electronic computer was built during the second world war, a mere fifty years ago. For many years after that, the computer was remote from the public, for its application required considerable skill in both programming and operation. Now, with the advent of computer games, spreadsheets, and other software packages, and of user-friendly interfaces with its users, the computer is accessible to the average member of the public. It is at last seen for what it always has been — a tool. But it is a tool of remarkable power and versatility, being serviceable and economic from the smallest to the largest applications and across the widest range of tasks. It has not merely been useful; it has influenced, and indeed altered, the nature of tasks and the structure of the working environment.

By the 1960s, computers were being used for process control. Not only were they being applied to control functions traditionally provided both electromechanically and manually, but also to new control functions. Since then, they have increasingly been used in safety-critical applications i.e., those where failure could result in death or injury to humans or damage to the environment. The major, well-publicised examples of these include chemical processes, railway signalling, nuclear power station protection systems, and fly-by-wire aircraft. However, safety-critical applications extend to almost every sector of industry. In medicine they include patient monitoring and intensive care units; in road transport they include the control of traffic lights and of many aspects of both commercial and domestic vehicles — for example, anti-lock braking and anti-jackknife systems.

In almost all cases, the application of computers improves reliability and safety. However, it is not easy (or, in many cases, even possible) to prove this other than in retrospect, and this takes too long and may be costly. The old electromechanical systems were easy to understand, but the ease of change of software leads to the introduction of new functions and, therefore, to increased complexity, and this necessarily mitigates against safety and encourages the feeling that the accident could have been avoided by not using software.

So why are computers used? First, as observed above, because they have the potential to increase safety, and, second, because both suppliers and users want the power and flexibility which they offer. For a supplier, the computer is the cheapest and most obvious basis of a control system — when a competitor offers a new facility, the time during which he has an advantage is reduced if it can be introduced in software. Users too want the convenience of computers. They want the rapid addition of new facilities to gain advantage over, or to reduce the advantage of, their competitors. The advance of technology will not be halted, and the first choice of the designer is the computer.

The advantages of software are not necessarily compatible with safety. The onus is therefore on the suppliers of safety-critical systems to do two things. The first is to employ the best safety and software technologies and standards in the development of their systems; the second is to convince others — principally the licensing authorities, the public and the press — that it is right to use software. As in these days the use of software is almost taken for granted, it goes unquestioned in most cases; but when a major accident occurs, such as a plane or rail crash or an explosion in a chemical plant, questions are raised. Frequently the blame is placed on those in charge at the time of the accident — the operators. However, the software, like the hardware, is created by humans whose characteristics and fallibilities are similar to those of the system operators. Formal proof of safety is impossible, and, in any case, absolute safety is impossible to achieve. Even proof of correctness does not guarantee safety, for correctness implies conformance with the specification, and the specification itself may be deficient. Thus, convincing others that it is right to use software in safety-critical systems cannot be achieved by proof of safety. Suppliers are most likely to achieve credibility for their systems by using the best technologies and standards and by presenting well-argued safety cases (see Chapter 12 for a discussion of safety cases).

Designing and building a safe system is one thing; another is using it safely. In both of these activities, which may be said to be the two sides of the system-safety coin, the human being is implicated, and human beings are notoriously unreliable. Computers are vastly superior in tasks which require speed, repetition, or the adherence to well-defined rules. As most control functions involve these categories of tasks, computers are likely to improve both reliability and safety. This is borne out by the fact that almost all accidents are wholly or partially caused by human beings.

Computers, however, are not universally superior to human beings. Humans are adaptable, and they are superior in flexibility and coping with unforeseen situations. In emergencies, these attributes are often crucial, not only to avoiding a catastrophe, but also in recovering the situation and minimising the consequences after an accident has occurred. Conversely, human actions can worsen an emergency situation.

So, in general, safety will not be achieved by omitting either computers or humans from the system. Computers offer safe control; humans are unpredictable and can introduce errors and breaches in safety; yet, humans may offer the only hope of retrieving the situation in emergencies. In many situations, the safest option is likely to be a combination of human and computer control. It is therefore the business of designers not only to plan safety-critical systems so that tasks are allocated to humans and computers according to their relative competence in dealing with them, but also to design the human-computer interface optimally.

The allocation of functions between humans and machines therefore forms a critical part of the design process. Nevertheless, human factors specialists are far too infrequently consulted during task design, with the consequence that human and machine responsibilities have already been allocated by the time human factors issues are formally considered. A result of this is that one of the operator's greatest assets — that of adaptability — is frequently used to facilitate the day-to-day system operation rather than as 'spare capacity' to enable safe and efficient operation in the case of unfamiliar or emergency situations.

Yet, the relationship between person and machine cannot be reduced to the mere allocation of tasks. Even when the human operator has been freed to concentrate on the more demanding and complex cognitive tasks that the operation of sophisticated systems requires, there is still the need for human monitoring, supervision and maintenance of the system. Lisanne Bainbridge [Bainbridge 87] has suggested that automation may in fact expand rather than eliminate the problems encountered by the human operator and that, the more advanced the control system, the greater the importance of the operator's contribution to its correct functioning. Thus, it is important to adopt a holistic approach to system design, aiming at a symbiotic relationship between the people in the system and the tools which they use. This involves consideration not only of the tasks that operators will perform and the physical environment in which they will perform them, but also of the social, organisational and cultural context which should be seen not only as the environment within which the system functions but also as an integral part of the functioning of the system.

The chapters of Part 2 of this book consider human factors in the design of safety-critical systems and those in Part 3 address their social, organisational and cultural context. This chapter sets the scene by introducing safety-critical systems.

1.2   SAFETY AND SAFETY-CRITICAL SYSTEMS

1.2.1   Some Fundamentals

Safety is not a physical entity, but a state. It is a state in which human life and well-being and the environment are not endangered. Perfect safety is impossible to achieve and, even if it were achieved, it would be impossible to prove that it had been achieved. Thus, when we speak of a safe state, we refer to a situation in which the risk is perceived to be acceptably low — though it may also include situations in which an actual risk is unperceived. What is acceptable depends on a number of factors, including public attitude (the subject of 'tolerable risk' is addressed in Section 1.4 below). Further, our attitudes are informed by familiarity; although we are surrounded by risk throughout our lives, in most circumstances we are unaware of it.

A safety-critical system is one which may affect safety — that is, it may present a hazard to human life or well-being, or the environment. It is often said that a safety-critical system is one whose operation may affect safety, but it is important to note that a system may affect safety during non-operation (e.g., an aircraft), maintenance (e.g., an alarm system), commissioning (e.g., a chemical plant), decommissioning (e.g., a nuclear power station), or indeed any part of its active life cycle. It is therefore necessary to determine while the system is being planned how it could affect safety — and this may be achieved by carrying out risk and safety analyses (see Section 1.5 below).

Typically, a safety-critical system has a number of components, one or more of which may be a computer. In the present context, it is assumed that a computer is involved. In general, it may be expected that the total system consists of three components: the plant, or equipment under control (EUC), one or more computers, or programmable electronic systems (PES), and a human component (see Figure 1.1). The second component, the PES, may be for controlling the EUC or for monitoring its state and providing protection if a dangerous state is entered. In either case, the fabric of the computer or computers involved comprises the hardware and the functionality is effected by software.

Frequently, only the first two mentioned components are perceived as comprising the system, with the human component being seen as part of the environment. Even when the human component is considered in the system's description, its scope is often limited to the operator. Yet, as well as normal operation, human participation in the system's operation includes policy making, management, operational scheduling, maintenance and, crucially, abnormal operation. Further, as accidents (catastrophic or not) can almost always be traced to a human origin (in operation, design, or management decision), it seems important for this to be considered seriously, not only in descriptions of the system, but also in the planning of its management, operation, adjustment and maintenance; indeed, in advance of that, in the strategic plans for its use and in its development and installation. Moreover, human intervention in the case of the failure of other components of the

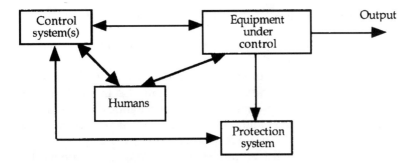

Figure 1.1 *The total system*

system (hardware or software) can be critical to the aversion of a catastrophe.

A hazard introduces risk. It imposes a threat, but a threat which may not be triggered or which may be consciously avoided. On a golf course, a bunker is a hazard. In the current context, however, the word 'hazard' is used to imply not a sporting difficulty or benign impediment but a threat to safety. Normally, a hazard is directly posed by the equipment under control, and not by the computer *per se*. A software-based control system, or function within a system, may cause the equipment under control to switch to an unsafe mode, and a protection system, or function, may fail to avert a hazard, but safety is never wholly a software issue. Yet, from the point of view of the world outside the system, what matters is whether there is or is not a breach of safety, not which part of the system causes the damage. Safety is an issue which concerns the total system and not merely one part of it. Thus, each part of the system — the hardware, software and human components — may be referred to as being 'safety-critical'. Leveson, in an instructive paper [Leveson 86], defines safety-critical software functions as 'those that can directly or indirectly cause or allow a hazardous system state to exist'. Here we see a recognition that it is the system which poses the hazard, together with an acknowledgement that the software influences the overall system. Leveson goes on to define safety-critical software as 'software that contains safety-critical functions'.

As yet, the human component of safety-critical systems seems not to be widely referred to as safety-critical, but it is important to recognise that it often is. The human component should be the subject of hazard analysis and careful, verified design to the same degree as the other components of the system. That it often is not is the cause of many otherwise avoidable accidents. It should be added, however, that the hazard analysis of humans, in a given context, is a great deal more difficult than that of machines. The options open to machines are limited to those embodied in their design — intentionally or unintentionally. On the other hand, people have freedom of choice and, while this is a strength in that it provides them with their power of judgement, it also renders their behaviour unpredictable and its analysis more complex and difficult.

1.2.2 Modes of Operation

A system may be in one of a number of modes, ranging from normal operation to shut-down. The modes available are dependent on the given system, because any number of modes of degraded operation may be defined. However, the general case is shown in Figure 1.2. In this figure, the modes are represented by the boxes, and the possibilities for transitions between the modes by the lines and arrows. It is seen that a direct transition does not exist from every mode to every other.

For a safety-critical system, all modes are not necessarily safe modes. Perhaps it should reasonably be assumed that 'normal operation' is safe. But what in any given case is 'normal'? Normal operation gives the impression of the system operating as it was specified to do. But although the specification of an aircraft must include unusual operations, is a passenger aircraft which is being put into a dive operating normally? Or what about a nuclear power station which is operating normally but whose alarm systems have been disconnected? We may deduce that operation which is normal to the system may in some circumstances be inherently unsafe, due, for example, to bad management or to the possibility of rash or careless human intervention. 'Normal operation' should, therefore, more correctly be stated as 'normal operation under normal conditions', with the characteristics and limits of both 'normal operation' and 'normal conditions' being carefully defined. In other words, safety is dependent on conditions and context. The harm that may ensue from an incident is a function of the environment in which the safety-critical system is placed and the conditions in which it operates.

Figure 1.3 (from Chapter 1 of [Redmill 89]) shows the shift of a system from a safe to an unsafe state. There is a point at which circumstances should suggest that there is an increased danger of a disaster occurring. These circumstances may take the form of a failure in the system, as in the figure, or some other form, such as one or more defined variables (for example, temperature and pressure) exceeding their accepted values. At this point, it is expected that some part of the system, designed or trained for the purpose, will detect the danger and switch the equipment under control into a safe

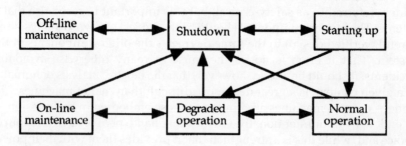

Figure 1.2 *Modes of system operation (generalised)*

mode. A subsystem which performs these monitoring and switchover functions automatically is often referred to as a 'protection system' and, in modern systems, it may itself be a computer (see Section 1.6.3). On the other hand, people are often used to monitor systems and ensure that they remain within safe operating tolerances — and it should be observed that, because it can be a boring task, it is something which they are 'bad' at.

There should always be a defined safe mode to which a system may be switched in the event that normal safe operation cannot be maintained, or in the event of failure. For some systems, the 'fail-safe' mode is 'off'; in others, it is a mode of degraded operation — in which case, returning the system to a safe mode may involve compromising operational availability or reliability. Most systems are never so safe as when they are off, and this may lead to a trade-off between availability and safety. There are other systems, such as aircraft and missiles in flight, in which safety is not achieved by shut-down — indeed, shut-down would bring such systems into a highly unsafe state — so one of their primary requirements is to continue to be safe even in the presence of failures, i.e., they must 'fail-operational'. To achieve the levels of safety necessary for public confidence, they must exhibit fault avoidance in their development, fault tolerance in their construction (see Section 1.6.2 below), proven safety procedures in their operation, and sound management throughout their life cycle.

But what about the safe modes of people? While safe modes can be defined, people, with their freedom to choose their behaviour, may not select them — for a number of reasons, such as poor interface design (perhaps leading to the misinterpretation of data — see Chapters 4 and 6), a lack of training (perhaps resulting in erroneous judgement — see Chapter 7), operator failure (perhaps because of a lack of concentration — see Chapter 2), a belief that a different course of action is more appropriate (see Chapter 11 for a discussion of 'violations'), or malicious intent. One solution which designers may attempt to achieve is to design the human out of the system, employing only machines which are both more predictable and more reliable. But, while this may be technically possible for system operation, it is almost certainly neither cost effective nor desirable in most cases [Bainbridge 87]. In any case, the human cannot be avoided in policy making and management. Further, a machine is only as predictable and reliable as its design and construction permit, so this solution demands a great deal of confidence in the designer, who is subject to error — and whose human involvement has not yet been designed out of the system!

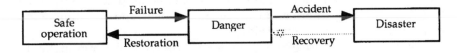

Figure 1.3 *The effect of the system on safety*

1.2.3   Computer Involvement

While the reference 'systems involving computers' is simple to make, it does not represent either the role that the computer is expected to play or the manner in which it interacts with the other components of the system, including the human component. Yet, the role of the computer in achieving or maintaining safety may be active or passive, causative or curative, interventionist or merely supportive. The way in which the computer is put to use determines the nature of the interface with users, and, frequently, even the roles and types of the users themselves (see Chapter 6 for a treatment of the factors to be considered in interface design). Further, the manner of the computer's intended application is likely to determine what design problems need to be solved, and it may even affect the design process itself. To make the point, five forms of computer involvement are briefly introduced. The extent to which the computer affects safety, the manner in which it does so, and the nature of its interaction with human beings are different in each case.

First, the computer may be used as a continuous controller, as in manufacturing process control. Here, the equipment under control cannot carry out its normal function without the involvement of the computer, and may malfunction as a result of errors in the software or hardware.

Second, there is the computer as mediator, as in a monitoring and protection system. Here, the computer is not necessary to the normal functioning of the plant. It monitors defined parameters related to the plant or its output (for example, temperature and pressure) and takes action if any of these parameters strays beyond defined limits. The limits may be predefined levels or they may be evaluated at the time in accordance with complex criteria; the actions taken may follow simple preprogrammed instructions or they may be determined in real time, having consideration to the status of one or more variables.

Third, the computer may be a repository and analyser of data and a source of analysed information. Examples of this are the storage and analysis of data from the equipment under control, the presentation of analysed data to operators (such as aircraft pilots), and the storage of information such as maintenance and fault histories, or operational or emergency data. In these cases, the computer may not be essential to normal operation, but it has the potential to compromise safety and, if poorly designed, will do so.

Fourth, the computer may be used as a 'third party intermediary', such as a telephone exchange control system. In such a case, it is not directly involved in either the operation or the safety of the plant, but it has the potential to make an existing unsafe situation worse as it may be essential to the recovery of the situation (for example, the telephone exchange is essential to communicating information about an incident and to summoning maintenance, medical, or other aid).

Fifth, there is the use of the computer as a design tool. For example, an architect's software package (apparently simple and sold over the counter) which contains a bug has the potential to compromise the design and thus the safety of a bridge or a building. Even the common spreadsheet may be used in safety-critical applications, for example, when it is the basis of modelling (for analysis or comparison) various hazardous situations.

1.3   THE HUMAN ELEMENT IN FAILURE

On 6 March 1987, the roll-on-roll-of ferry, *Herald of Free Enterprise*, sank off Zeebrugge with the loss of 188 lives [Steel 87]. It was the responsibility of a member of the crew, the assistant bosun, to close the bow doors prior to the ship leaving port and he did not discharge that duty.

On 12 December 1988, a train crash at Clapham Junction, in which 35 people died, was attributed to faulty refurbishment of the track [Hidden 89].

The immediate cause of each of these accidents was judged to be human error, through action or inaction. However, behind each there are further questions which suggest that the final single human action was not isolated but the end of a chain of human factors. In such a chain, the errors which form the links may occur in quick succession or with a delay between them. In the latter case, the first error creates a dangerous situation, latent until the occurrence of the next error — the situation often referred to as 'an accident waiting to happen'.

In the Zeebrugge instance, why was this safety-critical task of closing the doors not checked by an officer? Why had the designers of the ship not installed a monitor on the doors and an indicator on the bridge so that the captain could ensure that the doors were closed before putting to sea? Indeed, representation had been made by the captains of such ships to the directors of the company involved to install indicators, but they had been ignored.

In the case of the Clapham Junction train crash, it seems that the maintenance man in question had been allowed to do so much overtime work, and to work for such long continuous periods, that tiredness played a part in his error. Why did the management procedures not preclude this? Or, if they did, why were they not followed? The Hidden inquiry [Hidden 89] pointed to 'poor management' and said that 'the appearance of a proper regard for safety was not the reality'.

These are two cases that demonstrate the human involvement in accidents, but an investigation of almost any other accident would make the same point — the fire at King's Cross Underground Station, the explosion at the nuclear power station at Chernobyl in the USSR, almost any plane crash. If there is not an obvious immediate human causative action, there is usually a traceable human contributory factor such as a management decision, inadequate maintenance, a failure to replace ageing equipment, or a lack of safety procedures. It is said that 80% of all accidents have a human cause. Perhaps all but 'natural disasters' do, the deciding factor being only the number of steps which we retrace along the causative chain.

This is due both to the freedom of choice with which humans are endowed and the inadequacies which are inherent in them. Of course, freedom of choice is not an inadequacy; indeed, in some situations it is a positive attribute, but it causes problems when, for example, choice is made capriciously or when it does not coincide with the system's design constraints. One significant inadequacy which repeatedly is a component of accidents is our susceptibility to a loss of concentration — through boredom, disinterest, distraction, or an attempt to do two or more things at once. The mechanisms used to control individuals, both in constraining their freedom of choice and in overcoming their inadequacies, tend to be organisational, and these are discussed in Chapter 9. It is more problematic to devise mechanisms to overcome the inadequacies of organisations, though the development of a safety culture (see Section 1.8) is effective in optimising behaviour within an organisation.

A system throughout its life cycle depends to a great extent for its success and safety not only on human dependability but also, and in many instances more importantly, on human intelligence, management, breadth of vision, political bias, economic involvement, environmental awareness, competence, professionalism and integrity. Whereas there has been a great deal of research by human factors experts into the human-computer interface and the operator's role, as evinced by the remaining chapters in Part 1 of this book and those in Part 2, the considerable human factors involvement in the development process has hardly been addressed. Yet, as Reason says, 'Rather than being the main instigators of an accident, operators tend to be the inheritors of system defects created by poor design, incorrect installation, faulty maintenance and bad management decisions. Their part is usually that of adding the final garnish to a lethal brew whose ingredients have already been long in the cooking' [Reason 90, p. 173].

1.4   A CONSIDERATION OF RISK

As pointed out in Section 1.2.2 above, there is in most systems a trade-off between safety and availability or between safety and reliability. A few years ago (and even today in many industries), a system which needed to be safe would simply be designed to be as reliable as possible, with the fall-back being 'shutdown'. These days, safety-critical systems have penetrated so many industries, and competition is so intense in many of them, that the combination of safety and availability is not only desirable but also necessary. Plant operators want both safety and availability because shutdown loses production. It can also be expensive, as in the case of a nuclear power station. System designers and suppliers want safety without compromising availability because this would give them an advantage over their competitors. Further, in some cases, the trade-off between safety and availability does not exist, such as in an aircraft in flight.

Now, it is recognised that safety is not necessarily implicit in reliability, and there is a greater emphasis on safety engineering. Simultaneously, safety technology has advanced towards seeking a greater assessment and understanding of the risk involved in any given case. Then, before a system is built and deployed, rational decisions can be taken on the extent to which its design, and that of its environment and its operational procedures, should counteract the risks. Working Groups 9 and 10 of the International Electrotechnical Committee's Sub-Committee 65A have put forward the draft of an intended standard [IEC 95] which provides guidelines on the assessment of risk and offers a 'safety life cycle' to define the safety measures to be taken at the various stages of a system's life cycle.

In any given case, risk is a function of two variables. The first is the probability of a hazard developing into an accident; the second is the expected consequence of the accident if it did occur. One proposed definition of risk is that it is equal to the product of the probability and the consequence of a given scenario. If these two variables are quantified, risk could be deduced in numeric terms, which would be convenient for arriving at relevant decisions. Unfortunately, numeric values for a hazard's consequences are not always obvious, particularly as many consequences are qualitative. Similarly, the probability of an accident is seldom accurately known, particularly as the knowledge would almost certainly need to be based on the previous monitoring of the occurrence of several such accidents. Nevertheless, identifying the two components of risk has at least four significant advantages:

- It focuses attention on identifying the risk and what must be done to reduce it — and risk reduction can be achieved without knowing the precise value of the risk itself;
- It allows a risk value to be calculated when numeric values are available;
- Attributed numeric values, even if their bounds of accuracy are known to be wide, can provide a meaningful comparison between two or more options for taking countermeasures;
- It encourages decisions to be made about what level of risk is acceptable in a given circumstance — the concept of 'tolerable' risk.

While numeric values are useful, qualitative risk assessment is a normal function of everyday life — in crossing the road, driving a vehicle, deciding where to live. And assessing industrial risk is very much in conformity with common practice in our daily lives. If our perception of a risk is too great, we refuse to take it; if it is negligible, we take it unthinkingly; if it lies between these two extremes, we consider whether and how we can reduce it. Moreover, risk is balanced against perceived benefits: the greater the prize, the greater the risk we are prepared to take in an attempt to secure it.

And so it is with industrial systems, except that the decisions should not be taken unconsciously. If the risk associated with a safety hazard is clearly too great, and it cannot be reduced to a tolerable level, it should be refused; if the system is clearly not safety-critical, no safety precautions need be taken. When the reality lies between these two extremes, consideration must be given to risk reduction, accident prevention, and, as a back-up, damage limitation. Damage limitation involves planning for and taking efficient and effective action when an accident has occurred with the intention of reducing its effect. It is discussed in Chapter 8.

If either the consequences or the possibility of occurrence is removed, the risk of a given accident occurring is obviated. As usual, however, cost enters the equation, and the risk is accepted if a reduction which is economic brings it within what is considered tolerable. It should be noted, however, that reducing costs has been shown in the enquiries of many disasters to have contributed to the creation of the conditions suitable for an accident — an example being the fire at King's Cross Underground on 18 November 1987 in which 31 people died [Fennell 88]. The human factor at the strategy or 'policy' level can determine whether technology becomes a safe advantage or an accident waiting to happen.

To aid the conscious assessment of risk, a simple matrix may be used as a tool (see Figure 1.4). It plots an accident's severity against its probability using, in each case, only three categories. Exact numeric values are not assumed. A qualitative assessment of the risk involved in any given case is a starting point; it shows where further, more detailed analysis is necessary (see [Redmill 97] for a more detailed description of the use of this tool).

In the UK, the 'ALARP (as low as reasonably practicable) Principle' [Royal Soc 92] is used in risk management. ALARP states that risk reduction measures should be taken unless and until their cost (in terms of finance, time, or convenience) is grossly disproportionate to the risk itself. Figure 1.5 shows an example of the use of the ALARP principle and tolerable risk. Whereas the matrix of Figure 1.4 is in qualitative terms, the units of Figure 1.5 are quantitative, specifically in terms of human fatalities. The diagonal lines define the areas of tolerability or otherwise of the risk. Such a figure could be plotted for any given circumstance, with the areas of tolerability being defined according to current opinion.

Note two things, however. The first is that the chosen risk-reduction measures may not bring the risk down to a tolerable level; the second is that what is considered to be an acceptable level of risk is a subjective judgement, dependent on various factors, a good many of them social. Through familiarity, perhaps because we think ourselves to be good drivers, no doubt because we depend on our cars and cannot bear the thought of their being a threat to us, and perhaps because the rate of attrition is low (though persistent), we accept a high annual death toll on the roads. Yet, we are shocked at the relatively rare aircraft crashes. Over several years, we became relatively comfortable with nuclear power stations, in spite of doubts about our ability to cope safely with their radioactive waste products; now, even though their level of risk has not increased, new attitudes have caused our confidence to decline. Our view of what is tolerable has changed.

Note, too, that the quantification of an accident's consequence or probability is, in most cases, to a great extent subjective. In many industries it is the industrialist desiring to take the risk (presumably because of the value of the potential pay-off) who determines that it lies within the realms of public safety, or whose agent determines this! The human factor is inescapable, and business interests have an influence on declared levels of safety and on the actual (if undeclared) risks that are taken. It is to avoid the final decisions on safety being taken by those with commercial interests that in some industries (such as civil aviation and nuclear) there is a statutory requirement to license plant before it can be brought into operation and there are regulatory authorities (see Section 1.7 below and also Chapter 12) to enforce this requirement.

The human factor is inescapable too in the unpredictability of the manner in which an operator may act in an emergency situation. The best laid plans may be thwarted by a decision taken (or not taken) in haste, rashness, overconfidence, overload, or panic. Indeed, the importance of action in abnormal situations is of such importance that Chapter 8 is devoted to it.

Accident prevention involves the steps taken to detect when the system begins to go into an unsafe state and to return it to a safe one. Considering Figure 1.3, the preventive functions should detect when the system moves towards 'danger' and prevent it entering 'disaster'. These measures should be analysed, specified, designed, built, tested, implemented, operated, maintained and managed to the highest standards. They may be implemented on one or more computer systems (control or protection) which are then designated as being safety-critical. Planning and implementing safety measures is so closely analogous to planning and implementating fault tolerance to achieve reliability and availability that an understanding of the principles of the latter, which are explained in [Lee 90], is recommended.

Yet, even these measures do not abolish risk, as perfect safety cannot be guaranteed. This is recognised in insurance, which is another means of risk,

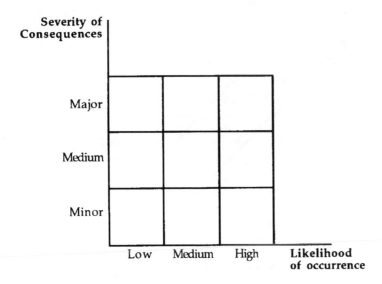

Figure 1.4 *Qualitative categorisation of risk*

or uncertainty, management. Any measure taken to improve safety has a cost. The decision of whether a given measure should be implemented is determined by considering its cost against the consequence of an accident and determining where on the graph of Figure 1.5 the risk lies. Unfortunately, some cynical plant manufacturers and operators, encouraged by their economists, consider only the financial costs likely to be incurred by them and disregard the costs which would fall on the wider community, thus denying responsibility for the environmental and long-term consequences of their actions. Others, more ethical, accept the costs of extra safety measures in order to protect the environment — but the extra costs which they incur can reduce their competitiveness. In the publicity following the Piper Alpha oil rig disaster in the North Sea in 1988 [Cullen 90], it was interesting to note the difference between the Norwegian and the British attitude to safety measures in the off-shore oil industry. The design of the Norwegian rigs was

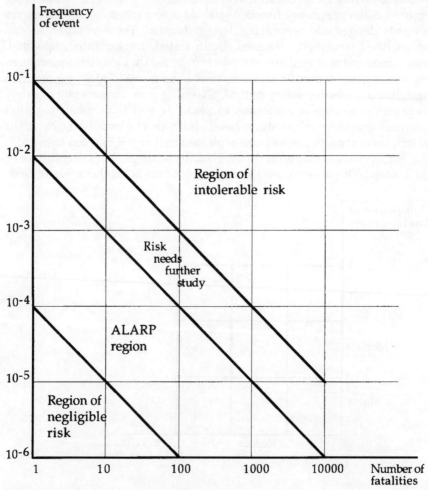

Figure 1.5 The ALARP principle and tolerable risk

safer but more expensive than those of the British, and the policy of the Norwegians was stated as being to put people before economics. This was not obvious in the British regulations. The current 'green' attitude and legislation promoted by it are raising the awareness of everyone, and it is to be hoped that those who take safety seriously, to the extent of increasing their costs, are not forced out of business by those whose products are cheaper at the expense of risk to life and the environment.

Having said this, however, it should be admitted that ultimate safety is not always the only criterion to be considered. In railway signalling, aviation, and nuclear systems, where failure could lead to multiple fatalities, the highest levels of reliability and safety are rightly demanded. But what about medical intensive care units and the monitoring of patients under anaesthetic? Here, computers have concentrated the outputs from several sources onto a single display and provided a more rapid scan of the monitored information than humans ever could. In doing so, they have improved the reliability and safety of the monitoring systems — which, in the main, are used in circumstances when only one person is at risk at any time. If a further improvement in safety could only be achieved at a cost which would place the systems beyond the reach of most hospitals, the question to be asked might be phrased as, 'Is the system safer than that which it has replaced?' rather than, 'Is the system as safe as it is possible to make it?'

Here we are referring to the system. In much of the discussion on risk, however, the reference was to a given accident. It should therefore be pointed out that the safety of a system may depend not on a single risk but on a number of risky scenarios. The risk involved in the operation of a system depends on all of these. The hazard implicit in each scenario needs to be identified and assessed, the risk of each hazard maturing into an accident needs to be analysed, and countermeasures against a number of scenarios may need to be defined, designed, implemented and tested. Thus, the risk involved in a system is broader than that involved in a single incident. Chapter 3 provides a discussion on the evaluation of the human element of such risk.

1.5　HAZARD AND RISK ANALYSIS

A significant problem in computer system development, with far-reaching consequences, is determining and specifying the system requirements. Indeed, the likelihood of accurately reflecting all the requirements in a specification is low, and, in most cases, specifications are poor. Far more reworking of software is carried out because it is the wrong software than because the software does not meet its specification. Given that a system's design is based on its specification, then if the specification is incorrect, incomplete or ambiguous, the appropriateness of the system to its intended purpose is certain to be reduced and its effectiveness jeopardised.

Similarly, the determination and specification of safety requirements is not trivial, as explained in Chapter 5. It is based on an assessment of risks involved, and these in turn are a function of the hazards — which must be identified and analysed. It has already been observed that risk is dependent on context, so analysis needs to consider the working environment as well as the modes of operation of the system.

In safety engineering, a number of techniques have been developed to identify and analyse the hazards and, thus, to assess the associated risks. An assessment of an extensive variety of these has been carried out by the European Workshop on Industrial Computer Systems, Technical Committee No. 7 (EWICS TC7) in its *Techniques Directory* [Bishop 90]. In the end, all the techniques seek to identify the reasons for taking safety-related action. Then, from an understanding of the risks, the safety requirements are deduced, these being the requirements for preventing the occurrence of hazards or reducing their frequency, and for minimising their effects if they do occur. Examples of frequently used techniques are given below.

In *HAZOP Studies*, the potential hazards are examined and their possible causes and consequences determined. This is done in formal meetings by a team carefully chosen to represent the various disciplines concerned in the production of the system, and led by a trained specialist in the method. Detailed examinations of the various parts of the system are carried out by posing a set of hazard-related questions based on the use of 'guide words', which therefore form an integral part of the method. These guide the study to consider 'what if' the guide word applies. For example, one guide word is 'more', and examination is made of what could occur if more of each component of a process were present. If the result could be hazardous, further questions like 'How could it occur?', 'Under what conditions?', 'What are its consequences?', and 'How could it be avoided?' are asked. This is a powerful method, not only for identifying system hazards, but also for determining what variables should be monitored by the protection system in order to provide the final safeguard against breaches of safety.

Fault Tree Analysis commences with an undesirable event. From this, it works backwards, seeking first the immediate cause and then earlier causes and combinations of causes. A graphical tree structure is produced, using logical symbols, such as AND and OR, to show combinations and alternatives and to demonstrate the relationships between the elements which compose the tree. While this method is powerful, the trees produced can rapidly become extremely complex.

Event Tree Analysis works forward from events which are chosen for analysis. The various outcomes of an event are identified and recorded. As these outcomes are themselves events which are causes of further events, a tree evolves until the final events in the chains are identified. Among these, there may be one or more hazardous events.

Failure Mode and Effects Analysis (FMEA) analyses failures of systems or components, or of functions, for their effects. Each component or function is considered in detail, its failure modes defined, and the effects of failure analysed. In computer systems, it is not possible to consider the failure of each line of code, so functions must be considered.

In *Failure Mode, Effect and Criticality Analysis (FMECA)*, a le
is added to each identified failure to indicate its effect on safet,
determined in various ways by different organisations and ca.
hazard's consequence or probability of occurrence, or both
analysis is valuable in deciding on options for risk-reduction
their costs.

From the above summary, it may seem that the techniques are complementary and that, by combining them judiciously, a consistent set of links in a causative chain from component to hazard may be deduced. This is not so in practice. The techniques were developed independently, without an intention to standardise, or even to define, the boundaries between them. Nor is there consistency in the terms used in the various techniques. There is a need for a great deal of work in placing the techniques relative to each other, adjusting their interfaces, and deriving a set of terms which is consistent across all of them. Then there will be a further need to evolve the techniques so that they are as suitable for safety analysis for software systems as they have been for plant in other industries.

There is also a need to evolve the methods so that they inherently take account of human factors in systems, rather than being aimed primarily at technical issues. Not only should there be documented procedures within the methods for seeking human causes of failures and accidents, but it should become standard practice for a human-factors specialist to be on the team involved in any safety analysis.

It will be apparent that the safety analysis methods described above are intensely person-dependent. Many of them require careful repetitive work and the documentation of meticulous detail — the very tasks in which people are error-prone — so there is scope for the semi-automation of safety analysis techniques. Indeed, some computer-based tools already exist and others are being produced for the support of the techniques.

The evidence suggests that many accidents are caused not by a single failure but by a number of failures — which may be related or independent of each other. This does not mean that there is necessarily the coincidence of several failures at the same moment; rather that the system was not vulnerable to a single failure and tolerated it until another occurred. A problem for the safety analyst therefore lies in identifying not only hazardous single failures, but also hazardous combinations of failures. To achieve this exhaustively is almost impossible. Indeed, it is particularly difficult when the human element is considered, for the potential for a catastrophe may be created by high-level company policy — for example, the decision to store toxic or flammable chemicals permanently rather than bring them to site only when they are needed; the decision to shorten an oil tanker's journey by choosing a more reef-laden and stormy sea route. Such decisions are usually taken for economic reasons — economics often neglecting the potential cost of damage to anything, such as flora and fauna, which does not have a visible price tag. This introduces the subject of ethics into safety-critical system design and operation, and it is reassuring to note that the Engineering Council's new Code of Professional Practice, *Engineers and Risk Issues* [EC 92], is concerned with ethics in the engineering profession.

Once the hazards have been identified, their probabilities may be deduced (perhaps by 'educated' guesses) and a risk analysis carried out. The safety requirements are then determined, based on the need to eliminate or reduce the identified causes or consequences of hazards. When the safety requirements have been identified and isolated, a safety plan is drawn up to show how they will be satisfied in the design, implementation and testing of the system. A number of the safety functions may be applied structurally, for example, by reinforced or fire-proofed buildings; others may be included in manual procedures (often these include strategies and plans for damage limitation to minimise the effect of an accident after the event); and typically, these days, many are built into the computer systems which are part of the total system.

It should also be noted, however, that hazard and safety analyses should not only be carried out at the requirements stage of a system's life cycle. They are also appropriate at other times, for example every few years during the system's operational life. Changes to the system or its environment can alter the hazards involved in operation, and these new and changed hazards should be identified, assessed, and the resulting risks reduced.

1.6 DESIGNING FOR SAFETY

1.6.1 Principles

Designing for safety must commence with an analysis of the plant, or equipment under control, as outlined in the previous section. If there is no hazard, safety precautions are unnecessary; when a hazard exists, the safety precautions to be designed into the system need to be commensurate with it.

The safety requirements, for the countermeasures to remove or mitigate the identified risks, are documented. They should be defined in such a way that the extent to which they are met is measurable, that their designs are easily verifiable, and that, when the system has been developed, they themselves can be validated. Some interesting and informative observations on managing requirements have been made by Dobson and Strens [Dobson 93].

By defining the safety requirements independently of the functional requirements of the system, particular attention can be given to meeting them in the design. Relatively costly methods or components, not justifiable for normal operational functions, may be employed in the design or construction of a safety-critical part of a system or of a subsystem intended for meeting the defined safety requirements. Thus, safety may not be limited by the practices used in the rest of the system, and the cost of the rest of the system is not aggravated by the imperatives of achieving safety. The independent identification and documentation of the safety aspects of the system allows a similar higher level of attention to be paid to such activities as verification and validation, project management and assessment. It also allows particular attention to be paid to management practices, such as the signing off of documents, which should then be readily auditable. For the reader who wishes to study the development process, McDermid offers an introduction [McDermid 93], and guidance on design for safety is offered by the EWICS TC7 in Chapter 1 of [Redmill 89].

Writing about the safety case (see Section 1.7 below and Chapter 12), Hawkesley suggested that there are three questions to be answered [Hawkesley 89]:

- What could go wrong?
- Why won't it?
- But what if it did?

The basis for a proof of safety being offered is that the first question has been thoroughly investigated in the safety analysis and that the other two questions have guided the design for safety. The following are a number of principles (but not an exhaustive list) which should influence the designer. For each principle, questions are raised which, typically, a designer should ask.

(i)   Risk reduction. Where is risk reduction necessary and what level of reduction is required? Can the hazard be eliminated? If not, can the required level of risk reduction be achieved by reducing the accident's effect or its probability of occurrence?

(ii)   Human error. Do the designers understand the nature of human error? What are the potential causes of human error in this system? How can the design minimise them? What mechanisms can be installed to provide, or maximise the potential for, recovery from error?

(iii)   Safe control. What hazards exist in controlling the plant? How can these be eliminated or avoided? What design characteristics (for example, redundancy) are necessary in the control system in order to ensure the specified level of safety?

(iv)   Protection. What safe and unsafe system modes exist? What variables need to be monitored in order to detect all possible excursions into unsafe modes? What actions are necessary in each case in order to ensure safety? How reliable does a protection system need to be in order to provide the required level of safety?

(v)   Emergencies. If all the above failed (even if the impossible occurred), what systems and procedures need to be in place in order to ensure that the effects of the accident are minimised, that communication is maintained with all persons involved, and that the public are adequately informed?

(vi)   Human factors. Which tasks are most suitable for humans and which for computers? What training and support will operators require in order to equip them for their tasks? What policies and management will be required in order to develop a safety culture?

(vii)   The impossible, the unthinkable. What have we thought of that we know to be impossible? What have we not yet thought of? Recognising that the impossible and unthinkable do occur, how could they occur? What must we do to preclude their occurrence? What must we do to minimise damage if they did occur? Often it takes an outsider to recognise those events which we assume are out of the question and to demonstrate how easily they could in fact occur, given the right circumstances.

1.6.2   Safety and Reliability

One view of achieving safety is via reliability, implying the assumption that if the system functions as it should it will remain safe. However, it is as well to recognise that, while normal operation should certainly be designed and planned to be safe, safety and reliability are not synonymous. Reliability is operation in conformity with specification, and the specification may not have taken account of all possible safety implications. A motor cycle may be specified and designed to travel at 170 kilometres per hour; but at that speed it may be highly vulnerable to a small undulation in the road and, therefore, unsafe, even when performing reliably. As already pointed out, safety is dependent on context. Safety is dependent on operation within the system's design constraints; it can never be absolute. It is imperative to carry out a safety analysis and determine the safety requirements independent of the functional specification.

Nevertheless, it is not unusual for designers to seek to achieve safety via reliability. This implies avoiding the introduction of faults into the system, designing the system so that it is tolerant of faults when they do occur, and having a mechanism for bringing the system to a safe mode in the event of safety being in any way threatened.

The first step in this process is *fault avoidance* (also referred to as *fault prevention*) in the design and construction of the system or sub-system. High-grade components, rigorous procedures, stringent, extensive and independent testing, and sound engineering and quality practices are all essential contributors to this.

While fault avoidance is an obvious and desirable goal, it is unlikely to be achieved entirely. In hardware, even if design were perfect, there is the certainty of random failures resulting from component wear. Moreover, even if complete fault avoidance were achieved, it could never be proved: finding faults proves that faults are present, but not finding faults does not prove their absence. This is especially so in software where, even in small systems, the huge number of possible logical paths through the system renders exhaustive testing in a finite time impossible.

The next step in designing for reliability is therefore to design and build *fault tolerance* into the system (see [Lee 90] for a thorough consideration of this subject). This is complementary to *fault avoidance* and usually has the objective that no single fault should cause a system failure. It includes hardware redundancy, both in the computer systems and the plant, and a great deal of diagnostic and maintenance software. Software redundancy usually takes the form of back-up versions of the same software. This can be reloaded in the event of corruption, and it can also be loaded into replacement hardware which is introduced automatically when a faulty item is switched out of service. It should be noted, however, that such duplicate software must necessarily suffer from the same faults as its original, so it cannot provide the solution to a software design problem. There are, however, instances of *n-version programming* — separate systems being independently designed and programmed, sometimes in different languages, to perform the same functions [Avizienis 85].

The final stage of designing for reliability, that of switching the system into a safe mode in the event of failure or any other threat to safety, has been briefly outlined in Section 1.2.2 above. It should be mentioned, however, that the logic for detecting the need to make the switch, and the mechanism for achieving it, need to be operational even in the event of failure. The detection and decision process must therefore be of both high reliability and high availability and, therefore, designed and constructed to the highest standards, with *fault avoidance* and *fault tolerance* being employed. In some systems, the monitoring and fail-safe functions are built into the normal control of the operational equipment; in best safety-critical practice, they comprise a separate 'protection' system.

1.6.3 Control and Protection Systems

If, as suggested in Section 1.2.2 above, a system is to be switched into a safe mode in the event of malfunction, there needs to be both a monitoring function designed to recognise when it is entering an unsafe state and a fail-safe switching function. Given the potential complexity of software systems, confidence in achieving safety is increased by separating the functions relating to safety protection from those concerned with control — leading to the design of control and protection computer systems which are distinct, not only functionally but also geographically (though they may be very close together).

The control system's main purpose is to direct the operation of the plant. This, of course, implies safe control, so a number of safety functions must be designed into this system. However, with an increase in functionality goes an increase in the number of logical paths through the software, a decrease in the proportion of those paths which can be tested in a finite time, an increase of the probable number of faults in the system, and an increase in the chance of malfunction.

A protection system's sole purpose should be to monitor the state of the equipment under control and to take action when safety is threatened. It is a sound principle that the protection system should be kept free of unnecessary functions and thus simple. The plant may fail, or the control system may malfunction, but the protection system should be available to detect these occurrences and take preventive measures to ensure that the system does not enter a non-safe mode or, if it does, to take corrective action to return it quickly to a safe mode. The safe mode may be 'shutdown', or it may be a degraded operational state.

In industries where safety-critical systems were employed prior to the introduction of computers, particularly those industries (such as nuclear) where safety certification or licensing is necessary, the use of independent protection systems is traditional. In these industries, protection systems remain the norm, even as computers replace electromechanical equipment. In other industries, where the first designs of safety-critical equipment have been computer-controlled, the separation of control and protection systems is not always observed. In some cases, reliance is placed wholly on reliability; in others, safety features, added in the control system, are taken to be sufficient: where monitoring and protection features are provided, they are built into the control system. System and software designers in these industries would do well to study the existing, proven safety technology. The human component in system design needs to be adequately educated and trained, and this means acquiring an understanding not only of software engineering, but of safety engineering as well.

One feature of some safety-critical systems is that of an automatic facility to preclude or limit human intervention for a given time in the event of certain types of failure. As explained in Chapter 8, human beings have a tendency to try to find confirmation of the first feasible solution they find to a problem, rather than to search systematically through possible alternatives. In limiting human intervention (say, for 30 minutes) the intention is to avoid hasty or rash human action. The system may be brought to a safe state by the automatic safety systems, but the interval also allows the human operator time to evaluate the situation and to gather information to plan an appropriate course of action.

Other solutions to the problem of human impulsive reaction to events include operator support by the use of decision aids, as discussed in Chapter 7, and the availability of teams of specialists who can provide expertise in the event of an emergency.

1.6.4   Software Engineering

An obvious starting point in the quest for safety in software-based systems is in the engineering and production of the software itself.

Software development is a new technology. Until the late 1960s, computer-system development was seen almost universally as consisting only of programming, and it was believed that, with just a little more research, programming could be made perfect. Then there would be no more software errors.

Things did not turn out like that. By the 1970s, it was recognised that the programmers were letting the side down. They sought 'clever', complex solutions rather than simple ones, their work was not submitted to independent testing and verification, and they coded before designing, or even specifying. Improvements in software development awaited the introduction and use of engineering and management principles.

Traditionally, in engineering, there has been a distinction between the engineer and the technician. The engineer is responsible for making decisions, for example, design decisions. The technician is responsible for implementing the engineer's decisions. In software development, the equivalent of the traditional engineer is the designer, that of the technician is the programmer or coder.

The introduction of true engineering principles and the achievement of real improvement in software development has been gradual and, even now, has not reached a culmination. One reason for this is that software engineering was at first seen largely as a technical issue; it was overlooked that techniques were already in the hands of the programmers. Later, as the emphasis on management increased, it was recognised that 'engineering' implies not more highly skilled technicians, but improved design and control, in the form of professionalism, discipline, coordination, and all those management attributes which hold teams together and make projects successful. Significant factors in the human element of any enterprise are making the right skills available and allocating tasks appropriately; and this in turn relies on the management function. Too often 'human factors' admits only the technicians and operators and overlooks or denies the importance of planning, decision making, supervision, and the way in which specifications (written or verbal) and designs are prepared and presented.

More recently, the extreme importance of management, not only of development projects, but indeed of departments and companies as a whole, has been acknowledged in the software domain. New management styles and an attention to quality have been advocated, and quality standards, such as ISO 9000 [ISO 94], are accepted as the minimum requirement for companies claiming to take a serious and competent approach. Indeed, there is now an additional ISO guideline [ISO 91] on applying a quality management system to software. The use of standards and guidelines in software engineering is also more prevalent than hitherto; their employment is fundamental to consistent good practice in the technical aspects of software development and to the facilitation of process assessment and quality assurance.

Yet, a study of several of the main software engineering standards in use has shown substantial weaknesses in many domains, particularly in measurement during the development process [Devine 93]. This being the case, it may be that improved software engineering requires not merely better adherence to current standards, but a new attitude to their use — that is, to use standards to provide initial guidance, but with a conscious intention to do better than they exhort.

There is also a great deal of work to be done in producing standards which are easily usable and which consistently achieve their aim, and the starting point may need to be a reform of the standardisation process itself.

Nevertheless, the foundation of reliable and safe software-based systems must be good software engineering principles and practices, including independent verification at each stage of the development life cycle and independent validation of the system. It does not take much investigation to show that a great deal of current software engineering is human-intensive: determining requirements, preparing specifications, carrying out design, programming, and testing. Where there is human involvement, there is the likelihood of human error. In safety-critical systems, human error and the resulting faults can lead to hazardous consequences. In traditional methods of software production, there were insufficient formal mechanisms and checks to offer a high level of confidence that given levels of safety had been achieved. However, the discipline of software engineering attempts to impose organisational mechanisms to overcome the inadequacies of individual humans (Chapter 9 discusses such organisational mechanisms).

In addition, the other essential supports of software engineering — high-level policy decisions, project management, and independent assessment of the processes and the products — lie almost entirely in the hands of humans. The report into the failures of the new London Ambulance Service's Computer Aided Dispatch System in October and November 1992 is one of the most instructive documents in this field [SWTRHA 93]. A system had been commissioned to automate the allocation of the most appropriate ambulance in London in response to each emergency call. On the system's introduction on Monday 26 October 1992, it failed to function satisfactorily and caused huge delays in the dispatch of ambulances. Finally, on 4 November, further problems resulted in its being withdrawn from service. The report reveals serious flaws in almost every aspect of the project, from the policy decisions of senior management to the technical development and testing of the product. Particularly interesting were the paragraphs of the report which revealed the remoteness of senior management from the project, the lack of confidence of project staff in their seniors, the failure of management to integrate their decisions with the technical realities of the project, and the breakdown in human relationships. The dangers of a project team being isolated from the rest of the business have been pointed out [Redmill 91].

A development project is never successful if it consists only of a technical development team; nor if management has failed to create harmonious human relationships. The lack of sound project management and a failure by senior management to identify strategic goals have caused numerous software development projects to founder. Software engineering is not merely a set of techniques and tools, but a discipline based on management, sound engineering practice, and control.

1.7   THE SAFETY CASE

However competent and conscientious the engineers and other development staff are, there always needs to be independent scrutiny and assurance that procedures and standards have been conformed to, tests have been conducted thoroughly, and quality standards and integrity levels have been met. Many believe that minimum requirements for assurance should call for the development department or company to be certified against the international standard ISO 9000 [ISO 94], and that there should be a number of independent assessments, namely: verification at every stage of development, validation of the system, quality assurance throughout development, certification of the development process, and certification of the system's conformity to the required standards and integrity levels. However, there is as yet no agreement on minimum requirements, and many (perhaps most) companies developing safety-critical systems do not comply with those mentioned above.

When a system is safety-critical, similar processes are necessary for ensuring that the safety requirements (as opposed to the functional requirements) of the system have been met. The documentation of the processes and their results go towards making up the 'safety case' or 'safety argument' — the formal documented argument of why the system is believed to be safe and therefore worthy of being brought into service. The safety case needs to demonstrate that risk analysis has been applied, that all foreseeable hazards have been identified, that the risk of these hazards has been removed or reduced to an acceptable level in the design, and that damage limitation resources and procedures are planned and provided for all predictable accidents. Such information is necessary to convince not only outsiders but also the organisation intending to operate the system of the system's safety; it would be irresponsible of an organisation to deploy the system if it were not convinced of its safety.

In many industries, there are independent or governmental regulatory authorities which must be satisfied as to the system's safety. In some industries, these bodies need to certify or license the system before it can be brought into service, and there the safety case is mandatory by law rather than merely to satisfy or protect the organisation deploying the system.

The safety case may be likened to a business case which is drawn up to convince a financier that all factors influencing a business have been identified, that all practicable measures are in place to take advantage of positive factors and to counteract negative ones, and that, as a consequence, the viability of a business is secure. The safety case is intended to provide the same assurance with respect to the safety of a system, and drawing it up is normally the responsibility of the operating company.

The safety case should first be submitted to a regulatory authority internal to the operating company. There is then an iterative loop between the internal regulatory authority on the one hand and the management of the system on the other. Once the system has been accepted as 'safe', it may be operated within defined limits and in accordance with documented procedures. Further, any change, other than the replacement of a part with its exact replica, must be approved by the internal regulatory authority (see Chapter 5 of [Redmill 89]).

For highly safety-critical systems, such as in nuclear power plants, certification or licensing of the system by external regulatory authorities is required before the system is allowed into service. Then there are defined aspects of the system which are subject to external regulation — other safety aspects being the responsibility of the internal regulatory authority. Some changes, deemed to be consequential, would need to be approved by the external regulatory body. An example of the principles on which a regulatory authority may base its assessments is given in *Safety Assessment Principles for Nuclear Plants* [HSE 92].

The formal relationships between the plant management and the internal and external regulatory authorities are shown in Figure 1.6. When communication is restricted to the formal channels, however, safety case approval can be a protracted and frustrating business. It is always advisable for the external safety assessors to be involved not only with the plant management, but also with the system developers, from the earliest possible stage of the project. Then, not only can the developers better understand the criteria for a successful assessment, but problems can be detected and resolved at minimum cost, with minimum delay, and with minimum antagonism, as the project proceeds.

In making the case for the system's safety, the safety case consists,

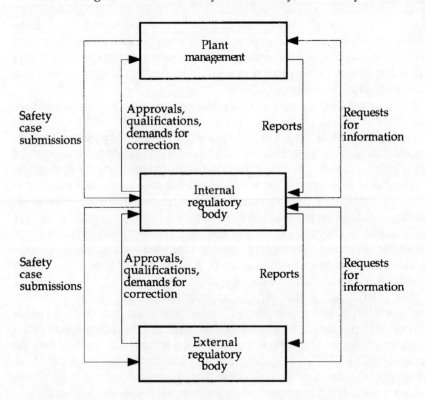

Figure 1.6 *Relationships between plant management and regulatory bodies*

typically, of a number of arguments, all converging on a single conclusion: 'The system meets its safety criteria and is therefore acceptably safe to be allowed to operate in accordance with its defined objectives and procedures.' Hawkesley defined the purpose of the safety case as being to answer three questions: 'What could go wrong?', 'Why won't it?' and, 'But what if it did?' [Hawkesley 89]. If the safety case demonstrates, by reasoned argument, that all foreseeable hazards have been anticipated, that there is confidence that the hazards will not be activated to become accidents, and that, even if they did, there are clearly defined systems in place to limit the damage, it will have performed its function. The regulatory bodies then need to determine whether the confidence claimed by the safety case that the system will be safe when it is brought into service is firmly based.

Given an adequate level of confidence, approval is given for the system to be brought into service. The level of confidence required may depend on the nature of the potential hazards, and there can be no certainty of absolute safety. It was mentioned above that risk and safety analyses should be carried out periodically throughout the life of the system, and these may show the assumptions made earlier to have changed or even to have been invalid. In either case, steps should then be taken to counter the newly discovered, or changed, hazards — by design, training, or security measures. Once changes have been made to the system, the safety case needs to be revised to demonstrate the effectiveness and efficiency of the new measures. Then, certification and licensing must also be reviewed. Thus, the safety case is a dynamic entity, reflecting the system and its changes, from its design, through its installation and operation, to its decommissioning.

Once a system has been certificated or licensed, there is a class of changes which must be approved not only by the internal, but also by the external, regulatory body. And once such a change has been made, the new system may not be returned to service without first being recertified. A guideline of good practice for maintaining and changing a software-based safety-critical system was produced by EWICS TC7 (Chapter 5 of [Redmill 89]).

If a safety case is concerned only with the technology of the system, it omits a major cause of accidents — the human factor. The possibility of human error in specification, design, policy making, management, operation and maintenance need all to be taken into account. Then there is the human aspect of the operational environment: security, access to the system by unqualified staff or other persons, sabotage, etc. Finally, there is the possibility of human error in developing the safety case itself. This likelihood can only be minimised by employing the most appropriate authors and rigorous quality assurance methods in the preparation of the document. These issues are explored in Chapter 12.

In some industries, for example the nuclear and civil aviation industries, a satisfactory safety case is a necessary prerequisite to the operation of a system. In many other industries, such as the medical and road transport industries, there is no requirement for a rigorous safety argument to be drawn up before a system can be brought into service. There is a need for stricter regulation in such industries.

1.8   SAFETY CULTURE

As pointed out in Chapter 9, the best chance of achieving safety occurs not simply when a special effort is made but when a pervasive safety culture exists in an organisation. If there is to be a safety culture, its pattern must be set by top management, it should be evident in policy decisions, and it should be propagated and maintained by all levels of management. Necessary ingredients of a safety culture are awareness, commitment, and competence, and the following brief discussion will be structured under these three headings. This does not, however, imply an attempt either to provide an exaustive description of safety culture or to limit its scope to these three subheadings — indeed, see also [Levene 97].

1.8.1   Awareness

Many manufacturers are not aware of the safety implications of their products. Many operators are not aware of the safety implications of their processes. Yet, awareness is the most crucial contributor to safe practice. Without awareness, designers are likely to design systems for functional performance only, without making provision for safety based on what could go wrong. Without awareness, policy makers and managers are likely to base their operations, including staffing, on economic functionality at the expense of safety — see, for example, the report on the enquiry into the King's Cross fire on the London Underground, in which 31 people died [Fennell 88]. Given such designs and policies, accidents are more likely to occur and, when they do, they are more likely to be catastrophic.

Awareness is also a necessary attribute of a true professional, who needs to be aware not only of the safety implications of his actions, but also of his own limitations. To assume responsibility beyond one's authority or field of competence, or to undertake tasks beyond one's ability, is at any time unprofessional; to do so when safety is at stake could set the scene for a disaster. The issue here is not merely commitment or competence, but, more fundamentally, perpetual awareness.

Recently, a number of learned, professional, and governmental organisations have taken initiatives which demonstrate the growing importance of safety-critical systems and the need for an increased awareness and understanding of risk. The Engineering Council's *Engineers and Risk Issues* [EC 92] is a code of practice which emphasises individual as well as corporate responsibility for the actions of engineers and managers. The Royal Society's *Risk: Analysis, Perception and Management* [Royal Soc 92] is a study group report which not only offers abundant immediate advice but which will also be an invaluable reference document for all engineers and managers involved in safety and risk issues. The Department of Trade and

Industry's initiatives have included the creation in 1991 of the Safety-Critical Systems Club whose purpose is to raise awareness and facilitate the transfer of information, technology, and current and emerging practices and standards. The club seeks to involve both engineers and managers in all sectors of the safety-critical community and, by doing so, to facilitate communication among researchers, the transfer of technology from researchers to users and feedback from users, and the communication of experience between users. The potential benefits are more effective research, a more rapid and effective transfer and use of technology, the identification of best practice, and the definition of requirements for education and training. As awareness is increased, all these benefits can be achieved, at least in some measure.

1.8.2 Commitment

The development of a culture can be directed and led, but it cannot be forced. Every organisation has its culture and, while this is demonstrated most obviously by the staff, it follows the pattern set by its senior management. Slovenly conduct and lack of attention to detail are most noticeable in the activities of engineers, technicians, operators, and other staff, but they reflect the lead set by those at the top. The changing of culture is a long-term endeavour and requires senior management to define goals and to commit and involve themselves in leading the changes.

Safety awareness is the first requisite of a safety culture. Then there must be the commitment to doing what awareness shows to be necessary. Threats to safety need to be recognised and, when they are, they should be removed. When there has been a breach of safety, there needs to be immediate recovery followed by an investigation. Nothing should be swept under the carpet. The investigation should provide feedback which, if the investigation has been thorough, is on at least two levels. First, there should be the information regarding the particular incident, and then the information regarding the context of the incident, and this should reveal the generic aspects — those circumstances which could lead to another similar incident. Analysis of all the information (see Chapter 10 for a description of a tool for the analysis of incidents) should be followed by the implementation of changes necessary to remove the possibility of a recurrence of the incident. These may take any of a number of forms — for example, design changes, procedural changes, training or retraining, or improvements in security. If action is not taken, the circumstances under which a major accident could occur begin to accumulate, and the probability of the accident occurring begins to rise. As has been shown by the investigations into so many accidents, it is the diversion of awareness from safety warnings and the lack of commitment to take action which made accidents inevitable.

Initiating investigations and taking corrective action in accordance with their results often requires courage and always requires leadership and commitment. It also requires competence, both in knowing how to act and in knowing when to seek advice or assistance or to turn the problem over to someone with more appropriate skills. It also costs; but, as in all quality and safety initiatives, wisely applied initial costs lead to long-term savings.

1.8.3   Competence

In a policy statement on safety-related computer systems, issued on 8 June 1993, the British Computer Society said, 'Safety-related computer systems must be developed and supported by suitably qualified personnel, working within the framework of an organisation which is accredited to international and national quality management standards.' It also made a further point: 'The quality of every safety-related computer system should be the responsibility of a named engineer who should be a Chartered Engineer, or a named individual whose competence and qualifications for such tasks have been registered with an appropriate professional body. This supervisor must have up-to-date training and qualifications in the relevant technologies.'

While there are as yet no agreed standards of competence in the field of safety-critical systems, and it is not yet accepted in industry that only Chartered Engineers should be entrusted with certain safety-critical responsibilities, it is agreed by the professional bodies that there is a need to specify competence levels for those working in various capacities on safety-critical systems. Competence, however, is not easily defined or verified, for it comprises a number of components, some referring to personal traits rather than tangible qualifications. The guidance for engineers on safety-related systems published by the Hazards Forum [Hazards Forum 95] points out that 'competence depends not only on the capability of an individual, but on the appropriateness of that capability to the particular task in hand'. This reminds us that an individual's competence needs to be considered in the context of the task in which he is involved and not in absolute terms. The professional brief goes on to offer the following definition of competence:

> Competence requires the possession of qualifications, experience, and qualities which include:
>
> - Such training as would ensure acquisition of the necessary knowledge of the field for the tasks which they are required to perform;
> - Adequate knowledge of the hazards and failures of the equipment for which they are responsible;
> - Knowledge and understanding of the working practices used in the organization for which they work;
> - An appreciation of their own limitations and constraints, whether of knowledge, experience, facilities, resources, etc., and a willingness to point these out.

The brief then adds that:

> Professionals with responsibility for design or for supervision of operators may be expected to have in addition:

- A detailed working knowledge of all statutory provisions, approved codes of practice, other codes of practice, guidance material and other information relevant to their work; an awareness of legislation and practices, other than these, which might affect their work; and a general knowledge of working practices in other establishments of a similar type;
- An awareness of current developments in their field of work.

When it comes to training for work with safety-critical systems, emphasis has traditionally been placed on the needs of operators. Now, with the recognition that managers can create the context for accidents, within which operators merely provide the final triggering actions, there is also the recognition of the need to train managers and to provide independent assessment of policies and decisions as well as of processes and products.

It should also be recognised that the 'operators' of safety-critical systems are not necessarily traditional computer operators: there is a new type of operator who has had to adjust to safety-critical system competence from a previous task of high responsibility. An example is the aircraft pilot who formerly was in control of a cockpit of electromechanical instruments and dial displays and is now reliant on computers for both the receipt of status data and the input of instructions to the aircraft's operational equipment.

A pilot requires certification of his competence to fly a particular aircraft. If there is a change to the aircraft, the pilot needs to be retrained and recertified, for, in an emergency, even a small change to a display could lead to hesitation or a wrong decision. But how can such competence be proved, or guaranteed? A pilot may be informed of the small change, but how much 'hands-on' practice will it require to ensure that in an emergency his reflex is not still conditioned to the previous situation? It is notable that a recommendation in the report of the crash at Kegworth, UK, on 8 January 1989, as a result of which 47 people died, was: 'The Civil Aviation Authority should review current airline transport pilot training requirements to ensure that pilots, who lack experience of electronic flight displays, are provided with familiarisation of such displays in a flight simulator, before flying public passenger aircraft that are so equipped' (recommendation 4.9 of [DOT 90]).

Perhaps there is a need for a Configuration Management System of certification. In the case of aircraft pilots, this would imply the keeping of a database which would define the precise nature of the certification of each pilot and the precise status of each aircraft. The database would be configured so that training records and system changes are coordinated. Resulting from this would be clear definitions of which aircraft types a given pilot is certified to fly and what further competence the pilot would need in order to fly other given aircraft. Then the pilot could not fly a changed plane unless he had undergone retraining. Any change to an aircraft or to a pilot's competence would be recorded in the database, and the new relationships between pilot and aircrafts would be defined. In such a system, the effect would be that

tion is carried out not merely on the aircraft or on the pilot, but on the
te safety-critical system — which includes the operator.

it happens, civil aviation is a well-regulated industry in which
certification is practised. However, it is not obvious that equally strict
regulations apply in other industries. In the medical industry, for instance,
there are computer-controlled intensive care units, surgical aids, patient
monitors, and record systems. Perhaps in most cases there is adequate
training of the surgeons and physicians who use them, but there have been
disturbing suggestions that surgeons, trained in the use of the scalpel, are
deploying 'keyhole' surgery without being trained in it — and that as a result
the safety of their patients has suffered. Is there a need for formal certification
of medical equipment and practitioners and a supporting Configuration
Management System?

In some industries there is a move towards the certification of competence.
This is not widespread, however, and the difficulty of making it so is
increased as the number of industries into which safety-critical systems are
introduced increases. There is a need for universal standardisation. The
professional bodies have made a start, with reports on education and
training by the Institution of Electrical Engineers [IEE 92] and the British
Computer Society [BCS 90], but there is a long way to go.

1.9   THE RIGHT QUESTIONS

So a safety-critical system is assessed as being safe. But what is the source of
our confidence? We know, or think we know, because the questions we have
asked about its safety have been answered satisfactorily. But what were the
questions, and who has asked them? It is always easier to find answers than
to pose the right questions, particularly when it is economically urgent for
the system to be brought into service.

The three questions referred to above [Hawkesley 89] are a starting
point: What could go wrong?, Why won't it?, But what if it did? But asking
them once is not enough; they need to be asked by everyone involved in every
activity throughout the life cycle of the system, for everyone has a different
viewpoint and every viewpoint reveals new evidence. Safety analysts must
ask the questions in preparing the safety requirements; designers must ask
them in planning risk reduction; assessors, operators, policy makers, security
planners and managers must ask them in the light of their respective
relationships with the system. A full guarantee of safety cannot be given; no
system is inherently safe. If it is designed to be safe, and the design has been
extended to its human components, the risk of a safety breach is diminished.
But then, if the system is operated outside of its design limits, or in the
absence of a safety culture, the risk is increased. The awareness to ask the
right questions, the commitment to seek the answers and not to disregard
them, and the competence to judge when to act and when to seek assistance,
add up to the essence of a safety culture. Awareness, commitment and

competence need to be present in all who are concerned with safety-critical systems. The responsibility for safety lies with everyone.

ACKNOWLEDGEMENTS

Thanks to Tom Anderson and Barry Hebbron for their criticism of drafts of this chapter.

REFERENCES

[Avizienis 85] Avizienis A: *The N-Version Approach to Fault-Tolerant Software*. IEEE Transactions on Software Engineering, SE-11 (12), December 1985

[Bainbridge 87] Bainbridge L: *The Ironies of Automation*. Chapter 24 in Rasmussen J, Duncan K and Leplat J (eds): *New Technology and Human Error*. John Wiley & Sons, 1987

[BCS 90] British Computer Society: *BCS Safety Critical Systems Group Report of the Education Working Party*. BCS Report, September 1990

[Bishop 90] Bishop P G (ed.): *Dependability of Critical Computer Systems — 3*. Elsevier Science Publishers, London, 1989

[Cullen 90] Cullen The Hon. Lord: *The Public Enquiry into the Piper Alpha Disaster*. HMSO, London, 1990

[Devine 93] Devine C, Fenton N and Page S: *Deficiencies in Existing Software Engineering Standards as Exposed by "SMARTIE"*. Chapter 19 in Redmill F and Anderson T (eds): *Safety-Critical Systems — Current Issues, Techniques and Standards*. Chapman and Hall, London, 1993

[Dobson 93] Dobson J and Strens R: *A Methodology for Requirements Management Applied to Safety Requirements*. Chapter 8 in Redmill F and Anderson T (eds): *Safety-Critical Systems — Current Issues, Techniques and Standards*. Chapman and Hall, London, 1993

[DOT 90] Department of Transport: *Report on the Accident to Boeing 737-400 G-OBME near Kegworth, Leicestershire on 8 January 1989*. DOT Air Accident Investigation Report 4/90 (1990)

[EC 92] The Engineering Council: *Engineers and Risk Issues*. The Engineering Council, London, October 1992

[Fennell 88] Fennell D: *Investigation into the King's Cross Underground Fire*. HMSO, London, 1988

[Hawkesley 89] Hawkesley J L: *A View from ICI*. Part of Chapter 7 of Lees F P and Ang M L (eds): *Safety Cases within the Control of Industrial Major Accident Hazards (CIMAH) Regulations 1984*. Butterworth, 1989

[Hazards Forum 95] *Safety-related Systems — Guidance for Engineers*. Hazards Forum, London, 1995

[Hidden 89]   Hidden A: *Investigation into the Clapham Junction Railway Accident*. HMSO, London, 1989

[HSE 92]   Health and Safety Executive:*Safety Assessment Principles for Nuclear Plants*. HMSO, London, 1992

[IEC 95]   *Draft IEC 1508 — Functional Safety: Safety-related Systems* (in 7 parts). International Electrotechnical Commission, June 1995

[IEE 92]   Institution of Electrical Engineers: *Educational and Training Requirements for Safety Critical Systems*. Public Affairs Board Report No. 12, January 1992

[ISO 91]   International Standards Organisation:*ISO 9000-3: Guidelines for the Application of ISO 9001 to the Development, Supply and Maintenance of Software*. ISO, 1991

[ISO 94]   International Standards Organisation:*ISO 9001: Quality Systems — Model for Quality Assurance in Design, Development, Production, Installation and Servicing*. ISO, 1994

[Lee 90]   Lee P A and Anderson T: *Fault Tolerance: Principles and Practice*. Springer-Verlag, 1990

[Levene 97]   Levene T: *Getting the Culture Right*. Chapter 2 in Redmill F and Dale C (eds.): *Life Cycle Management for Dependability*. Springer-Verlag, London, 1997

[Leveson 86]   Leveson N G:*Software Safety: Why, What, and How*. Computing Surveys, Vol. 18, No. 2, June 1986

[McDermid 93] McDermid J:*Issues in the Development of Safety-Critical Systems*. In Redmill F and Anderson T (eds): *Safety-Critical Systems — Current Issues, Techniques and Standards*. Chapman and Hall, London, 1993

[Reason 90]   Reason J: *Human Error*. Cambridge University Press, 1990

[Redmill 88]   Redmill F J, Johnson E A and Runge B: *Document Quality — Inspection*. British Telecommunications Engineering, Vol. 6, Jan. 1988

[Redmill 89]   Redmill F J (ed.): *Dependability of Critical Computer Systems — 2*. Elsevier Science Publishers, London, 1989

[Redmill 91]   Redmill F: *Project Management in Context — An Environmental Model*. UKCMG Annual Conference, Birmingham, May 1991

[Redmill 97]   Redmill F: *Practical Risk Management*. Chapter 8 in Redmill F and Dale C (eds): *Life Cycle Management for Dependability*. Springer-Verlag, London, 1997

[Royal Soc 92]   Royal Society: *Risk: Analysis, Perception and Management*. Report of a Royal Society Group, The Royal Society, London, 1992

[Steel 87]   Steel D: *Formal Investigation into the MV Herald of Free Enterprise Ferry Disaster*. HMSO, London, 1987

[SWTRHA 93]   South West Thames Regional Health Authority:*Report of the Inquiry into the London Ambulance Service*. South West Thames Regional Health Authority, February 1993

2

The causes
of human error

2.1   THE AIM OF THIS CHAPTER

This chapter provides an introduction to recent scientific thinking about human error. It also outlines the practical applications of this research through human reliability assessment. The chapter is neither a theoretical treatise nor an exposition of all available practical methods for error prediction and reduction. Rather, the aim is to provide the building blocks of understanding for readers who are approaching the topic for the first time and for those who need a quick overview of the field and of recent developments. For the reader who wishes to obtain deeper insights, there are now a number of textbooks that cover the topic of human error in some depth (see, for example, [Rasmussen et al 87, Reason 90, Glendon and McKenna 95]).

This chapter offers a broad introductory summary, many of the issues introduced here being treated in more detail in later chapters of this book. It is divided into three main parts which cover:

- Human error and major accidents — where 'human error' has contributed to a serious accident resulting in significant loss of life, injuries, loss of plant, or environmental damage;
- Understanding the causes of human error — what we currently know about why people make errors;
- Practical uses of theories of human error — how we can apply our

knowledge of error causation to eliminate, reduce or control human error in safety-critical contexts.

2.2   HUMAN ERROR AND MAJOR ACCIDENTS

2.2.1   Recent Disasters Involving Human Error

The human being has a considerable influence on the safety and reliability of safety-critical systems. As examples of this, a selection of major accidents in which human error has played a significant role, in a variety of industrial sectors, are outlined below.

(a)   In the nuclear industry

Three Mile Island (1979) — This classic example of incomplete and misleading feedback to operators occurred in the instrumentation of the power-operated relief valve. The indicator light in the control room illuminated when a signal was sent to the solenoid to cause it to close the valve. The light did not monitor the actual position of the valve, and the operator's interpretation depended on his remembering that the light might not necessarily mean the valve had closed. In the incident, the failure of the valve mechanism was not recognised for some time.

Chernobyl (April 1986) — Chernobyl's 1000 MW Reactor No. 4 exploded, releasing radioactivity over much of Europe. The death toll and environmental effects are still mounting. Although much debated since the accident, a Soviet investigative team admitted 'deliberate, systematic and numerous violations' of safety procedures by operators.

(b)   In the off-shore and petrochemical industries

Union Carbide, Bhopal (December 1984) — The plant released a cloud of toxic methyl isocyanate. The death toll was 2500 and over a quarter of the city's population was affected by the gas. The leak was caused by a discharge of water into a storage tank. This resulted from a combination of operator error, poor maintenance, failed safety systems and poor safety management.

Piper Alpha (July 1988) — One hundred and sixty-seven people died in the North Sea when a major explosion, resulting in a fire, occurred on Occidental's platform. Lord Cullen's inquiry [Cullen 90] revealed a variety of technical and organisational causes of the accident. Reviewing the evidence, Pate-Cornell [Pate-Cornell 93] concluded that the precursors to the accidents were 'rooted in the culture, structure and procedures of Occidental Petroleum... The maintenance error that eventually led to the initial leak was the result of inexperience, poor maintenance procedures, and deficient learning mechanisms.'

(c) In the transport sector

Space Shuttle Challenger (January 1986) — An explosion shortly after lift-off killed all the astronauts on board. An 'O ring' seal on one of the solid rocket boosters split after take-off, releasing a jet of ignited fuel. The history of how that type of defective O ring was in the rocket is a complicated story of organisational mindset, conflicting scheduling of safety goals, and the effects of fatigue on decision making.

Herald of Free Enterprise Ferry (March 1987) — The roll-on-roll-of ferry sank in shallow water off Zeebrugge, Belgium, killing 189 passengers and crew. The inquiry [Sheen 87] highlighted the commercial pressures in the competitive ferry business and the friction between ship and shore management that led to safety lessons not being heeded. The inquiry also reported that the Townsend Thorensen company 'was infected with the disease of sloppiness'.

King's Cross Underground Fire (November 1987) — A fire at the King's Cross Underground Station in London killed 31 people. The Fennell inquiry [Fennell 88] concluded that London Underground's senior management had 'narrow horizons and a dangerous, blinkered self-sufficiency'.

Clapham Junction (December 1988) — Thirty-five people died and 500 were injured in a triple train crash near Clapham Junction Station, London. The immediate cause of the accident was a wrong-side signal failure caused by a technician failing to remove a wire. The Hidden inquiry [Hidden 89] highlighted 'bad workmanship, poor supervision and poor management'. It was pointed out that 'the appearance of a proper regard for safety was not the reality'.

Kegworth M1 Air Crash (January 1989) —British Midland Flight BD 092 crash-landed on the M1 motorway close to Kegworth village in Leicestershire. A total of 47 people died and many others were seriously injured. The pilots probably misinterpreted their instruments and shut down the wrong engine.

2.2.2 Accident Costs and Causes

It is clear from the above examples that humans can cause or contribute to incidents and major accidents, resulting in loss of life or injury, loss of plant and production, and environmental damage and pollution. Such events are expensive in human and business costs and the reputations of the companies concerned [HSE 1993].

Human failures are typically cited as causal factors in between 30% and 80% of accidents — see, for example, [Glendon and McKenna 95]. The level cited varies between industrial sectors. For example, the shipping industry recently named human failure as a major cause of over 80% of accidents

[Donaldson 94]. This level of citation is perhaps not surprising as the human is involved in all stages of the life cycle of a plant, from design and construction through to operations, management, maintenance and, ultimately, decommissioning. There are indications that in certain industries, such as petrochemicals, the level of accidents attributed to human error is increasing as hardware reliability improves.

2.2.3   Human Involvement in Major Incidents

Humans can cause or contribute to incidents in four ways, not all of them with negative consequences.

(a)   Initiating event

An accident may be the result of a direct or an indirect failure. A direct failure has an immediate consequence. For example, the Kegworth air crash in January 1989 almost certainly occurred because the pilot shut down the wrong engine — with immediate effect. On the other hand, the impact of an indirect failure may not be felt for some time and may be difficult to trace. For example, in August 1989 on London's River Thames, the *Marchioness* pleasure vessel was hit by the dredger *Bow Belle* because there had been a failure to maintain a proper look-out on the dredger.

(b)   Escalation or control

Humans can make disastrous decisions even when they are aware of the danger. Remember the captain of the *Titanic* who carried on even when he knew icebergs were present? Humans can also misinterpret a situation and act in an inappropriate way. For example, there was a fatal crush at a football match in Hillsborough, Sheffield, in 1989. The evidence indicates that the police officer in charge believed he was coping with rioting fans and not with people struggling for their lives. His actions in not opening the gates to relieve the crush further exacerbated the disaster.

Fortunately, humans can also intervene to mitigate potential disasters. Incidents of this type are not well reported, but many companies have their own anecdotes based on human recovery of high-potential incidents. For example, in a loss-of-coolant accident in the nuclear industry, the operators showed lateral thinking by using a hosepipe to introduce water into the loop when all other means had failed. This on-the-spot quick thinking was later formalised as an alternative emergency procedure.

(c)   Emergency evacuation, escape and rescue

The degree of loss of life can be reduced by the effective emergency response of operators and crew. Emergency planning and response, including appropriate training (see Chapter 7 for a discussion of this), can have a significant impact on rescue efforts. Poor response can lead to delays and further loss of life, while effective response and even acts of heroism can save many lives. Emergencies off-shore and at sea are particularly relevant examples of this. Rapid rescue of people in the water can significantly reduce the final death toll of sea incidents.

(d)   Latent failure

As we shall see later in this chapter, management, particularly senior management, can lay the foundations for accidents through the decisions they take, either because the decisions have the potential to cause failure and, perhaps, a breach of safety, or because they create the conditions in which more junior staff are likely to fail or commit violations (see Chapter 11). Management decisions (which may not only relate to safety matters), if flawed, are usually termed 'latent failures'. Safety management systems (see Chapters 9 and 12) are the means by which such failures are commonly monitored and managed. In the King's Cross, Clapham and Zeebrugge disasters, management failures became the focus of the follow-up investigations, and since then much research effort has been directed into the impact of organisational factors on safety.

2.3   WHAT CAUSES HUMAN ERRORS?

In this section we take a brief look at various paradigms for understanding and reducing human error. Key concepts and issues that are the current focus for behavioural scientists are presented. As this review is only an introduction to the topic, interested readers are encouraged to look further at the texts which are referred to in the chapter.

2.3.1   Human Error Causation Paradigms

Throughout Western history, philosophers and scientists have questioned why humans err. However, the psychological study of human error, with which this chapter is mainly concerned, has only emerged during this century. In recent years interest in the causes of human error in large-scale industrial systems has been fuelled by the study of major accidents such as those outlined earlier.

Looking at the different perspectives on the role of human failures in accident causation, various views have fallen in and out of fashion. From the history of recent practical and scientific thinking on human error, a number of *error paradigms* can be distinguished. (Here a paradigm is defined as a collection of shared concepts, perceptions and practices that forms a particular view of reality, and which guides understanding, collective actions and research — see [Kuhn 62] for further information on paradigms in scientific thinking.) This century, there have been four error paradigms, as follows:

(i)   *The Engineering Error Paradigm* — which focuses on the technical aspects of a system. Human factors that are targeted by this approach typically include the design of the man-machine and human-computer interfaces and issues surrounding plant automation.

(ii)   *The Individual Error Paradigm* — which considers both motivational and personality issues as well as accident proneness of individuals.

(iii)   *The Cognitive Error Paradigm* —which has focused on the psychological or information processing causes of human errors. This error paradigm covers both skill and decision-making errors and considers the capabilities and limitations of the individual human information processing system.

(iv)   *The Organisational Error Paradigm* — with its focus on management decision making, safety management and issues such as safety culture, participation, competence, control and communication.

The shifts between error paradigms have tended to relate to alternative regulatory (and perhaps legal) views of accident causation and prevention. In particular, during the 1980s there was a discernible move away from blaming accidents on individuals and their errors and towards emphasising management decisions and actions. Through the 'organisational disasters' such as King's Cross, Zeebrugge and Clapham Junction, management decisions and management liability issues became prominent.

In conjunction with the shift towards corporate responsibility for safety (and the possibility of 'corporate manslaughter' charges on the occurrence of a serious accident), there has been a move from prescriptive safety legislation towards the encouragement of self-regulation. This change encourages companies to promote issues such as empowerment, worker participation in safety matters, and a positive safety culture.

We will now look at some of these error paradigms in more detail.

2.3.2   The Engineering Error Paradigm

The main characteristic of the Engineering Error Paradigm is the recognition that the human is typically an unreliable 'component' of the total system. The paradigm views humans as though they are almost equivalent to hardware components. The extreme approach proposed for 'engineering out' human failures is to remove the 'human factor' entirely through automation. An alternative, more widespread and accepted, view is to improve human reliability through better ergonomic design of the workplace or the man-machine interface (see Chapter 6 for a consideration of such design issues).

(a)   Automation

Designing the man 'out of the loop' through automation is often considered as a proposal for reducing risk. For example, an unmanned location, particularly at a hazardous installation, would obviously have a reduced risk of personnel fatalities. However, automation may not always remove human error and indeed may create new sources of human unreliability. In her description of the 'ironies of automation' in a now classic paper, Bainbridge included the following [Bainbridge 87]:

- The designer has more opportunity to introduce errors into the system during the design process;
- The operator still has to do those tasks which the designer has not been able to automate;
- The operator's role is changed to one of the supervisory monitoring of an automated system, but with reduced information about the system itself (humans are generally poor at monitoring tasks that require extended periods of vigilance);
- The operator may become de-skilled and yet be expected to intervene when the automated system fails.

More recent research has looked at the 'glass cockpit syndrome' where an operator may find difficulty in understanding the internal functioning of a nearly fully automated system. In the 1988 incident in which the *USS Vincennes* erroneously shot down a civilian Airbus carrying 290 passengers from Southern Iran to Abu Dhabi, this syndrome was evident. The *Vincennes* had been fitted with a very sophisticated Tactical Information Co-ordinator (TIC) which gave sharp warnings that there was a hostile aircraft in close vicinity to the US warship. The captain also received a warning that the aircraft may be commercial, but being under considerable time pressure and considering the safety of his crew, he decided to accept the TIC's interpretation. In another US warship equipped with a less sophisticated system, the officers and crew felt more able to evaluate alternative scenarios. They realised that the aircraft was indeed a civilian airliner. The formal inquiry found the TIC technology to be a weak link in the chain of decision making and information processing. (Chapter 9 offers a more detailed discussion of this incident.)

(b)   Human-machine interface design

Here the view is that human errors often occur as a result of a human-machine mismatch between the demands of the task, the physical capabilities of the person, and the characteristics of the interface provided to enable the person to carry out the task. This model concentrates on the individual and his or her immediate work situation. Typically the focus is a single operator in a control room. This view promotes recognised solutions to human reliability, that is, ergonomic design changes, improved training, and the provision of job aids such as procedures (see Chapter 7 for a discussion of training and job aids). There is a clear focus on performance-influencing (or -shaping) factors and on their control. Proponents of this view of human error stress the need to consider human performance, including human reliability, early in the design process. Other chapters in this volume develop this argument in more detail. See, for example, Chapter 6 for further information on design issues in HCI and Chapter 8 for information on design in abnormal situations.

2.3.3 The Individual Error Paradigm

In the individual error paradigm, the focus is on a person's motivation and safety attributions. This paradigm has moved in and out of fashion this century, but is usually linked with the traditional safety model.

(a) The traditional safety model

The traditional view of human error is closely linked to the legal concept of negligence and blame. The fundamental belief is that errors are caused by a person 'not trying hard enough' or 'not paying sufficient attention' to the task. When an accident involving a human error occurs, an investigator with this view typically questions the motivation of the worker to carry out the system of work safely.

This view is related to the idea that the majority of accidents are caused by 'unsafe acts' as opposed to 'unsafe conditions' — see [Heinrich 50, Peterson 75]. A closely related view is the discredited concept of 'accident proneness' (see, for example, [Hale and Hale 70, Sass 87]).

Traditional 'solutions' to human error are often limited to disciplinary measures such as dismissal and suspension, but these can at best be partially successful in this case as they are suitable only for errors which have been intentionally committed.

(b) Risk perception, violations and safety culture

Recently research interest has been renewed in the individual error paradigm in risk perception and behaviour-based accident prevention methods.

Behaviour modification techniques, 'safety blitzes' and safety reward schemes, are increasingly used in industry to reduce the unsafe behaviours which are the precursors of accidents (especially those involving personal injury). Current research is also looking in more depth at perceptions of risk and safety-climate or safety-culture issues (see, for example, [ACSNI 93, Royal Soc 92]). We can expect to be presented with more insights and guidance in this area in the future.

A focus for researchers has been violations of rules or procedures, particularly non-malevolent violations such as not wearing personal protection equipment and habitual corner-cutting (see Chapter 11). These form one type of human failure that can be attributed to motivational causes. Factors which influence such behaviour include beliefs and attitudes (safety culture), risk-taking behaviour, and situational factors such as the conflict of productivity and safety goals. A taxonomy of different types of violations which emphasises the organisational pressures on individual workers is now emerging (see [Reason et al 95]).

2.3.4 The Cognitive Error Paradigm

The cognitive perspective analyses human errors in relation to the information-processing abilities, human error tendencies, and interaction with tasks and situations of the individual. Human errors are assumed to occur as a result of mismatches between the capabilities, physical and mental, of people and the demands of the job. Error reduction is considered to be dependent on establishing the cognitive 'root causes' of the error, for example, memory failure, attentional failure, information overload, and decision-making failure.

This view received much attention in the late 1970s and early 1980s, with a focus on everyday slips of action and memory lapses, so-called 'absentminded' behaviour (see [Norman 81, Reason and Mycielska 82, Reason and Lucas 84]). The researchers applied the findings and theories developed with data from non-industrial settings to safety-critical industries. The focus then turned towards theories of mistaken intentions and decision-making failures.

Key concepts and definitions of this error paradigm are presented in Section 2.4.

2.3.5 The Organisational Error Paradigm

From the middle of the 1980s, psychologists began to adopt a broader perspective, looking higher than the individual operator and up the organisational structure at the management and policy makers. While the catalyst for this change of error paradigm was the spate of major accidents around that time, the line of argument followed logically from the previous studies of action slips and memory lapses in safety-critical industries.

The idea emerged that human errors are caused by certain preconditions in the work context. These include such aspects as poorly designed procedures, unclear allocation of responsibilities, lack of knowledge or training, low morale, poor equipment design, time pressure, and so on. One key aspect, as illustrated in Figure 2.1 — from [Wagenaar et al 93] — is that such preconditions can be traced back to management and organisational policies and decisions. Figure 2.1 shows that the unsafe acts (human errors and procedure violations) which breach the system defences and barriers (shown as wavy lines in the diagram) and lead to accidents and incidents have their origins in management decisions. These decisions are fed through to general failure types such as poor maintenance planning, communication failures, and inappropriate workload levels, which are some of the immediate causes of unsafe acts.

Reason argues that organisational and management decision-making failures are 'organisational pathogens' and may contribute indirectly to accidents through latent failures [Reason 90]. Such pathogens are more likely to be present in complex, interactive, tightly coupled, opaque systems [Perrow 84]. Improving human reliability is a question of searching out such latent pathogens and correcting them. There are obvious parallels between this form of pathogen audit and safety management system audits.

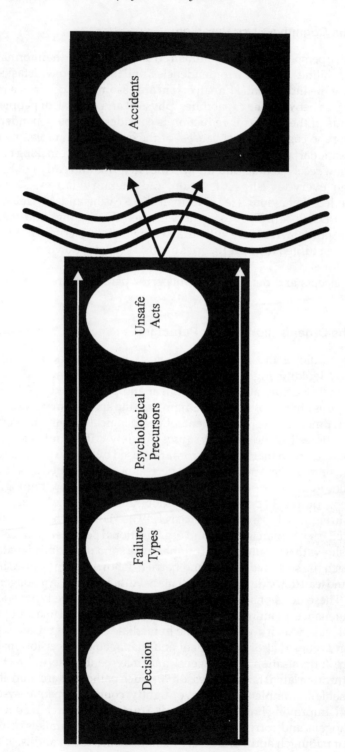

Figure 2.1 General accident scenario

2.4   SOME KEY CONCEPTS AND DISTINCTIONS

The aim of this section is briefly to introduce the key ideas that are often referred to by researchers and consultants looking at human reliability issues. The focus is on the cognitive error paradigm.

2.4.1   Working Definition of Human Error

Senders and Moray define 'error' in the following way [Senders and Moray 91, p. 25]: For most people error means that something has been done which:

- Was not intended by the actor;
- Was not desired by a set of rules or an external observer; or
- Led the task or system outside its acceptable limits.

The key elements of this definition are intention, deviation from some accepted standard, and undesirable consequences of the action or decision.

2.4.2   Latent and Active Failures

A distinction is usually made between active and latent human failures.

Active failures are usually made by 'front-line' operators and have an immediate effect. For example, by deviating from standard operating procedures humans can cause safety-critical systems to fail or increase their failure probability.

Latent failures are made by construction workers, maintenance staff, inspectors, plant management, and high-level decision makers. They can have adverse effects that may lie dormant for a long time, only becoming apparent when they combine with other factors to breach the system's defences (see Figure 2.1). Latent failures can create dependencies between redundant systems or even between physically unrelated systems. They can also have an impact on human reliability.

2.4.3   Human Recovery

Humans can intervene in an accident sequence to mitigate the effects of system failure. This can either be in a reactive mode (e.g., an operator finds a leak and takes action by switching to a standby system) or in a proactive mode through regular monitoring or checking.

2.4.4   Automatic or Conscious Performance

Humans need to respond to different situations, including routine, abnormal and emergency situations. We now know that according to the nature and familiarity of a task, our brains process the information required in different ways. The different modes of thinking predispose us to various causes of error. The two fundamental modes in which we interact with the world are 'conscious' and 'automatic'. Table 2.1 shows the characteristics of these two modes of thought. Conscious thought is used in novel situations and involves considerable conscious effort and attention. It is slow and is often seen as involving learning. The causes of error which occur with conscious thought are those of overload of information processing, lack of knowledge, and lack of awareness of the consequences of actions.

On the other hand, the automatic mode of thought is the hallmark of a skilled performer. It occurs with routine tasks in familiar environments. Skilled performance is fast and effortless, requiring little feedback and minimal conscious attention. The error causes of such automatic performance include the intrusion of stronger, more frequently used action sequences (e.g., pushing the more frequently used button 'A' when you actually intended to push button 'B' on this one occasion).

These two extremes of control have been related by Rasmussen to three different types of performance: skill-based, rule-based, and knowledge-based [Rasmussen 83]. These are three levels of cognitive control that depend on the degree of familiarity with the environment and the nature of the information used.

The skill-based level represents sensori-motor performance during activities which take place without conscious control. They are smooth, automated and highly integrated patterns of behaviour. The types of errors that can occur here are the slips and lapses of the skilled person performing a familiar task. Think about the types of errors you typically make when driving (assuming you are an experienced driver), which are losing-place errors (omitting or repeating a step in a familiar sequence due to a memory lapse typically caused by distraction or preoccupation), and habit intrusions.

Rule-based behaviour typically occurs in a familiar situation and involves stored rules or procedures. Rules are of the form 'if <situation> then <diagnosis>' or 'if <situation> then <action>'. The general types of errors that occur at the rule-based level involve the tendency to use familiar

Table 2.1 Modes of interacting with the world

Conscious	Automatic
Unskilled or occasional user	Skilled, regular user
Novel environment	Familiar environment
Slow	Fast
Effortful	Effortless
Reuires considerable feedback	Requires little feedback
Causes of error	**Causes of error**
Overload	Strong habit intrusions
Lack of knowledge of modes of use	Manual variability
Lack of awareness of consequences	Frequently used rules used inappropriately

solutions even when these are not the most convenient or efficient.

Knowledge-based behaviour occurs in unfamiliar circumstances. Here it is necessary consciously to formulate a goal and then develop a plan to achieve that goal. Planning or problem solving typically involves reasoning from first principles or using analogies rather than employing existing rules for diagnosis or action. Mistakes occurring at the knowledge-based level generally consist of misdiagnoses and miscalculations.

2.4.5   Slips and Mistakes

A distinction is made between slips and mistakes as follows:

- Slips and lapses are actions which deviate from the intended plan — i.e., plan satisfactory, execution of plan faulty;
- Mistakes occur when a plan deviates from some adequate path towards a desired goal — i.e., plan faulty, execution of plan satisfactory.

Action slips and memory lapses together constitute 'absentmindedness' and are unintentional errors. Such errors can increase in frequency when an individual is under stress (see [Reason 88]) and it appears that some people are generally more absentminded than others [Broadbent et al 82].

2.4.6   Errors and Violations

Human failures may be divided into two broad categories: errors and violations. Errors are defined as the failure of planned actions to achieve their desired goals. They are unintentional failures, that is, the slips, lapses and mistakes referred to earlier. Violations are generally intentional infringements of safe working practices (see Chapter 11 for a thorough discussion of violations).

2.4.7   Classification of Human Error

Using these insights into human cognitive behaviour, researchers have been able to develop a classification of unsafe acts, as in Table 2.2, which is based on [Reason 90].

For each group of human failures, different error-reduction strategies are appropriate. For example, exhorting a person to 'try harder' may prove moderately effective for routine violation prevention but will not reduce unintentional slips. Redesigning the man-machine interface to prevent inadvertent operation of the wrong switch will reduce slips but will not prevent exceptional violations.

It is interesting to relate the immediate human causes of some of the major accidents involving human error to this classification (see Table 2.3).

Of course, this is only one classification of error and there are many others that have been devised for specific theoretical and practical purposes. Most classifications involve several levels of analysis (for example, *what*

Table 2.2 *Classification of unsafe acts*

Unintentional human errors
a)   Slips and lapses due to failures in
     skill-based performance
b)   Mistakes due to failures in rule- or
     knowledge-based performance

Intentional human violations
a)   Routine violations
b)   Exceptional violations

Table 2.3 *Illustrating the classification of unsafe acts*

Unintentional error	
Skill-based action slips and memory lapses	M1 air crash Clapham Junction train crash
Rule- or knowledge-based mistaken decisions	Challenger USS Vincennes Hillsborough
Intentional violations Routine violations	King's Cross fire Herald of Free Enterprise
Exceptional violations	Chernobyl

happened, *how* did it happen, and *why* did it happen). All such classifications are ways of analysing the scope and nature of human error. Senders and Moray list three broad categories of taxonomies of human error as follows [Senders and Moray 91]:

(a)   Phenomenological taxonomies which describe errors superficially in terms that refer to observable events. Typical categories include omissions, substitutions and repetitions. This type of classification is used to establish *what* happened, and an example is given in Table 2.4. Classifications of this form are widely used in human reliability assessments.

(b)   Cognitive mechanism taxonomies which classify errors according to the stages of human information processing at which they occur. For example, perceptual errors, memory lapses and attentional errors. Table 2.5 gives an illustrative example using part of the framework of [Rasmussen et al 81]. Such taxonomies classify errors at the *how* level and are increasingly used in post-accident investigations.

Table 2.4 *Example of a classification at the 'what' level*

Action or check made too early or too late
Action or check omitted or partially omitted
Action too much or too little
Action too long or too short
Action in wrong direction
Right action or check on wrong object
Wrong action or check on right object
Information not obtained
Wrong information obtained
Misalignment

(c)   Taxonomies for biases or deep-rooted tendencies that errors are thought to reveal. Examples of such biases include the confirmation, availability and recency biases (see Table 2.6 — and see also [Reason 90] for a discussion of these and other underlying biases). Such taxonomies are used to classify errors at the *why* level. Currently these tend to be research tools.

Table 2.5 *Example of a classification at the 'how' level*

Discrimination
Stereotype fixation
Familiar short-cut
Stereotype takeover
Familiar pattern not recognised

Input information processing
Information not received
Misinterpretation
Assumption

Recall
Forget isolated act
Mistake alternatives
Other slip of memory

Inference
Condition or side effect not considered

Physical coordination
Motor variability
Spatial disorientation

Table 2.6 Classifying at the 'why' level (performance error-shaping factors)

Skill-based	Recency and frequency of use
	Environmental control signals
	Shared schema properties
	Concurrent plans
Rule-based	Mindset
	Availability
	Matching bias
	Overconfidence
	Oversimplification
Knowledge-based	Selectivity
	Working memory overload
	Out-of-sight-out-of-mind
	Thematic vagabonding and encystment
	Memory cueing and reasoning by analogy
	Matching bias
	Incomplete or incorrect mental model

2.4.8   Types and Tokens: Mapping Individual Errors onto Management Failures

One recent distinction that is likely to receive further attention in the next few years is that between so-called 'types' and 'tokens'. Types are defined as 'general classes of organisational and managerial failures', while tokens are 'more specific failures relating to individuals at the human-system interface' [Reason 93]. The distinction between them has emerged out of research on human error and management failures in the petrochemical industry. However, the distinction has far wider application than the chemical processing sector. The models that emerge from this notion will be important because they represent a serious attempt to map individual human errors onto general management and organisational failures. Currently fashionable ideas such as organisational culture, management decision-making failures, maintenance failures and individual violations, are all found in this research.

Organisational and management factors (types) include such problems [Reason 93] as:

- Hardware defects;
- Design failures;
- Poor maintenance procedures;
- Poor operating procedures;
- Error-enforcing conditions;
- Poor housekeeping;
- System goals incompatible with safety;
- Organisational failures;

- Communication failures;
- Inadequate training;
- Inadequate defences.

Types and tokens are subdivided as follows.

Types can be source types (associated with the decision-making failures of policy makers) or function types (line management decision-making failures).

Tokens are divided into condition tokens (psychological and situational precursors of unsafe acts) and unsafe acts (slips and lapses, mistakes and violations). Condition tokens cover performance-influencing factors such as the man-machine interface and the immediate work environment, social factors such as climate, group norms and attitudes, and cognitive or information-processing factors such as attentional capacity, memory load and knowledge.

A summary of these concepts is given in Figure 2.2. A more detailed listing of performance-influencing factors (which are sometimes referred to as performance-shaping factors) is provided in Figure 2.3 and in Table 2.7. (The list of performance-influencing factors is the author's own and is drawn from a wide variety of human factors sources.)

Table 2.7 *Typical performance-influencing factors*

Task demands and characteristics
Frequency, workload, critical nature, duration, interaction with other tasks, perceptual, physical, memory, attention, vigilance

Instructions and procedures
Accuracy, sufficiency, clarity, level of detail, meaning, readability, ease of use, applicability, format, selection and location, revision

Environment
Temperature, humidity, noise, vibration, lighting, work space, movement restriction, control of environment

Displays and controls
Compatibility, ease of operation, reliability, feedback, sufficiency, location, readability, identification, distinctiveness

Stresses
Time pressure, workload, fatigue, high-risk environment, monotony, isolation, distractions, shift work incentives

Individual
Capacities, training and experience, skills and knowledge, personality, physical condition, attitudes, motivation, risk perception

Socio-technical
Manning, work hours and breaks, resource availability, social pressures, conflicts, team structure, communication, roles and responsibilities, rewards and benefits, attitude to safety

---

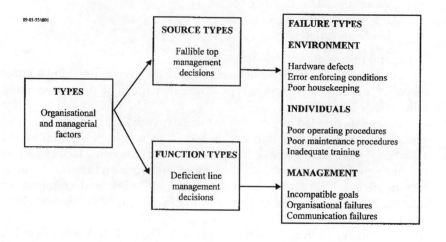

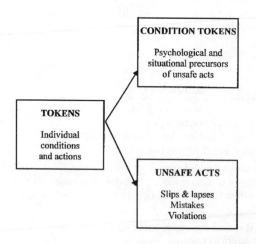

Figure 2.2 *Types, tokens and failure types*

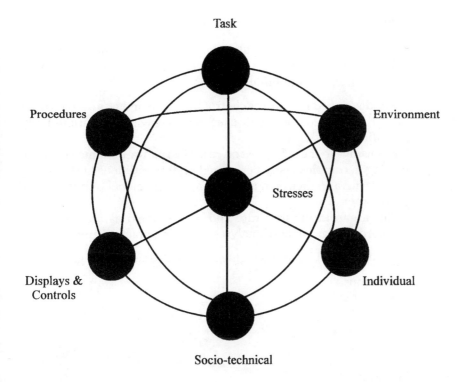

Task

Procedures

Environment

Stresses

Displays &
Controls

Individual

Socio-technical

Figure 2.3 *Performance-influencing factors*

2.5   COGNITIVE MODELS OF HUMAN ERROR

There are many cognitive models of human error. In this chapter we only present two very briefly. These are framework models of human error, so-called because they offer a broad view of the possible information-processing mechanisms underlying human error. Framework models are global rather than focused and have breadth of coverage rather than a depth of explanation for one information-processing function. This means that they are virtually impossible to disprove by conventional hypothesis generation and testing. However, their breadth enables practical insights to be developed and also, perhaps an undervalued commodity, it facilitates interdisciplinary communication on human error.

The two classic and now widely cited framework models of human error which are briefly outlined here are the 'step-ladder' model [Rasmussen 83] and the Generic Error Modelling System (or GEMS) proposed by Reason [Reason 87, Reason 90]

2.5.1   Rasmussen's 'Step-ladder' Decision Model

This model is the basis of the skill, rule, knowledge distinction already described in Section 2.4. It is important for several reasons, which were summarised by [Reason and Embrey 85]:

- It was developed for process control situations and was derived from the verbal protocols of electrical technicians engaged in various fault-identification tasks;
- It is sensitive to the reflexive and recursive nature of human decision making in complex tasks, performed under conditions of stress;
- It is a general rather than a situation-specific model of information processing;
- It fills a gap in the literature and was adopted as the theoretical standard for a number of analyses involving practical applications.

The step-ladder model identifies eight stages of decision making:

- Activation;
- Observation;
- Identification;
- Interpretation;
- Evaluation;
- Goal selection;
- Procedure selection;
- Procedure execution.

These stages are not represented in a linear fashion but in the form shown in Figure 2.4, which is a simplified representation of Rasmussen's original diagram. From this figure it can be seen that Rasmussen has focused on the short cuts that human decision makers take in real-life situations. Such short cuts are represented as transverse links between stages of decision making. Reason and Embrey point out that the most common error patterns involve using rule-based routines when knowledge-based operations are demanded by the novelty of the situation [Reason and Embrey 85]. In the diagram this is shown by a possible linkage between observation of the system state and automatic (skill-based) selection of remedial procedures.

Although superficially this is a simple model it has been applied with some success by a number of researchers, one application involving the modelling of critical decision making in nuclear power plant control rooms under emergency conditions [Pew et al 81]. In this study, the step-ladder model was also used to develop a technique for establishing priorities among proposed control room improvements.

2.5.2   The Generic Error Modelling System

The Generic Error Modelling System (GEMS) [Reason 87] was aimed at providing a context-free framework to be used to analyse the more predictable

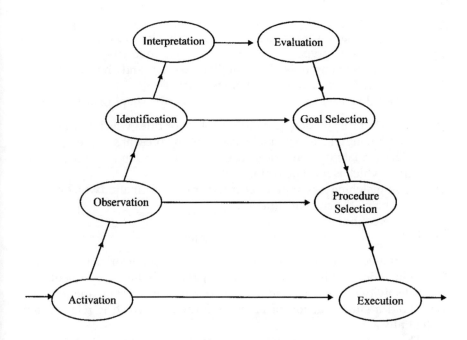

Figure 2.4 *Simplified 'ladder' model*

types of human error. GEMS is based on Rasmussen's skill, rule, knowledge taxonomy. However, it also includes many aspects of cognition derived from other theories of human information processing. In particular, GEMS builds on the theories of Norman and Reason and Mycielska on the processes underlying slips and lapses [Norman 81, Reason and Mycielska 82]. It also includes elements of problem-solving theories and incorporates the research on heuristics and biases which have an impact on the processing of rule-based information (e.g., [Tversky and Kahneman 74]).

Reason distinguishes between the three basic error types of skill-based slips, rule-based mistakes and knowledge-based mistakes using five dimensions: activity, control, focus of attention, forms, and detection. Figure 2.5 shows a summary of his criteria. By combining these distinctions with known psychological mechanisms underlying error, it was possible for Reason to identify error-shaping factors at each level of performance. Table 2.6 has already summarised these factors.

At the skill-based level there are four main error-shaping factors which are taken from the research into slips and lapses. These cover the recency and frequency of use of schemata (postulated high-level knowledge structures), the presence of environmental control signals which may trigger schemata, the overlap between schemata, and the holding of more than one plan or intention at once. The effect of these error-shaping factors is to produce a number of familiar error forms such as strong habit intrusions (sometimes referred to as stereotype takeovers), omissions following interruptions, perceptual confusions (mistaking similar objects), and the forgetting of intentions (the 'what have I come here for?' lapse).

At the rule-based level, the five identified error-shaping factors are:

- Mindset (the tendency to retrieve plans or rules which have worked before but which may not be useful in the present situation);
- The availability heuristic (retrieval of information is biased towards that which is most activated);
- The matching bias (the tendency to retrieve information which appears to match the current situation);
- Overconfidence;
- The tendency to oversimplify a problem.

At the knowledge-based level there are a number of error-shaping factors which include selectivity and thematic vagabonding and encystment.

Selectivity implies that the working database or working memory will only hold a small fraction of all the information which is relevant and potentially available.

Thematic vagabonding and encystment refer to two cognitive strategies observed during emergency planning and decision making. In thematic vagabonding, people tend to flit between issues quickly, treating each rather superficially. In encystment, certain topics are dwelt on to excess while other important topics are ignored.

Error tendencies at the knowledge-based level generally have their

Factors	Skill-based slips	Rule-based mistakes	Knowledge-based mistakes
Activity	Routine actions	Problem solving	
Control	Mainly automatic processors (Schemata)	(rules)	Resource-limited conscious processes
Focus of attention	On something other than present task	Directed at problem-related issues	
Forms	Largely predictable 'strong-but-wrong' error forms (actions)	(rules)	Variable
Detection	Usually fairly rapid	Difficult, and often only achieved with help from others	

Figure 2.5 *The three basic error types*

origins in the limitations of the cognitive system. These include the limited capacity of working memory and attention and the difficulty of retrieving relevant information from long-term memory. Simon expresses it succinctly: 'The human information processor is always struggling with the limits of his own processing and storage capabilities in the face of a wealth of information to be processed and stored' [Simon 72].

In the GEMS model, cognitive operations are divided into two types, those which precede the actor's detection of the existence of a problem (skill-based) and those which follow such detection (rule- and knowledge-based). The slips that occur at the skill-based level are assumed to be primarily due to monitoring failures while Reason groups mistakes under the general heading of problem-solving failures.

The major advantage of the GEMS model is that it allows the cognitive processes that operate at each of the three levels of performance to be specified. The failure modes which various processes may lead to can also be identified to allow prediction of the types of errors which are likely to arise. This specification of the processes of human error means that we are able to speculate about why a failure arose as opposed only to identifying what error has occurred.

The GEMS model continues to be developed. The interested reader is referred to [Reason 90] for a fuller description of it.

2.6   PRACTICAL USE OF MODELS OF ERROR CAUSATION

2.6.1   Overview

This chapter has summarised the major insights of the past decade or so into the mechanisms underlying human failures. It has concentrated on the cognitive errors which can lead not only to everyday, usually inconsequential, slips and lapses, but also to significant incidents in safety-critical industries. Have these research findings and the models and classifications of human error paid off? Are they used in industry to reduce, control or eliminate opportunities for human error?

As yet the answer to these questions must be 'not entirely'. Over the past decade there has been increasing interest in the application of theoretical ideas concerning human error. There are some examples of accident and near-miss investigation systems which apply ideas about the causes of human error (see Section 2.6.2). Indeed, a model for such investigations is described in detail in Chapter 10. In addition, awareness of the different types and causes of human error has increased in many industrial sectors. However, penetration of these ideas is slow and time consuming, and more effort is required to convert them into usable tools with clear and quantifiable benefits.

In the remainder of this chapter we look at some of the main areas where knowledge of the causes of human error can be applied in the life cycle of a system. There are four relevant channels:

- During design;
- During post accident and incident investigation;
- During human reliability assessment;
- During human factors auditing.

The influence of human error modelling on design, particularly human-computer interface design, is increasing. Since design issues are covered in detail in other chapters of this book, principally in Chapters 1, 6 and 8, this issue is not further dealt with here.

Human factors auditing is a novel area which offers scope for reduction of human errors in an existing plant. Such audits should be built on both practical and theoretical insights into the causes of human error and the factors that increase their likelihood (see [Hudson et al 94] for one relevant example). In the remainder of this chapter we consider in more depth only accident investigation and human reliability assessment.

2.6.2   Accident and Incident Investigation

Serious incidents have to be investigated to satisfy regulatory and insurance requirements. However, the main philosophy behind accident and incident investigation is to learn the lessons from past incidents in order to implement effective error-reduction strategies to guard against future failure. Chapter 10 discusses the approaches that can be taken to learning from incidents. With the predominance of human error as a cited cause of incidents and accidents, such investigations need to take account of what we know of the origins of human error. This issue is addressed through 'root cause analysis'.

'Root cause analysis' is a concept that is open to differing interpretations and subject to fads and fashions. What is clear is that by understanding root causes one should be able to devise effective remedial actions. For root cause analysis to be effective for human error, it must consider a range of factors, including individual, task, team, supervisory, management, and policy influences. It should identify performance-influencing factors (see Figure 2.3) and it may need to search for the underlying psychological causes. The analysis of root causes provides the insights needed to develop effective error-control strategies.

In practice, there are two approaches to human error root cause analysis. The first, which can be referred to as 'Leave it to the investigator', relies on the accumulated experience and knowledge of the investigator. Little guidance is given and few analytical techniques are provided. This approach appears to be quite common and results in incident reports of variable quality. Its main advantage is that it avoids standard solutions and the need for training in specific techniques. The major disadvantage is that few root causes tend to be identified and these may not result in the development of appropriate error-control strategies.

The second approach is to provide a tool box of techniques for incident investigators. This method standardises the quality of incident reports and potentially may increase the number of root causes found and human error reduction strategies developed for each incident. The tools need to be easy to use and supported by a training programme. They may need to be tailored to a specific industrial context. Examples of human error root cause analysis methods include:

- MORT (Management Oversight Risk Tree) [Johnson 73];
- Root Cause Tree [Paradies et al 93];
- TRIPOD [Shell 93];
- Tailored classifications.

MORT and the Root Cause Tree are both generic tools. The first is the basis of the technique described in Chapter 10, and the Root Cause Tree is described in the recently published 'Human Performance Investigation Process' [Paradies et al 93]. Both methods originate in the nuclear industry. TRIPOD is an approach to safety based on a model of human error that is used in one petrochemical company to investigate accidents and to monitor the potential for future human-related accidents. Many organisations have their own tailored classification of causes that reflect the specific context of their industries (see [Taylor and Lucas 91] for an example from the railway industry). Increasingly such classifications may include insights from human error models.

2.6.3 Human Reliability Assessment

Human reliability assessment (HRA) is a practical set of tools for assessing the risks attributable to human error in high-risk industries. The method originated in the nuclear industry and received substantial impetus following the Three Mile Island accident in 1979. It is also increasingly used in the military, petrochemical (especially for off-shore installations) and transport sectors. A human reliability assessment may be incorporated into a safety case in the nuclear, off-shore and transport sectors as part of a quantitative risk assessment (QRA).

HRA consists of three main stages:

- Human error identification (HEI) — identifying what errors can occur;
- Human error quantification (HEQ) — determining how likely the errors are;
- Human error reduction (HER) — improving human reliability by reducing the error likelihood.

A thorough description of the HRA process and the tools and techniques commonly used in it are given in [Kirwan 94] (but see also [Embrey 94]), and further details of human error quantification are given in Chapter 3 of this volume.

It is in the human error identification stage that insights from research into human error causation can be incorporated. Originally HEI took a secondary role to HEQ, and analysts would rely on their own judgement or a classification of observable error forms such as omission or commission errors. More recently, there has been an increased focus on the HEI stage, and new methods to assist the analyst have been developed. Methods such as the SHERPA and HRMS techniques and MURPHY diagrams (see [Kirwan 94]) aim to improve the identification of external (observable) forms of error and also psychological error mechanisms and performance-shaping factors. Such techniques are receiving increasing attention and acceptance, and it is likely that the development and validation of these HEI methods will continue.

2.7   CONCLUSIONS

The aim of this chapter has been to provide the reader with an overview of our current knowledge about human error modelling. This is an area of research which will continue to be important and from which further insights will emerge in the years ahead. While there are now a number of error-causation theories, these tend to be broad framework models and need to receive more validation and further development.

The practical use of human error causation models in industry is increasing and we can expect this to continue in the future. Undoubtedly the practical use of such ideas is 'horses for courses', and it will be important for practitioners to address issues such as perceived validity and applicability, the resources and training required, assessment of the benefits of their use, and the view of human error held within an organisation.

The future is likely to see further linking of the organisational influences on human error causation with existing safety management approaches. There is also likely to be more focus on the practical solutions to human error which organisations can implement. These may consider individual, task, team and organisational perspectives [HSE 89]. Finally we can expect to see substantial attention given to the role of attitudes, job satisfaction and risk perception on human errors in safety-critical industries.

ACKNOWLEDGEMENTS

The views expressed in this chapter are the author's and do not necessarily reflect those of the Health and Safety Executive, or any other body.

REFERENCES

[ACSNI 93] ACSNI: *Human Factors Study Group. Third Report: Organising for Safety.* London: HMSO, 1993

[Bainbridge 87] Bainbridge L:*Ironies of Automation*. In Rasmussen J, Duncan K and Leplat J (eds):*New Technology and Human Error*. John Wiley & Sons, Chichester, UK, 1987

[Broadbent 82] Broadbent D E, Cooper P F, Fitzgerald P and Parkes K R: *The Cognitive Failures Questionnaire (CFQ) and its Correlates*. British Journal of Clinical Psychology, 19, 177-188, 1982

[Cullen 90] Cullen, Lord D: *The Public Inquiry into the Piper Alpha Disaster*. HMSO, London, 1990

[Donaldson 94] Donaldson, Lord: *Safer ships, cleaner seas* (Report of Lord Donaldson's inquiry into the prevention of pollution from merchant shipping). HMSO, London, 1994

[Embrey 94] Embrey D E: *Guidelines for Preventing Human Error in the Chemical Processing Industry*. Centre for Chemical Process Safety, American Institute of Chemical Engineers, New York, 1994

[Fennel 88] Fennel D: *Investigation into the King's Cross Underground Fire*. HMSO, London, 1988

[Glendon and McKenna 95] Glendon A I and McKenna E F: *Human Safety and Risk Management*. Chapman and Hall, London, 1995

[Hale and Hale 70 Hale A R and Hale M:*Accidents in Perspective*. Occupational Psychology, 44, 115-121, 1970

[HSE 89] Health and Safety Executive: *Human Factors in Industrial Safety*. (HS(G)48), HMSO, London, 1989

[HSE 93] Health and Safety Executive: *The Costs of Accidents at Work*. (HS(G)96), HMSO, London, 1993

[Heinrich 50] Heinrich H W: *Industrial Accident Prevention: A Scientific Approach*. (3rd Edition) McGraw Hill, New York, 1950

[Hidden 89] Hidden A: *Investigation into the Clapham Junction Railway Accident*. HMSO, London, 1989

[Hudson et al 94] Hudson P T W, Reason J T, Wagenaar W A, Bentley P D, Primrose M and Visser J P: *Tripod Delta: Proactive Approach to Enhanced Safety*. Journal of Petroleum Technology, 46, 58-62, 1994

[Johnson 73] Johnson W G: *MORT: The Management Oversight and Risk Tree*. US Atomic Energy Commission SAN 821-2, Government Printing Office, Washington DC, 1973

[Kirwan 94] Kirwan B: *A Guide to Practical Human Reliability Assessment*. Taylor and Francis, London, 1994

[Kuhn 62] Kuhn T: *The Structure of Scientific Revolutions*. University of Chicago Press, 1962

[Norman 81] Norman D A: *Categorisation of Action Slips*. Psychological Review, 88(1), 1-15, 1981

[Paradies 93] Paradies M, Ungar L, Haas P and Terranova M: *The Development* of NRC's Human Performance Investigation Process (HPIP). NUREG/CR-5455, SI-92-101, 1993

[Pate-Cornell 93] Pate-Cornell M E: *Learning from the Piper Alpha Accident: A Post-mortem Analysis of Technical and Organisational Factors.* Risk Analysis, 13, 215-232, 1993

[Perrow 84] Perrow C: *Normal Accidents: Living with High-Risk Technologies.* Basic Books, New York, 1984

[Peterson 75] Peterson D: *Safety Management: A Human Approach.* Aloray, Englewood Cliffs, New Jersey, 1975

[Pew et al 81] Pew R W, Miller D C and Feeher C E: *Evaluation of Proposed Control Room Improvements through Analysis of Critical Operator Decisions.* Report prepared for the Electric Power Research Institute, NP-1982, 891, 1981

[Rasmussen et al 81] Rasmussen J, Pedersen O M, Carnino A, Griffon M, Mancini G and Gagnolet P: *Classification System for Reporting Events Involving Human Malfunctions.* Report RISO-M-2240, DK-4000, Roskilde, Riso National Laboratories, Denmark, 1981

[Rasmussen 83] Rasmussen J: *Skills, Rules and Knowledge: Signals, Signs and Symbols and Other Distinctions in Human Performance Models.* IEEE Transactions on Systems, Man and Cybernetics, SMC-13 (3), 257-266, 1983

[Rasmussen et al 87] Rasmussen J, Duncan K and Leplat J (eds): *New Technology and Human Error.* John Wiley & Sons, 1987

[Reason and Mycielska 82] Reason J and Mykielska K: *Absent-Minded? The Psychology of Mental Lapses and Everyday Errors.* Prentice Hall, Englewood Cliffs, New Jersey, 1982

[Reason and Lucas 84] Reason J and Lucas D: *Absent-mindedness in Shops: Its Correlates and Consequences.* British Journal of Clinical Psychology, 23, 121-131, 1984

[Reason and Embrey 85] Reason J and Embrey D: *Human Factors Principles Relevant to the Modelling of Human Errors in Abnormal Conditions of Nuclear Power Plants and Major Hazardous Installations.* Unpublished report: Human Reliability Associates, Parbold, Lancs, UK, 1985

[Reason 87] Reason J: *Generic Error Modelling System (GEMS): A Cognitive Framework for Locating Common Human Error Forms.* In Rasmussen J, Duncan K and Leplat J (eds): op. cit., 1987

[Reason 88] Reason J: *Stress and Cognitive Failure.* In Fisher S and Reason J (eds): *Handbook of Life Stress, Cognition and Health.* John Wiley & Sons, Chichester, 1988

[Reason 90] Reason J: *Human Error.* Cambridge University Press, 1990

[Reason 93] Reason J: *Managing the Management Risk: New Approaches to Organisational Safety.* In Wilpert W and Qvale T (eds) *Reliability and Safety in Hazardous Work Systems: Approaches to Analysis and Design.* Lawrence Erlbaum Associates, USA, 1993

[Reason et al 95] Reason J, Parker D, Lawton R and Pollock C: *Organisational Controls and the Varieties of Rule-related Behaviour.* ESRC Conference on Risk in Organisational Settings, York, May 1995

[Royal Soc 92] The Royal Society: *Risk Analysis, Perception, Management.* The Royal Society, London, 1992

[Sass 87] Sass R: *Accident causation as victim-blaming.* Proceedings of the Annual Conference of the Human Factors Association of Canada, October, 1987

[Senders and Moray 91] Senders J W and Moray N P: *Human Error: Cause, Prediction and Reduction.* Lawrence Erlbaum Associates, New Jersey, 1991

[Sheen 87] Sheen, Lord Justice: *MV Herald of Free Enterprise, Report of Court No 8074 Formal Investigation.* HMSO, London, 1987

[Shell 93] Shell Intl. Petroleum Mij. BV: *The Tripod Manual, Second Edition.* EP 93-2800, The Hague, 1993

[Simon 72] Simon H A: *On the Development of the Processor.* In Diggory S F (ed): *Information Processing in Children.* Academic Press, New York, 1972

[Taylor and Lucas 91] Taylor R K and Lucas D A: *Signals Passed at Danger: Near Miss Reporting from a Railway Perspective.* In van der Schaaf T W, Lucas D A and Hale A R (eds): *Near Miss Reporting as a Safety Tool.* Butterworth Heinemann, 1991

[Tversky and Kahneman 74] Tversky A and Kahneman D: *Judgement under Uncertainty: Heuristics and Biases. Science,* 185, 1124-1131, 1974

[Wagenaar et al 93] Wagenaar W A, Souverijn A M and Hudson P T W (1993) *Safety Management in Intensive Care Wards.* In Wilpert W and Qvale T (eds) *Reliability and Safety in Hazardous Work Systems: Approaches to Analysis and Design.* Lawrence Erlbaum Associates, USA, 1993

3

Human reliability assessment: methods and techniques

3.1 INTRODUCTION

A key issue in the safety assessment of systems based on modern technology is the evaluation of the role of Human Factors (HF) in the evolution of an incident. This has been proven numerically by incident statistics derived from a variety of domains, such as avionics [Poucet 88, IAEA 89], nuclear power production [Dougherty 88] and the petrochemical industry [Embrey 92]. The impact of human factors is exacerbated by the fact that the quality and reliability of hardware components have vastly improved over the last few decades and the role of operators has quite substantially changed. Indeed, the function of plant operators has been modified from that of manipulators and controllers of physical phenomena to supervisors and monitors of automatic processes governed by computerised tools. This has increased the probability of operators misunderstanding their new roles in the management of the plant and, thus, increased the likelihood of error.

The results are a significant reduction in the number of component failures and an increase in the instances of erroneous human actions. Currently, human failures account for about 70-80% of total incidents. Moreover, the tasks and goals demanded of operators, although of a supervisory nature, are also increasing in complexity and importance. Consequently, the accidents caused by erroneous human behaviour often have greater potential consequences for both the plant and the environment than was previously the case.

This chapter covers human error as it results from the decision-making process within the plant. Human errors resulting from the behaviour of people not directly related to the management, operation and maintenance of the plant are not considered — other than peripherally, when necessary. Examples of human errors included here are those made during maintenance, those due to mismanagement, and the obvious errors made during the planning and execution of a control procedure. Examples of human errors not considered are those in the design of components, voluntary damage or sabotage, and the mistakes made at the level of strategic planning and policy making. Chapter 2 provides a more detailed consideration of the intrinsic nature of human error.

The safety study of a plant or system is usually performed either by analysing its 'deterministic' behaviour (during predefined incident conditions called 'design basis incidents') or by evaluating the frequency and risks associated with hazardous events, using probabilistic approaches.

Probabilistic analysis has become of great importance for safety assessment, and large-scale studies covering the whole plant, including its control processes and human factors, are carried out. Such plant analyses are called 'probabilistic risk assessment' (PRA), or 'probabilistic safety assessment' (PSA) studies. In this chapter the latter term will be used as it is more recent and has been adopted in almost all technological domains. Historically, the major effort in a PSA study [US-NRC 75] has been dedicated to the evaluation of the safety and reliability of the hardware components of the system, including the controls and interfaces.

In this chapter, the terms 'system reliability' and 'systemic reliability' will be used to identify the hardware elements of the study, and 'human reliability' and 'human factors analysis' to identify the human factors aspects of a PSA.

The techniques most commonly adopted to perform PSA studies are based on the construction of 'event trees' and 'fault trees', which are graphical representations of accident scenarios. As pointed out in Section 1.5 of Chapter 1, event trees and fault trees provide complementary evaluations, namely:

- The event tree analysis starts with an initiating event and depicts the accident in terms of the logical branches generated by the binary alternatives of successes or failures of the major protection and safety systems which are intended to intervene during the accident evolution. The aim of the analysis is to define the consequences in terms of failure or success in controlling the accident.
- The fault tree method starts with the failure of a system and develops a set of logical combinations of malfunctions of subsystems, or of related components, which may lead to that failure. In this way it is possible to identify the set of elementary components which, by

failing, contribute to the overall system malfunction. These are then called 'cut sets'.

In a PSA, the combination of the event tree and the fault tree analyses, performed for all initiating events and failure modes, provides an extensive description of the possible accident paths (i.e., the origins of failure and their possible outcomes). The definitions of the failure probabilities of the elementary components of a system, and the use of computer programs for reliability evaluation and fault tree construction, generate the probabilities associated with the accidents defined for study. These probabilities, combined with the likely consequences of the accidents, give a measure of the safety (or risk) associated with the plant under study (for a discussion of risk, see Section 1.4 of Chapter 1).

This chapter provides an outline of methods and techniques used for human reliability assessment (HRA) in the context of safety analysis and, in particular, in PSA studies. Since the original work of Swain and Guttmann [Swain 83], a number of HRA methods have been developed [Apostolakis 90, Dougherty 91, Parry 91, Roth 91] with the dual aims of:

- Enhancing the scope of HRA, for example by including new aspects of human behaviour connected with the changing role of operators;
- Maintaining the basic connections with PSA approaches for systems reliability assessment.

The resulting techniques have matched the corresponding system reliability methods, to which they have been coupled, in accuracy and formalism. Thus, although at the level of research, a number of innovative human factors (HF) methods are currently being developed and tested [Cacciabue 91, Dougherty 93, Hollnagel 93, Macwan 93], this chapter will focus only on a number of well-established and applied approaches.

First, a number of criteria for categorising the HF methods will be considered on the basis of the current operational working environment requirements and operator tasks. Subsequently, a methodology for carrying out an HF study is explained and a number of methods are described in some detail, with discussions of the advantages and drawbacks of each approach in comparison with the criteria previously developed. This approach aims to provide the safety analyst with information to allow him to identify the HF method best suited to the analysis to be carried out. In conclusion, the issues of on-going research and their potential new findings and outcomes will briefly be reviewed.

3.2   CRITERIA FOR CLASSIFYING HUMAN FACTORS METHODS

Having defined the role and relevance of HRA analyses for safety assessment studies, it is useful to classify the methods according to certain differentiating criteria.

In some reviews of HRA methods [Poucet 88, IAEA 89] a broad subdivision is made between qualitative and quantitative approaches, to distinguish between the techniques that aim at explaining the events from those that try to quantify them. However, the safety analyst constantly faces the problem of assessing the risk of inappropriate behaviour from a numerical as well as a logical viewpoint. This implies that a combination of the two approaches is required. Therefore, this chapter examines only quantitative methods which are supported by a sound qualitative background analysis of the nature of the human error.

In other reviews of HRA methods, the focus has been placed on the mathematical formulation of the models [Watson 86, Humphreys 88, Cacciabue 88] and on the correspondence between the human reliability technique and the system reliability technique. The present review of techniques is based on three differentiating criteria associated with the functions of operators.

(i)   The technique's ability to account for both cognitive and behavioural human activities;

(ii)   Its ability to include both dynamic and static human-machine interaction;

(iii) The source of data for quantifying probabilities within the technique; data are sparce and may come from databases, expert judgement, or direct observation.

This choice of criteria will now be explained.

3.2.1   Cognitive and Behaviouristic Models

The continued increase in automation has led to operators' roles becoming more supervisory and less the direct manipulation of controls. Their tasks have become more cognitive (e.g., they include more reasoning, decision making, and management of interfaces). Moreover, plant control commonly requires data input using computer keyboards, joysticks or other devices (see Chapter 6 for a discussion of the relationships between tasks and interface design).

Thus, the types of error to be analysed in HRA include those which arise from these new characteristics of operators' tasks. They occur mainly at the cognitive level, but may cause, eventually, an inappropriate action. Therefore, an HF study has to consider the errors of cognition as well as erroneous operator actions. The latter are accounted for by 'behaviouristic' models, because they only consider the external and visible part of human behaviour, while the former require models of cognition, to take into consideration the mental processes as well as their external manifestations [Cacciabue 92].

The first criterion for differentiation needs to be the ability of the method to handle both the behaviouristic and cognitive aspects of human factors. An example of the behaviouristic aspect is the omission of a step or an entire task during the execution of a procedure. An example of the cognitive aspect is issuing the wrong command or consciously selecting the wrong control. These are, respectively, errors of omission and commission.

3.2.2   Dynamic and Static Interactions

The second criterion distinguishes between the HRA methods which analyse dynamic situations and those which only address static representations of human-machine interactions.

In systemic PSA, where the focus is on the behaviour of the technical aspects of systems, only the consequences of single component failures are considered. The overall plant layout is subdivided into a set of independent subsystems as the basis for the construction of the fault tree and event tree diagrams; this leads to substantial problems when dependencies among subsystems are identified or envisaged. This is particularly the case when the human element of the system is considered. Studies of human-machine interaction indicate that human error and the associated sequence of human actions are likely to be the result of direct interaction with the plant's control systems and a consequence of the plant's physical behaviour [Cacciabue 93a]. Consequently, it is very difficult to analyse human errors by means of the usual fault tree and event tree approaches. Indeed, the interdependence of human behaviour and machine response is a complex phenomenon, and a detailed insight into it requires a dynamic simulation of the human-machine system. Fortunately, this level of complex analysis is not always necessary and, usually, a quasi-static representation of operator-plant behaviour is sufficient for a reliable formulation of the safety case. (The safety case was introduced in Section 1.7 of Chapter 1 and is discussed in detail, with particular emphasis on human factors, in Chapter 12.)

3.2.3   Data Sources

The availability of detailed and sound data on human errors is a necessary precondition for the performance of a credible human reliability evaluation [Lucas 89]. The third criterion therefore concerns the source of the data which forms the input to the analysis method. The human factors analyst must be able to identify the degree of accuracy required and, consequently, the most appropriate type of data and its source, for any given safety analysis. In principle, there exist three different approaches to obtaining human error data:

(i)   Databases: the use of existing data sources concerning past events in a variety of plants which bear similarities to the one under study;

(ii)   Direct observation: the collection of data on human behaviour by observing the actual control of the plant in normal conditions, by collecting the data during training sessions, or even by laboratory experiments;

(iii)   The use of expert judgement.

3.3   METHODOLOGIES FOR HUMAN RELIABILITY ASSESSMENT

Prior to reviewing the methods available for evaluating and quantifying the human error probability (HEP), it is important to understand the methodological approach by describing the steps involved in carrying out an HRA. A methodology is a body of methods and techniques organised in order to achieve a given result. By applying a methodology for human reliability assessment, the safety analyst can properly structure and perform three essential steps, namely:

(i)   The analysis of the working environment under study;
(ii)   The identification of the potential human errors;
(iii)   The quantification (by a probabilistic measure) of potential human errors and their consequences.

There are a limited number of methodologies for HRA and they bear a strong similarity to each other [Hannaman 84a, Embrey 86]. Therefore, it is sufficient to describe only one, and then focus our attention on the methods and techniques which apply to each step of the methodology. The SHARP (Systematic Human Action Reliability Procedure) methodology of Hannaman and Spurgin [Hannaman 84a] has been selected as the most representative and complete example.

3.3.1   Systematic Human Action Reliability Procedure

SHARP is a framework for systematically incorporating human-machine interactions into a PSA study. It helps to define the types of interactions that are important to risk analysis and enables the analyst to incorporate them into the PSA. SHARP is applied in seven distinct steps (see Figure 3.1):

(i)   Identifying human interactions (Definition);
(ii)   Stating key assumptions (Screening);
(iii)   Identifying key interactions (Breakdown);
(iv)   Describing in detail key interactions (Representation);
(v)   Integrating the key interactions with the hardware description (Impact Assessment);
(vi)   Quantifying the impacts (Quantification);
(vii)   Documenting the results (Documentation).

In Step 1, *Definition*, the basic logic trees developed by the safety analysts from the functional description of the plant are enhanced. At this level, the expected control actions on the system are identified in order to ensure that all different types of human interaction are adequately considered in the study.

In Step 2, *Screening*, the logic trees, with the human interactions integrated, are screened so that only the most important interactions are selected for further analysis.

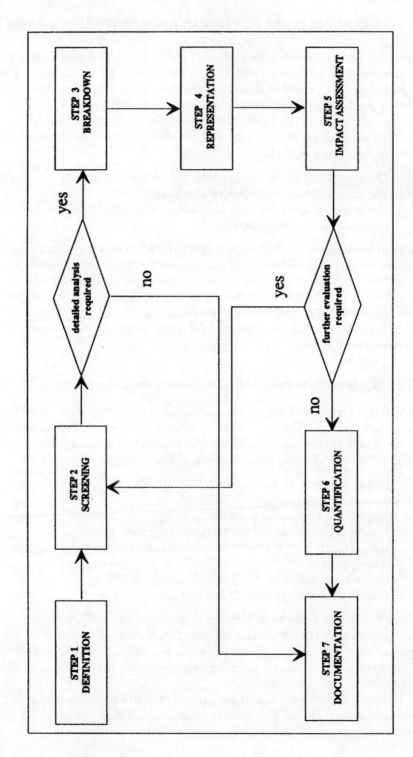

Figure 3.1 SHARP Methodology

In Step 3, *Breakdown*, each key interaction is decomposed into tasks and sub-tasks in order to define the factors necessary for accurate modelling.

In Step 4, *Representation*, these key interactions are explicitly modelled to define the alternative options that the operator may choose to achieve the interactions. In this way, it is possible to identify additional significant human actions that have an impact on the system's logic trees (but which have not been identified through the logic trees themselves).

In Step 5, *Impact assessment*, the detailed interactions identified in the previous step are now incorporated to update the logic trees. This allows the impact of significant human actions to be examined.

In Step 6, *Quantification*, probabilities are allocated to the human actions, and these are incorporated into the PSA study accident-sequence quantification. This allows the influence of human reliability on the consequences of an event to be assessed. Estimations of uncertainty are carried out, based on the reliability of the data used.

In Step 7, *Documentation*, the results of the study are documented.

Overall, the SHARP methodology integrates a comprehensive and well-structured sequence of models and methods, which may be applied at the various stages of the human reliability assessment. The quantification methods described in the following sections represent possible approaches for use in the context of the SHARP framework.

3.4 METHODS AND TECHNIQUES FOR HUMAN RELIABILITY ANALYSIS

This section presents the most commonly used HRA methods and techniques. The level of accuracy and the scope of the analysis performed by these methods varies quite substantially. Some focus only on one specific step of the overall methodology, while others cover several steps (and are, in a certain sense, smaller methodologies themselves). The most important steps of the methodology are the quantification of the human error probabilities, the representation of the human-machine interaction, and the evaluation of the probability of the success or failure of the sequence of actions under study. Consequently, all the methods described delineate an engineering solution to one or more of these problems.

Each of the selected methods will be reviewed and critically surveyed with reference to the scope of the method and to the criteria defined in Section 3.2. A detailed description of the statistical and mathematical formulations adopted by each method is not given as these can be found in the relevant references. The following HRA methods are reviewed:

(i)   Technique for Human Error Rate Prediction (THERP);
(ii) Operator Action Tree (OAT);
(iii) Maintenance Personnel Performance Simulation (MAPPS);
(iv) Absolute Probability Judgement (APJ);

(v)   Paired Comparison (PC);
(vi)   Tecnica Empirica Stima Errori Operatori (TESEO);
(vii)   Success Likelihood Index Methodology (SLIM);
(viii) Influence Diagram Approach (IDA);
(ix)   Human Cognitive Reliability Correlation (HCR).

The first three methods cover more than one step of the methodological framework described in Section 3.3, while the last six techniques are specifically dedicated to the evaluation of data on human error probabilities. These latter methods use an approach based primarily on expert opinion elicitation or judgement. This is coupled with the statistical treatment of the information retrieved from the different sources.

Following the review of the methods, a table provides, for each method, a summary of the steps of the SHARP methodology which it covers and the degree of its compliance with the classification criteria.

3.4.1   Technique for Human Error Rate Prediction

The Technique for Human Error Rate Prediction (THERP) [Swain 83, Bell 81] is the most widely used technique. It is a hybrid approach because it models human errors using probability trees and models of dependence as well as considering the performance-shaping factors (PSFs) affecting the operators' actions. The technique is linked to a database of human error probabilities (HEPs) which is described in Chapter 20 of the THERP handbook [Swain 83], and contains data derived from a mixture of objective field observations and judgements by the authors of the technique. The database, coupled with the method's engineering approach and with the fact that THERP was the first method to be accepted and used in the field, accounts for its popularity. The technique is carried out in four phases [Bell 81], each of which requires the completion of well-defined steps (see Figure 3.2).

The following description of THERP concentrates on its approach to the quantitative assessment of the human error probabilities and their combinations in the tree (Phases 3 and 4), as these are peculiar to the method.

Once the system failures of interest, i.e., the failures which may induce and influence human errors, have been identified, a detailed task analysis needs to be performed, followed by the identification of human errors. The implicit classification of the THERP database is of two basic error types:

(i)   Errors of omission, by which a step or an entire task is omitted;
(ii)   Errors of commission, which entail selection errors (such as issuing the wrong command or selecting the wrong control), sequence errors (such as executing a step too early in an operating procedure), and qualitative errors (such as performing too many repetitions of a particular task).

In order to describe and analyse these human errors, THERP uses the human reliability analysis event tree (HRA-ET) as its basic tool. By the use of HRA-

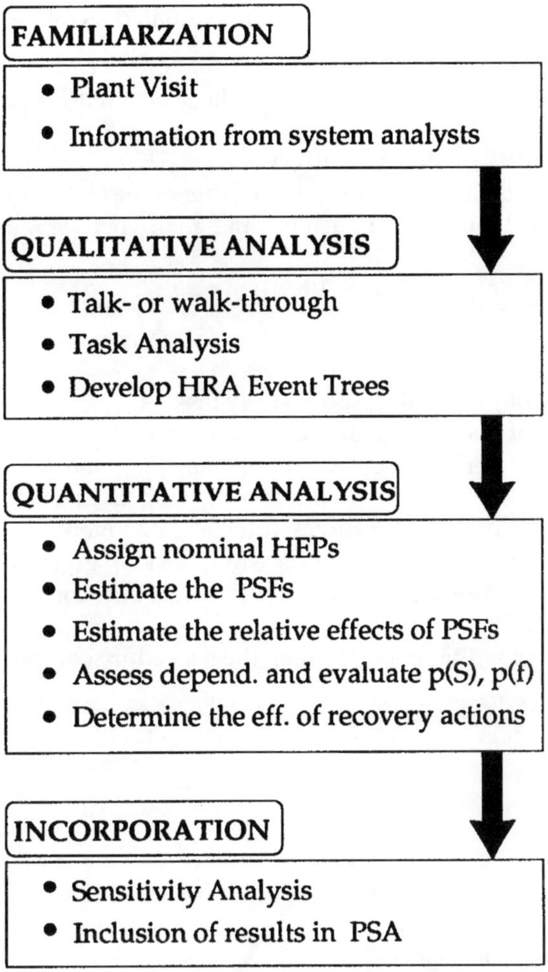

FAMILIARZATION
- Plant Visit
- Information from system analysts

QUALITATIVE ANALYSIS
- Talk- or walk-through
- Task Analysis
- Develop HRA Event Trees

QUANTITATIVE ANALYSIS
- Assign nominal HEPs
- Estimate the PSFs
- Estimate the relative effects of PSFs
- Assess depend. and evaluate p(S), p(f)
- Determine the eff. of recovery actions

INCORPORATION
- Sensitivity Analysis
- Inclusion of results in PSA

Figure 3.2 *Outline of THERP procedure for HRA*

ET, a graphical description of the procedural steps in a task are set out in a logical framework, in which, at each node of the event tree, there is a *binary decision point*, representing the failure or the success of the current action. Hence, these trees are consistent with conventional event trees, they can be evaluated in the formal mathematical sense, and, consequently, once the success or failure probability of each particular task step in a procedure is estimated, the overall reliability of the action under study can be calculated.

This decomposition approach is very similar to the engineering risk assessment of hardware components. In Figure 3.3, a schematic representation of an HRA-ET is shown. The success or failure probability of the mission depends on whether the two tasks 'A' and 'B' are to be carried out in series or are alternatives (in parallel). The HRA-ET is thus evaluated as follows:

(i)   If the procedure is of the type 'Series', both tasks have to be carried out successfully in order to achieve the goal and the $p(S)$ and $p(F)$ are:
$$p(S) = a(b/a)$$
$$p(F) = 1 - a(b/a) = a(B/a) + A(b/A) + A(B/A)$$

(ii)   If the procedure is of the type 'Parallel', then either of the tasks can be performed successfully in order to achieve the goal. Then, $p(S)$ and $p(F)$ are:
$$p(S) = 1 - A(B/A) = a(b/a) + a(B/a) + A(b/A)$$
$$p(F) = A(B/A)$$

Where:

a	is the probability of performing successfully task 'A'
A	is the probability of performing unsuccessfully task 'A'
b/a	is the probability of performing successfully task 'B' given a
B/a	is the probability of performing unsuccessfully task 'B' given a
b/A	is the probability of performing successfully task 'B' given A
B/A	is the probability of performing unsuccessfully task 'B' given A
$p(S)$	is the probability of performing successfully the procedure composed of task 'A' and/or 'B'
$p(F)$	is the probability of failing in performing the procedure composed of task 'A' and/or 'B'

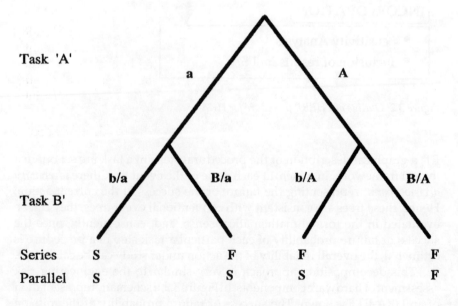

Figure 3.3 Human reliability analysis event tree for series and parallel systems

The failure probabilities are calculated by combining the data contained in the database of the technique's handbook and the data derived from recorded incidents, trials from simulators, and subjective judgements. However, the database itself is the primary source of information, comprising data tables as well as performance models and guidelines for the conversion of independent failure probabilities into conditional failure probabilities. The method of calculating probabilities using the database requires a sequence of steps which can be summarised as follows:

(i)   The definition of the errors, using tables corresponding to generic human error probability guidelines;

(ii) The identification of the performance-shaping factors (PSFs) which can affect human performance, and the definition of the effect on the human error probabilities of differences existing between the PSFs described in the THERP handbook and those existing in the actual task under evaluation;

(iii) A laborious step of correction and refinement, related to the dependencies of various tasks, where five levels of dependency are envisaged, going from 'zero dependence' to 'complete dependence';

(iv) A synthesis step requiring the probabilistic evaluation of the HRA. This uses conventional probability mathematics to combine the final values of the HEPs identified for each task step.

A large amount of work is involved in the performance of a detailed THERP analysis. Moreover, THERP is quite a rigid tool, i.e., it is not particularly sensitive to many possible performance-shaping factors. Nor does it attempt to identify underlying psychological aspects of a particular error. Furthermore, it does not allow the evaluation of errors concerning diagnosis or high-level decision making. However, it provides a structured, logical and well-documented record of factors and errors considered in the human reliability assessment, which makes it extremely useful for reference and as a data source, as well as for sensitivity analysis. Its compatibility with classical fault tree and event tree methods allows THERP to be integrated easily into probabilistic safety assessments.

In summary, THERP is a method which attempts not only to account for the complexity of the interaction between humans and machines, but also to achieve consistency with the methods of assessing system reliability.

The human reference model tries, through the dependencies and the performance-shaping factors, to consider some degree of cognition. Dynamic interactions are only superficially considered, although a complete and accurate static approach is performed.

The THERP handbook's database still represents a valid collection of data which is widely used by many safety analysts as a general reference for HF studies.

3.4.2   Operator Action Tree

The Operator Action Tree (OAT) [Wreathall 82] is a model of the sequence of actions needed to accomplish a function or task. It is applied to accident scenarios and the tasks related to them.

OAT focuses on decision making and is represented in Figure 3.4. It identifies alternative actions on the basis of ambiguities which become evident during operation or operator interpretations associated with observations, diagnoses, and selection of required responses. OAT allows the analyst to identify and display various types of decision strategies that may influence the sequence of events in an accident — for example, choosing a rarely used procedure which places strain on the equipment rather than the normal procedure or a standard operation. The method thus accounts for operator misinterpretation at various stages in the execution of a task, and models various operator decision paths. The assessor can determine whether or not these paths are key sequences of events in terms of the consequences for plant and environment. The method also incorporates the evaluation of the time available to the operator to allow successful task execution.

OAT contains a time failure, or a non-response probability relationship, used to quantify operator errors. The correlations for time failure are

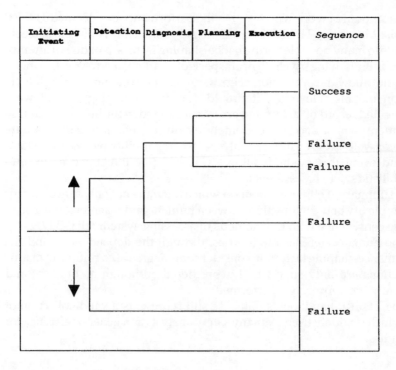

Figure 3.4 Basic operator action tree

represented as families of curves (Figure 3.5), and the reference model of operator behaviour is based on the paradigm 'skill-, rule- or knowledge-based behaviour', as defined by Rasmussen [Rasmussen 83] and discussed in Chapter 2. In this model, the human activities are subdivided into three main classes: the close coupling between sensory input and response action (skill), the execution of a set of known rules or procedures that have to be consciously recalled or checked before implementation (rule), and the analytical process of resolving successfully a problem in a complex unfamiliar environment, requiring laborious and difficult mindwork (knowledge).

Finally, it is worth pointing out that, although the OAT representation shown does not contain recovery actions, such an extension to the method may be appropriate, for example, by developing further branches which include recovery actions and the evaluation of timescales.

The OAT method tries to include in its framework both the cognitive elements and the physical manifestations of operator tasks. The interactions between the operator and the system under control are not directly simulated, so only static effects can be considered, even if recovery actions and 'jumps' from one branch to another are allowed in the event tree-like representation of the sequence. The method could thus be classified as a quasi-static approach.

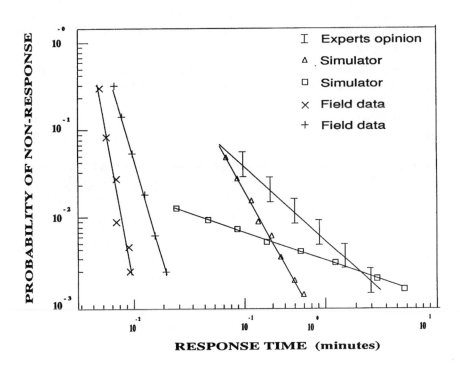

Figure 3.5 *Examples of non-response probability vs response time*

The data adopted in OAT use the correlations derived from experiments, field studies and simulator training results, and it is the responsibility of the safety analyst to apply the most relevant data from the families of curves.

3.4.3   MAPPS

Maintenance Personnel Performance Simulation (MAPPS) [Siegel 84] is a model-based method. It allows quantitative analysis of the effects of varying a set of conditions, represented by model inputs, on a second set of conditions or analytical results.

The analyst can systematically vary, both individually and in any combination, a variety of conditions, such as task variables, levels of operator ability, environmental and situation conditions, as well as cognitive human factors variables. The objective of the analysis is to yield sub-tasks, shifts, iterations and summary data, which show the effect of the introduced variations on the simulated task performance. The results can be used for defining trade-offs and regulatory decisions.

The direct method of MAPPS provides the capability to simulate:

- Corrective as well as preventive maintenance tasks;
- Contractor as well as 'in-house' maintenance;
- Maintenance conducted by personnel with any combination of skills and job titles, working in any usual conditions;
- Special sub-task types, such as decision making and troubleshooting, as well as sub-tasks of normal actions.

The model also allows for the inclusion of customised task analysis data for each simulated maintenance task, calculated values for average sub-task duration and success probabilities, use of default data when selected inputs are not defined by the user, possibility of using a Monte Carlo approach for introducing chance elements, rotation of personnel, and interactive input-output facilities.

The data adopted in MAPPS are drawn from existing collections or are formulated by the safety analyst on the basis of personal experience and ability. From a formal viewpoint, MAPPS does not contain a statistical treatment of data provided by different expert judgements.

MAPPS thus provides a model for a range of users, from the power plant architects and engineers to the plant maintenance management experts, as well as regulatory bodies. However, the application of the method requires the existence of a procedure unbounded by time constraints and man-machine interactions. Therefore, MAPPS is less applicable to incident analysis which has dynamic characteristics. MAPPS is oriented towards behavioural models, with little consideration of the dependence of the interactions of operators with the system under control.

3.4.4   Absolute Probability Judgement

Absolute Probability Judgement (APJ) [Seaver 82], also known as Direct Numerical Estimation, is the most direct approach to the quantification of Human Error Probabilities (HEPs).

APJ relies on the utilisation of 'experts' to estimate HEPs, based on their knowledge and experience. The rationale for using expert judgement in human reliability assessment is that there exists little or no relevant human error probability data, while there exist experts who have experience and appropriate knowledge that can be translated into quantitative estimates of the probability of occurrence of an event.

There are two forms of APJ, namely the Group APJ and the Single Expert Method. Most research involves the use of group methods, since it is seldom the case that a single expert has enough relevant information and expertise to estimate human reliability accurately. In the group methods, the individual knowledge and opinions are aggregated either mathematically or by bringing the judges to some form of consensus agreement.

The overall APJ procedure comprises the following steps:

(i)   Select subject matter experts;
(ii)   Prepare task statements;
(iii)   Prepare response booklets;
(iv)   Develop instructions for subjects;
(v)   Obtain judgements;
(vi)   Calculate inter-judge consistency;
(vii)   Aggregate individual estimates;
(viii) Estimate uncertainty bounds.

While steps (i) to (iv) are straightforward, steps (v) to (viii) require a deeper analysis. For step (v), there exist four major methods of obtaining experts' judgements:

- *Aggregated individual method:* In this approach experts do not meet, but make estimates individually. The estimates are then combined statistically by taking the geometric mean of the individual estimates.
- *Delphi method:* This is another anonymous method, in which experts make individual assessment. However, having made them they can review and reassess their individual estimates on the basis of the other estimates which are shown to everyone. The revised estimates are then statistically combined as above.
- *Nominal group technique:* This is similar to Delphi, except that some limited discussion is allowed between experts for clarification purposes only.
- *Consensus group method:* Each member contributes to the discussion, but the group must arrive at an estimate on which all members agree.

When non-group consensus methods are used, a statistical procedure for calculating the inter-judge consistency (step vi) is applied, based on a technique of analysis of variance (ANOVA). If the agreement between judges is adequate, the aggregation of the individual estimates is done (step vii) by taking the geometric mean of the single estimates. Finally, the uncertainty bound is estimated (step viii) using the statistical concepts of standard error and confidence range.

The APJ technique presents two main disadvantages, mainly linked to the fact that it relies entirely on expert judgement. These are the biases which can result from personality conflict and problems within the group, and the tendency to 'guessing' to which experts may be prone. On the other hand, the technique has been shown to provide accurate estimates in other fields, and is relatively quickly used.

APJ is useful for collecting data on cognitive behaviour and cognitive errors. However, it lacks a model of reference, either behavioural or cognitive, which is the consideration for interaction between humans and machines.

3.4.5   Paired Comparison

The Paired Comparison (PC) technique [Thurstone 80; Hunns 80] also aims at the definition of HEPs by expert judgement. Although PC is based on expert judgement, it does not require any quantitative assessment, but rather the experts are asked to compare a set or pair of tasks for which HEPs are required.

For each pair, the expert must decide which has the highest likelihood of error. These comparative judgements are elicited from a number of experts, and a scaling of the tasks is developed in terms of their relative likelihood of error. Two or more tasks with known HEPs are then used to calibrate the scaling, on the basis of a logarithmic transformation, in order to derive HEP estimates.

The complete PC procedure for estimating HEPs is based on 16 steps:

(i)   Define tasks;
(ii)   Incorporate calibration tasks;
(iii)   Select expert judges;
(iv)   Prepare exercise;
(v)   Brief experts;
(vi)   Carry out paired comparison;
(vii)   Derive row frequency matrix;
(viii) Derive proportion matrix;
(ix)   Derive transformation X-matrix;
(x)   Derive column difference Z-matrix;
(xi)   Calculate scale values;
(xii)   Estimate calibration points;
(xiii) Transform scale values into probabilities;
(xiv)   Determine within-judge consistency;
(xv)   Determine inter-judge consistency;
(xvi)   Estimate uncertainty bounds.

The details of the various steps will not be discussed here. However, the disadvantages and advantages of the method are worth some attention.

The PC method relies on three main assumptions, all of which can be violated in a human reliability assessment. These are: the independence of each comparison made, the homogeneity of the events or tasks to be performed, and that the event to be considered is not overly complex.

PC was originally developed to consider relatively simple perceptual phenomena, usually considered as 'uni-dimensional' events. When experts are asked to compare two complex operator error scenarios, such as during nuclear power plant emergencies, a simple comparison may prove difficult due to the multidimensional nature of the two situations. Similarly, when the tasks to be performed are not homogeneous, i.e., when they are not all maintenance tasks or emergency actions, then the comparison process is very difficult and the analyst may be induced into error. The independence criterion aims to identify tasks for comparison which do not distort the results by creating a bias for the expert (for example, by being somehow related to previously examined cases).

The advantages of PC are based mainly on the relative simplicity required at the level of probability evaluations. Indeed, the technique offers the possibility of rapidly arriving at the probability estimates simply by using the statistical methods applied to the PC estimations and to the two calibration points. Moreover, even without calibration, the technique provides a useful means of deriving a scaling and thus indicating the relative importance of different human errors. This rank ordering can be used also by other quantitative methods.

Finally, experts do not have to carry out the comparison as a group, and this eliminates the psychological drawbacks of the APJ method. In terms of compliance with the outlined criteria, PC contains only data estimation based on expert judgement.

3.4.6   TESEO

The TESEO model (Tecnica Empirica Stima Errori Operatori) [Bello 80] predicts human reliability as a function of five factors which are assumed to be the major determinants of operator performance, namely:

(i)   The type of activity carried out (K_1);

(ii)   The time available to carry out the activity $(K_2$, called 'temporary stress factor');

(iii)   The characteristics of the operator (K_3);

(iv)   The emotional state of the operator $(K_4$, called 'activity's anxiety factor');

(v)   The environmental ergonomics $(K_5$, called 'activity's ergonomic factor').

HEP (human error probability) $= K_1\,K_2\,K_3\,K_4\,K_5$

The value of each factor can be determined from standard tables.

There is a lack of theoretical foundation associated with this technique, especially in comparison with the previously discussed methods. Moreover, the model assumes that the five factors are always adequate for the assessment of the human performance, and no theoretical justification is given for the multiplicative relationship between the factors and the HEP. Indeed, it is stated by the authors that TESEO is based on a totally empirical formula which, in principle, does not need any deeper theoretical justification.

TESEO, as with all empirical correlations, is very easily applied and quick to use, for both error assessment and sensitivity analysis. It has an immediate appeal for performing a rapid assessment of different control room designs. Its most noticeable feature, i.e., its simplicity of application and the existence of a built-in database, should not bias the safety analyst towards its use, since the absence of a model of operator behaviour and the lack of consideration for the human-machine interaction make the model less attractive for detailed analysis.

3.4.7   Success Likelihood Index Method

The Success Likelihood Index Method (SLIM) [Embrey 84] originates from the field of decision analysis and is basically a method for quantifying a preference from a set of options. Its applicability to the assessment of human reliability derives from the consideration that human performance is affected by many factors and can only be quantified by calculating the additive effects of these performance-shaping factors on the human response.

The method is totally computerised and comprises two modules, namely: SLIM-MAUD (multi-attribute utility decomposition), for quantifying the effects of various factors on human reliability, using expert judgements for deriving the relative likelihood of success for a set of tasks; and SLIM-SARAH (systematic approach to the reliability assessment of humans), for transforming these relative likelihoods into absolute probabilities, using a logarithmic calibration relationship.

The SLIM procedure comprises 10 steps:

(i)   Definition of situations and subsets;
(ii)   Elicitation of PSFs;
(iii)   Rating the tasks on PSFs;
(iv)   Ideal point elicitation and scaling calculations;
(v)   Independence checks;
(vi)   Weighting procedures;
(vii)   Calculation of success likelihood indexes (SLI);
(viii) Conversion of the SLIs to probabilities;
(ix)   Uncertainty bound analysis;
(x)   Cost-effectiveness analyses.

The first eight steps are performed in the SLIM-MAUD module and the remaining two in SLIM-SARAH. The first step of SLIM-MAUD provides the judges with as much information as possible regarding the characteristics of the tasks. Then the judges are asked, during the interactive session at the computer, to identify and rate the performance-shaping factors relative to the tasks, on the basis of a linear scale from 1 to 9.

The judges have also to define the *ideal* point on the particular scale being used, and this allows MAUD to rescale all the other ratings. The rescaled ratings for each task on the various performance-shaping factors scales are subsequently multiplied by their respective relative importance weights, and these products are summed to give the overall SLI for each task. The weighting procedure is concerned with evaluating how much emphasis is to be given to each of the performance-shaping factors in terms of its effect on the likelihood of success. The SLIs for each task are calculated using the following formula:

$$SLI_j = \Re W_i R_{ij}$$

where:

SLI_j represents the Success Likelihood Index for task j ;
W_i is the normalised importance weight in the ith PSF;
R_{ij} is the scaled rating of task j on the ith PSF.

In order to transform these relative measures of likelihood of success into HEPs it is necessary to calibrate the SLI scale for each set of tasks considered. The relationship assumed in the SLIM method is:

$$\log HEP = aSLI + b$$

where a and b are constants.

The justifications for this formula are partly related to the PC technique, which also makes use of experts' judgements of likelihood of success, as well as of more theoretical works supporting the validity of the logarithmic relationship.

The extensive use of expert judgement and the logarithmic expression also represent the major drawback of the model, because of the amount of resources required for the initiation of a case. However, the methodology is based on a good theoretical background in decision theory and the easily managed interaction with the computer makes the technique highly visible and auditable. Moreover, once a detailed database has been established, the evaluations can be made easily and rapidly.

Considering the criteria of classification, the SLIM method shows the same characteristics as the other techniques based on expert judgement, though it is more easily applied by analysts who are less expert in statistics, as it is totally computerised. The method does not contain a complete methodology of application for HF analysis, it lacks a reference model of human behaviour, and it does not handle human-machine interaction. However, these drawbacks have been compensated for by including the method in a larger methodological approach, named SHERPA [Embrey 86]. In this, the data definition of SLIM is conjoined with the modelling aspects necessary for a complete reliability analysis.

3.4.8   Influence Diagram Approach

The Influence Diagram Approach (IDA) [Howard 80] originates from the field of decision analysis and is a graphical representation of the dependencies between the factors that influence human behaviour.

The objective of the technique is to assess the likelihood of occurrence of a particular human action, in relation to the combined influences of factors such as the quality of the information (external factors), the individual's motivation (internal factors), and the socio-technical environment (social factors). These influences act at a high level and are subject to other lower-level influences which are defined by an analysis of the work environment. These interrelations are expressed by the graphical representation of the influence diagrams. The numerical evaluation of the influences is made by expert judgements, which are then statistically combined to obtain direct estimates of error probabilities.

The IDA is based on 10 steps:

(i)   Description of the conditioning events;
(ii)   Refinement of the target event;
(iii) - (v)   Selection of a middle-level event and assessment of its influences;
(vi) - (vii)   Assessment of probabilities for the target event;
(viii)   Comparison of the results with holistic human error;
(ix)   Iteration to refine opinions;
(x)   Sensitivity analysis.

In the first step, the group of experts are asked to describe in general terms the work setting in which the target event may occur and the main influences which impact on success or failure in the scenario under study. A generic diagram is developed (Figure 3.6) and the influences of various factors on the scenario are represented, including the mutual influences between factors. Moreover, each influence is further broken down into a number of elements so as better to define the context of the factor. As an example, the factor 'stress' is broken down into 'shifts', 'time available' and 'operating objectives'.

In step 2, the target event is defined in such a way that no additional information is needed for confirming its occurrence or non-occurrence. Steps 3-5 are typical of the statistical methods based on the elicitation of expert opinions, in this case using production rules of the type 'if a and b then c, and on weighted combinations of estimates. In steps 6 and 7 the probabilities of the target events are assessed using the previously developed evaluations and the direct estimates of the success or failure probabilities of the target event expressed by the experts. In step 8 the unconditional probabilities and the weights of evidence are given to the experts for comparison with their own judgements. In step 9, the discrepancies are discussed and iterations take place until agreement is reached and no new observations are made

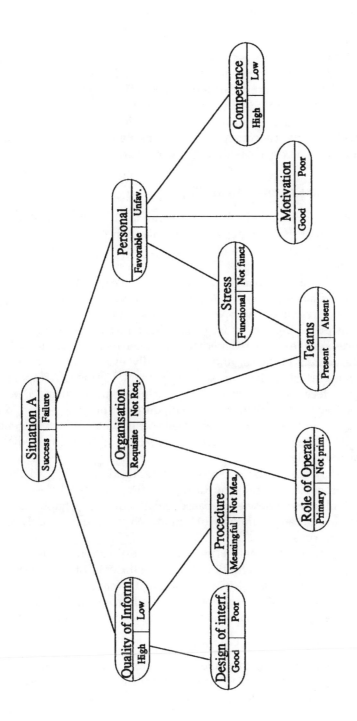

Figure 3.6 Generic influence diagram for human reliability assessment (from [Humphreys 88])

about the case under study. Step 10 evaluates the relevance of possible discrepancies still existing after this iteration process, by performing a sensitivity analysis on the variation of the probability estimate for the target event due to such discrepancies.

The IDA method shares the advantages and disadvantages of the other approaches based on expert elicitation, for example the high cost in terms of resources (time and experts). On the other hand, IDA exhibits some interesting features in its explicit modelling of the dependencies between factors, in the structured representation of the influences which is built by progressive refinement of levels of accuracy, and in the possibility of evaluating the effect of discrepancies across individual assessments (sensitivity analysis).

3.4.9   Human Cognitive Reliability

The Human Cognitive Reliability (HCR) model [Hannaman 84b] provides the evaluation of the time-dependent human non-response probability, in terms of key engineering and operational input parameters and their coefficients.

HCR is thus an approach mainly devoted to the quantification of human errors on the basis of key input parameters, which identify: the type of cognitive behaviour of the operator, the median response time, and the performance-shaping factors (see also Chapter 11). The median response time represents the estimated median time taken by the operational team to complete actions or tasks, and it is usually evaluated from simulator measurements, task analysis or expert judgement. The median response time assumes perfect task performance, i.e., excluding the occurrence of errors and the subsequent recoveries. However, the effect on team performance of operationally induced stress, equipment design, etc., which is generally expressed by means of PSFs, is accounted for by modifying the median time to perform the relevant task.

The behaviour of the operator is categorised according to the three cognitive processes identified by Rasmussen [Rasmussen 83; Rasmussen 86] as skill-, rule- and knowledge-based behaviour (see Chapter 2). These represent the reference paradigm of human behaviour most widely used for simulation purposes.

The results of the analysis of data and estimation of the task to be performed lead to the definition of the coefficients of a semi-empirical correlation. In this sense, HRC is more a mathematical correlation than a model. The shape of the curves, obtained from simulator data, are approximated by a three parameter Weibull distribution, given by:

$$P(t) = \exp^{-}(t/T_{1/2} - C_{\gamma j}\beta_j)C_{\eta j}$$

in which t represents the time taken by the crew to complete actions after a stimulus; $T_{1/2}$ is the median time to perform the task corrected by a performance-shaping factor; β_j is a shape parameter of the correlation for type j of cognitive process; $C_{\gamma j}$ and $C_{\eta j}$ are, respectively, a time delay parameter and a scale parameter, which are expressed in the form of fractions of $T_{1/2}$ for type j of cognitive process.

If T is the time window allowed for completing a task before a change in plant state, then $P(T)$ is the crew non-response probability. Figure 3.7 shows the normalised crew non-response curves corresponding to the three cognitive processes of skill-, rule- and knowledge-based behaviour. The probability of non-response, defined by this equation, becomes an element in the assessment of the overall non-success probability for input to PSA studies.

HCR, being a method based on expert judgement, suffers the same disadvantages of similar approaches that require the extensive and laborious activity of experts. However, it is a useful and easily applicable approach and does not need any specific mathematical or statistical elaboration before the evaluation of non-success probabilities. HCR represents a considerable effort of combination and interfacing of previously developed methods in related fields, namely: the use of time reliability curves, empirical correlations, and cognitive psychology approaches to human behaviour. HCR does not contain a simulation of the interactions between humans and systems, but it

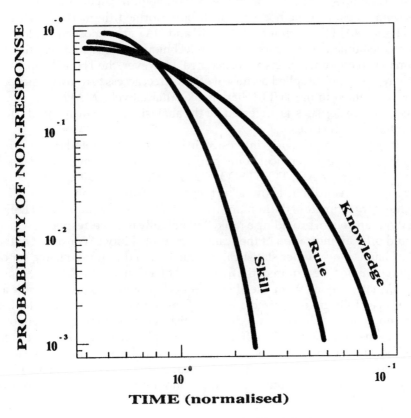

Figure 3.7 Normalised crew non-response curves for SRK behaviour

accounts for the key aspects of modern task allocation and demands through the data and the reference model of cognition.

3.5   COMPARISON OF METHODS

In order to obtain a comparative overview of the methods described in the previous sections, as well as a measure of their scope of application, Table 3.1 has been developed with reference to the criteria devised in Section 3.2 and the methodological framework of Section 3.3 (and reference should be made to these in interpreting it). Using the table, in conjunction with a consideration of the type of information available, in terms of data and statistical algorithms, the analyst can choose the most suitable method for the human factors analysis to be carried out.

3.5.1   Comparison of Methods on Methodological Aspects

In the fifth column of the table, the scope of applicability of each method is clearly identified with reference to the methodological framework of SHARP. If several steps of the methodology are to be performed, the methods THERP, OAT and MAPPS are preferred. THERP and OAT are almost equivalent as far as the human factors representation technique and the consideration of human behavioural aspects are concerned. However, the THERP approach is more commonly applied because it offers direct access to its own database — as contained in the THERP handbook. Alternatively, MAPPS, given its specific characteristics, is preferred for the analysis of human factors during maintenance processes.

If the user is primarily interested in achieving accurate human error probabilities, the preference should be placed on one of the more specialised methods, namely APJ, PC, TESEO, SLIM, IDA, or HCR. The differences existing between them allow a further level of distinction. Indeed, if a preliminary estimate of the human error probabilities is required, the TESEO method can help in providing a rapid, but not fully reliable, response, which would offer an indication of the scale of the probability of concern. On the other hand, if a sounder statistical formulation and a more reliable set of expert opinions are required, the methods APJ, PC, IDA and SLIM are more appropriate. Among them, the SLIM approach provides some advantage as it has been fully computerised and transformed into a user-friendly tool; it does not require a deep knowledge of statistics for combining the opinions of the experts. Finally, HCR offers better compliance of the data-selection method with the criteria for considering cognitive behaviour and the dynamic interaction typical of modern technological systems, even if it demands a very accurate data elicitation procedure and a cautious evaluation of the collected data.

Table 3.1 *Comparison between the HF methods*

	Beh.	Cog.	Sta.	Dyn.	Data base	Exp Jud.	Direct Ob.	Methodologic steps
Models			**Interaction**			**Data**		**Methodologic steps**
THERP	4	2	4	1	X		X	Breakd. — Repl Impact As. — C
OAT	4	2	4	2	X			Breakd. — Repl Impact As. — C
MAPPS	2	0	2	0	X		X	Breakd. — Repl Impact As. — C
APJ	n.a.		n.a.			X		Quant.
PC	n.a.		n.a.			X		Quant.
TESEO	0	2	1	0	X			Quant.
SLIM	n.a.		n.a.			X		Quant.
IDA	n.a.		n.a.			X		Quant.
HCR	3	4	2	1		X		Quant.

In the table above: X denotes the type of data used, 'n.a.' stands for 'not applicable', and 0-4 are the levels of detail of the approach, with 4 = full modelling, 0 = no modelling, and 1 = insufficient modelling

3.5.2   Comparison of Methods on the Selection Criteria

Some general remarks can be made concerning selection criteria adopted for the comparison of methods, namely: the modelling of human behaviour, the dynamic interactions of the human-machine systems, and the data sources.

The consideration of human behaviour has been included in all techniques, even if the need to interface the HF methods with the systemic approach of a PSA study has created a strong bias towards the quantification of action performance, mainly in terms of the success or failure of tasks, with less attention being paid to the causes of human error. The methods, which include some kind of simulation of operator activity, account properly for the human-machine interaction only if normative behaviour is considered, but are very superficial for the elements that concern cognitive processes. Although HCR, OAT and THERP attempt to include some of the aspects of cognitive behaviour, this subject is still a matter of research. Some techniques focusing on this issue have recently begun to appear in human factors analysis [Dougherty 93, Hollnagel 93, Cacciabue 93b, Macwan 93].

As far as the dependence of human errors on the dynamic evolution of the incident is concerned, the overall review is even less encouraging. Indeed, only a few methods make some reference to the dynamic aspects of the human-machine and human-human interactions. While the static dependencies are well considered in almost all approaches, the dynamic interplay of human behaviour with the surrounding environment is only examined in the OAT method and, to a minor extent, in THERP and HCR. This issue is strongly related to the ability to account for the reasoning and the decision-making processes typical of cognitive behaviour. Most of the above-mentioned recent techniques attempt to solve, in an integrated way, the two problems of cognitive behaviour and dynamic interaction. Solving these two issues represents a real breakthrough which will lead to the formal development and application of a body of new tools with respect to the methods and techniques described and discussed in this chapter.

With reference to data sources, the three types of database used for human reliability analysis exhibit very different characteristics and peculiarities. The data derived from direct observation of the work setting under study are usually held in confidential databases [Villemeur 86, Rouhet 89] and their use is specific and restricted to the industry or environment where the data collection has been carried out. When these data are available, the methods proposed by THERP or MAPPS can be applied for the assessment of the consequences of human error. On the other hand, the type of data contained in generic databases are for general application, as they represent the synthesis of experiences in different domains and they focus on generic errors, such as omissions in performing manual actions and in verification tasks during the execution of procedures. These databases can be applied in many similar contexts and are accepted and used by many analysts.

A classical example of this type of human error database is that associated with the THERP method, which is described in detail in Chapter 20 of the THERP handbook [Swain 83]. The use of expert judgement is also a valid and very helpful technique. It has been accurately developed for the statistical treatment of the data, while in some cases little attention has been paid to the connections with the remaining steps of an HF methodology. To be fully effective, this approach requires a substantial investment of resources for the collection of the data and its integration into a methodological framework.

In summary, the data from direct observation are very difficult to obtain, but they are very valuable in terms of content of information. The expert judgements are less costly to achieve, but they demand the collection of a considerable amount of data and an accurate statistical treatment before application. The use of existing databases offers a more rapid solution to the problem, but it results in greater uncertainty regarding the evaluation of the consequences. The safety analyst must make an appropriate trade-off between the advantages and drawbacks of each method before selecting the one to be applied to the case under study.

3.6   CONCLUSIONS

The overall conclusion to be drawn from examining the available human reliability assessment methods is that while a fair degree of accuracy and experience have been gained in the treatment of the data, only limited precision has been obtained in the modelling of human behaviour and in the consideration of the cognitive components of the root causes of human error. Moreover, the dynamic aspects of man-machine interactions are almost completely neglected.

There are two reasons for the deficiency in dynamic interaction analysis:

(i)   The intrinsic difficulty and complexity of the problem;
(ii)   The existence of an equivalent inadequacy in the systemic approaches to plant safety studies.

The solution to the first problem can be achieved by the developments of models of cognition based on new types of computer architectures and software tools. The second deficiency is methodological. The approaches to HF methods have traditionally been substantially static and there has been little impetus to develop dynamic HF methods. However, today, particularly with the increase in safety-critical systems, dynamic dependence is now recognised as highly relevant to system analysis. Consequently, 'dynamic reliability' (or 'dynamic PSA') is becoming a topic of research. This will certainly give new emphasis to the development of dynamic HF methods.

However, a substantial amount of work still remains to be done before such new methods mature from the stage of research to the formal application in safety assessment studies. Consequently, the approaches described in this chapter will remain, for some time, the tools adopted by analysts whenever a human error analysis has to be carried out in the context of an overall safety and reliability assessment.

REFERENCES

[Apostolakis 90]   Apostolakis G E, Mancini G, van Otterloo R W (eds): *Human Reliability Analysis*. Special volume, Reliability Engineering and System Safety, 29 No. 3, 1990

[Bell 81]   Bell B J and Swain A D: *A Procedure for Conducting a Human Reliability Analysis for Nuclear Power Plants*. NUREG/CR-2254, SAND81-1655, USNRC, Washington, US, 1981

[Bello 80]   Bello G C and Colombari V: *The Human Factors in Risk Analyses of Process Plants: The Control Room Operator Model, 'TESEO'*. Reliability

Engineering, 1, 3-14, 1980

[Cacciabue 88]   Cacciabue P C: *Evaluation of Human Factors and Man-Machine Interaction Problems in the Safety of Nuclear Power Plants*. Nuclear Engineering and Design, 109, 417-431, 1988

[Cacciabue 91]   Cacciabue P C and Vivalda C: *A Dynamic Methodology for Evaluating Human Error Probabilities*. In Apostolakis, GE (ed): *Proc. of the Int. Conf. on Probabilistic Safety Assessment and Management (PSAM) 4-7*, Beverly Hills, California. Elsevier, New York, NY, 507-512, 1991

[Cacciabue 92]   Cacciabue P C: *Cognitive Modelling: a Fundamental Issue for Human Reliability Assessment Methodology?* Reliability Engineering and System Safety, 38, 91-97, 1992

[Cacciabue 93a]   Cacciabue P C and Hollnagel E: *Human Models in Reliability and Safety Analysis of Interactive Systems*. Proc. of Int. ANS/ENS Topical Meeting on Probabilistic Safety Assessment, PSA 93, Clearwater Beach, FL, 26-29, January 1993. American Nuclear Society, La Grange Park, Ill, 25-31, 1993

[Cacciabue 93b]   Cacciabue P C and Cojazzi G: *A Human Factor Methodology for Safety Assessment Based on the DYLAM Approach*. Post SMiRT Seminar on 'Current Issues in Probabilistic Safety Assessment (PSA)', Heidelberg, Germany, 23-24 August 1993

[Dougherty 88]   Dougherty E M and Fragola J R: *Human Reliability Analysis. A System Engineering Approach with Nuclear Power Plant Applications*. John Wiley & Sons, New York, 1988

[Dougherty 91]   Dougherty E M: *Issues of Human Reliability in Risk Analysis*. In Apostolakis, G E (ed): *Proc. of the Int. Conf. on Probabilistic Safety Assessment and Management (PSAM) 4-7*, Beverly Hills, California. Elsevier, New York, NY, 699-704, 1991

[Dougherty 93]   Dougherty E: *Context and Human Reliability Analysis*. Reliability Engineering and System Safety, 41, 25-48, 1993

[Embrey 84]   Embrey D E, Humphreys P C, Rosa E A, Kirwan B and Rea K: *SLIM-MAUD: An Approach to Assessing Human Error Probabilities Using Structured Expert Judgement*. NUREG/CR-3518, USNRC, Washington, US, 1984

[Embrey 86]   Embrey D E: *SHERPA: A Systematic Human Error Reduction and Prediction Approach*. Proceedings of the International Topical Meeting on Advances in Human Factors in Nuclear Power Plants, Knoxville, US, April 21-25. American Nuclear Society (ANS), 184-193, 1986

[Embrey 92]   Embrey D E: *Managing Human Error in the Chemical Process Industry*. Proceedings of Int. Conf. on Hazard Identification and Risk Analysis, Human Factors and Human Reliability in Process Industry. Orlando Florida, January 15-17, Amer. Inst. of Chem. Eng., New York, 399-413, 1992

[Howard 80]   Howard R and Matheson J G: *Influence Diagrams*. SRI

International, Menlo Park, CA, US, 1980

[Hannaman 84a]   Hannaman G W and Spurgin A J: *Systematic Human Action Reliability Procedure (SHARP)*. EPRI NP-3583, Project 2170-3, Interim Report, NUS Corporation, San Diego, CA, US, 1984

[Hannaman 84b]   Hannaman G W, Spurgin A J and Lukic Y D: *Human Cognitive Reliability Model for PRA Analysis*. NUS-4531, NUS Corporation, San Diego, CA, US, 1984

[Hollnagel 93]   Hollnagel E: *Reliability of Cognition: Foundations of Human Reliability Analysis*. Academic Press, London, 1993

[Humphreys 88]   Humphreys P (ed): *Human Reliability Assessors Guide*. United Kingdom Atomic Energy Authority, RTS88/95Q, 1988

[Hunns 80]   Hunns D and Daniels B K: *The Methods of Paired Comparison and the Results of the Paired Comparisons Consensus Exercise*. Proceedings of the 6th Advances in Reliability Technology Symposium, NCSR R23, Culcheth, UK, 1980

[IAEA 89]   International Atomic Energy Agency: *Models and Data Requirements for Human Reliability Analysis*. Report of consultants meeting. IAEA-TECDOC-499. Laxenburg, 1989

[Lucas 89]   Lucas D A and Embrey D E: *Human Reliability Data Collection for Qualitative Modelling and Quantitative Assessment*. In Colombari V (ed): *Proceedings of 6th EUREDATA Conference on Reliability Data Collection and Use in Risk and Availability Assessment*. March 15-17, Siena, Italy. Springer-Verlang, Berlin, 358-370, 1989

[Macwan 93]   Macwan A P and Mosleh A: *A Simulation Based Approach to Modeling Errors of Commission during Nuclear Power Plant Accidents: Application to PRA*. Proc. of Int. ANS/ENS Topical Meeting on Probabilistic Safety Assessment, PSA 93, Clearwater Beach, FL, Jan. 26-29, 1993. American Nuclear Society, La Grange Park, Ill, 40-46, 1993

[Parry 91]   Parry G W, Singh A, Spurgin A, Moieni P and Beare A: *An Approach to the Analysis of Operating Crew Responses Using Simulator Exercises for Use in PSAs*. OECD/BMU Workshop on Special Issues of Level 1 PSA, Cologne, Germany, 28 May 1991

[Poucet 88]   Poucet A: *Survey of Methods used to Assess Human Reliability in the Human Factors Reliability Benchmark Exercise*. In Apostolakis G E, Kafka P and Mancini G (eds): *Accident Sequence Modelling: Human Actions, System Response, Intelligent Decision Support*. Elsevier Applied Science, London, UK, 257-268, 1988

[Rasmussen 83]   Rasmussen J: *Skills, Rules and Knowledge: signals, signs and symbols; and other distinctions in human performance model*. IEEE Transactions on Systems, Man, and Cybernetics, *IEEE-SMC*, 13, 3, 257-267, 1983

[Rasmussen 86] Rasmussen J: *Information processes and human-machine interaction. An approach to cognitive engineering*. North Holland, Oxford, 1986

[Roth 91]   Roth E M, People H E Jr. and Woods D D: *Cognitive Environment Simulation: a Tool for Modelling Operator Cognitive Performance during Emergencies.* In Apostolakis, G E (ed): *Proc. of the Int. Conf. on Probabilistic Safety Assessment and Management (PSAM)* 4-7, February 1991, Beverly Hills, California. Elsevier, New York, NY, 959-964, 1991

[Rouhet 89]   Rouhet J C, Francois N, Wanner J C and Charpentier M: *Human Error in Aviation Operations: a contribution for human behaviour modelling.* Proceedings of Second European Meeting on Cognitive Science Approaches to Process Control, Siena, Italy, October, 1989. CEC-JRC, Ispra, 361-370, 1989

[Seaver 82]   Seaver D A and Stillwell W G: *Procedures for using Expert Judgement to Estimate Human Error Probabilities in Nuclear Power Plant Operations.* NUREG/CR-2743, USNRC, 1982

[Siegel 84]   Siegel A, Bartter W D, Wolf J J, Knee H E and Haas P M: *Maintenance Personnel Performance Simulation (MAPPS). Model: Description of Model Content, Structure and Sensitivity Testing.* NUREG/CR-3626, Vols 1 and 2, USNRC, 1984

[Swain 83]   Swain A D and Guttman H E: *Handbook of Human Reliability Analysis with Emphasis on Nuclear Power Plant Applications.* NUREG/CR-1278. SAND 80-0200 RX, AN. Final Report, 1983

[Thurstone 80]   Thurstone L L: *A Low of Comparative Judgement.* Psychological Review, 34, 273-286, 1980

[US-NRC 75]   U.S. Nuclear Regulatory Commission: *Reactor Safety Study: An Assessment of Accident Risks in U.S. Commercial Nuclear Power Plants.* WASH-1400 (NUREG-75/014), Washington, US, 1975

[Villemeur 86]   Villemeur A, Monseron-Dupin F and Bouissou M: *A Human Factors Data Bank for French Nuclear Power Plants.* Proceedings of the Intern. Top. Meeting on Advances in Human Factors in Nuclear Power Systems, Knoxville, Tennessee, April 1986. American Nuclear Society, La Grange Park, Ill, 368-373, 1986

[Watson 86]   Watson I A: *Human Factors in Reliability and Risk Assessment.* In Amendola A and Saiz de Bustamante A (eds): *Reliability Engineering,* proceedings of the Ispra Course held at the Escuela Tecnica Superior de Ingenieros Navales, Madrid, 22-26 September 1986. D. Reidel, Dordrecht, NL, 1986

[Wreathall 82]   Wreathall J W: *Operator Action Tree, An Approach to Quantifying Operator Error Probability During Accident Sequences.* NUS Report 4159, NUS Corporation, Gaithersberg, Maryland, July 1982

2

Human-computer interaction

4

Introduction to HCI in safety-critical systems

4.1   WHAT MAKES HCI DESIGN SO IMPORTANT?

Why do people make mistakes when operating safety-critical systems? What is it that makes someone misread a display, ignore a warning signal, or press the wrong button? To the world outside the situation, it can look like dangerous stupidity or gross negligence. Yet close inspection of cases of 'human error' often reveals that the problem was linked more closely to design than operation — that something in the design of the human interface to the system was at fault, something that could have been foreseen and that could have been designed in a safer way.

The idea that good design of the user interface can make systems safer is not new; it has existed under the banner of 'ergonomics' for many decades. Human-computer interaction (HCI) is the discipline that applies ergonomic principles to the design of user interactions with computers. In the early days of the discipline, in the 1960s and 1970s, the emphasis was all on 'ease of use' and this is still by far the major concern of HCI practitioners and researchers. Computers have always been difficult to use, but it has become clear that careful design which takes into account the human need for consistency and clear and timely feedback, and avoids unnecessary ambiguity, memory load and complexity, can dramatically improve the effectiveness and efficiency of the user interaction. Chapter 6 gives a fuller discussion of the ways in which this can be practically achieved.

Fuelled in part by the enormous increase in the study of human error within psychology in the last two decades, as discussed in Chapter 2, an awareness of the fundamental causes of human error in the operation of safety-critical systems has developed. It has become clear that error arises not out of random or arbitrary processes but as a predictable consequence of basic and normally useful psychological mechanisms. As James Reason puts it, 'Correct performance and systematic errors are two sides of the same coin' (see [Reason 90] p. 2). This realisation gives us the hope that systems can be designed in ways that make appropriate allowance for the psychology of the people who operate them. However, in the real and messy world of engineering, theory and abstraction are of limited use. Practical and cost-effective techniques for developing user interfaces to computers need to be found which will give us an acceptable level of assurance that the resulting software will be safe to operate. When the user interaction with a safety-critical system goes wrong, the result can be catastrophic. There are many infamous examples.

On 20 January 1992, an Airbus A320 crashed into a hill near Strasbourg. The proposed reason for the disaster is that the crew instructed the aircraft to descend at a rate of 3300 feet per minute rather than the much gentler descent they really wanted, which was at an angle to the horizontal of 3.3°. What appears to have been the reason for this error is that the autopilot display on the aircraft had two modes. In one, in which climbs and descents were displayed in degrees, a 3.3° descent was shown as 'minus 3.3'. In the other, in which climbs and descents were displayed in units of 100 feet per minute, a 3300 feet/minute descent was shown as 'minus 33'. To the crew, the two displayed figures were almost identical. Only a knowledge of the current mode would easily distinguish their crucially different significance. In this case, the crew apparently made what is known in HCI as a 'mode error' and got it wrong — with tragic consequences.

It is common to speak of such mistakes as 'human error', as though the human operator were somehow negligent or incapable. Yet problems like mode errors are human-performance hazards that are well known and well understood and which can easily be avoided by good design.

Mistakes like the one just described, which have immediate consequences, are known as 'active' errors. There are others which have no consequences for some time or whose effects are only slowly recognised. These are known as 'latent' errors [Reason 90]. In the Three Mile Island disaster, in which a nuclear reactor was seriously damaged after a series of accidents led to a loss of radioactive coolant, one of the several errors that compounded the original problem was that operators failed to notice a warning light which indicated that a valve was closed when it should have been open. In fact, this was not surprising since the light was obscured by a maintenance tag. We could say that the latent error was the earlier sloppy placing of this tag. Yet we could also criticise the maintenance procedures that allowed such a mistake. We could criticise the operating procedures for not demanding a check of the light. We could even criticise the human interface designers for creating a panel with the potential for warning lights to be obscured (given the maintenance procedures). Latent errors such as this one, which are often ascribed to maintenance or management problems, are an indication that the human interface to a system depends for its efficacy on a wide range of system factors not obviously related to it.

The ease of use of a computer system is a complex property, relying on the quality of the user interface design, the amount of training and other support required to operate it, the quality of on-line and off-line documentation, and the skills and abilities of the operators. A highly skilled operator with good on-line support may be able to do just as well with a poorly designed interface as a less well-skilled operator with no on-line support but who has good training and a well-designed interface. The issues to be considered in supporting operation are considered in chapter 7. Many would accept that the safety of a system relies on its ease of use. Yet if we accept this, we have to reach some rather surprising conclusions, for example, that an organisation's recruitment policy and the manner in which it is operated affect the ease of use and, therefore, the safety of its systems. Given this, these factors should influence decisions about training and user interface development.

Many do not find it easy to accept that the user interface to a system is not just a skin on the surface but goes very deeply into the system. In fact, the user interface is any part of the system that affects any user or through which any user can affect the system. Thus, it is not just the form of the display and the entry of information and instructions that concern the human interface designer, it is also the content and the temporal and semantic structuring of that information. This means that almost all aspects of a system, from sensor accuracy to data processing and database design, are important from a human factors perspective because it is important to have the right kinds of information available for diagnosis or decisionmaking — or whatever the user's task requires — from moment to moment.

The purpose of this chapter is to provide an introduction to HCI in safety-critical systems. Chapter 6 will look more closely at the specific methods and techniques that can be used to improve safety through improving the quality of the user interface. Here, the purpose is to survey, generally, the issues involved in building user interfaces and to explore, in particular, the reasons why creating good HCI is so difficult; also to examine the kinds of approach to system development that address the problems arising from the difficulty.

4.2   WHAT MAKES HCI DESIGN SO DIFFICULT?

4.2.1   Difficulties in HCI Design

How often do you use a piece of software and think, 'Gosh, that's so easy to use!'? Sadly, the chances are that it isn't very often. Despite enormous strides in the past two decades in the development of user interfaces, we are still far from the stage where we can say that well-designed, usable software is being routinely produced by IT vendors and IT departments. For most of us who use computers, this means a daily struggle with complexity and inconsistency, additional stress and frustration, and a general lowering of productivity and quality in our work. For those who use computers to control and monitor safety-critical systems and processes, this daily struggle with unfriendly software could have far more serious consequences.

It may seem odd that I am insisting that HCI is difficult when we live in an age where graphical user interface (GUI) styles (such as Windows, Motif, Apple and Open Look) are being advertised as the simple answer to this very problem. Unfortunately, while the GUI, if properly used, can introduce a certain low-level consistency into screen design, it is far from being a panacea. Designing a good user interface involves a great deal more than simply assembling sets of standard components into more-or-less attractive arrangements. There are a number of sources of inherent difficulty in the process and it would be useful here to review the most important ones.

4.2.2   Users

Above all, users present the biggest problem for the interface designer; and the biggest problem with users is unpredictability. This stems from a number of sources but let me mention just two.

One major source of unpredictability in user behaviour is that there is a great deal of inherent variability between people — what the psychologists call 'individual differences'. Every single human trait, from reaction time to reading ability, from hearing acuity to height, and from memory to manual dexterity, can be very different from person to person. Permute these differences together and the result is the obvious (but almost always neglected) fact that no two users are the same. Yet, the problem of variability is by no means insoluble, as we can, to some extent, estimate its reasonable bounds.

The other major source of unpredictability is simple ignorance. We don't understand people. We can't say with certainty what an individual will do in any given set of circumstances; nor can we say, after the event, precisely why they did what they did. The science of psychology is still young (it is a mere 120 years since the first psychological laboratory was established) and still has a very long way to go. At its present stage of development, we are not able to model human behaviour in our safety analyses, in our designs, and in our test programmes, with anything but a tiny fraction of the degree of accuracy that we would like.

We will return to the subject of users in more detail in Section 4.6.

4.2.3   Tasks

People use computers for all kinds of tasks. As the use of computers becomes more and more pervasive in society, the range of tasks for which they are being used also increases. The nature of a particular user's task directly influences the requirements for support from the user interface. It only takes a moment's thought to see this. People's activities are structured in particular ways to suit the objectives they wish to achieve with the resources at their disposal. The types of activities involved determine the type of information that needs to be available or that needs to be captured, and the ordering of activities determines the sequencing of entering and displaying information at the user interface. It is not difficult to see the truth of this, but what seems to escape most system developers is its corollary: that, if task information is essential to the design of a user interface, then the tasks of the user must be analysed and designed before a satisfactory system can be built. Such task design is the essential requirement of what the user interface must be designed to support.

This means that developers must go through the long and difficult process of understanding exactly what it is that the end-users of a proposed computer system will be doing and why they will be doing it. Although this process is almost always absent in software development, without this understanding the designer is forced to make guesses about what the user needs and will, in many cases, get it wrong. Chapter 7 looks at the processes involved in identifying and analysing users' tasks.

4.2.4 Environments

Another source of variability for the interface designer is the environment in which the user will operate the computer. Here I am not referring only to the physical environment, but also to the social and organisational environments. Physically, we need to worry about work station ergonomics, ambient temperatures, noise levels and lighting levels (can the user reach the button, hear the warning tone, or see the state of a machine?). There may be a variety of stressors such as vibration, g-forces, and pressure. The user may be wearing protective clothing (can they operate the keyboard with gloves on? Can they see the overhead panel with a helmet on?) or they may be subject to environmental hazards.

Also under this heading, we include the organisational environment, looking at patterns of authority and responsibility — each of which can affect user action in safety-critical ways. The existence and knowledge of procedures is also important — especially procedures for correcting dangerous states of the system, possibly under circumstances of degraded physical and organisational conditions (for instance, the operations room is full of smoke and the operator's immediate supervisor is absent or unconscious).

We must also consider the social environment in which work takes place. The pattern of informal communication is often a considerable influence on the movement of information in an organisation. It is often through conversations over coffee or in slack periods that people learn about organisational changes or other environmental changes that might be relevant to their jobs. In this way they learn who is off sick or on holiday, they learn and rehearse new procedures by discussing them, and they learn about problems and practical remedies.

They also come to learn about their colleagues. They learn their strengths and weaknesses and they form bonds of trust and support. New systems that disrupt old social structures run the risk of weakening vital channels of communication and sources of knowledge, and of destroying valuable processes of learning and group cohesion.

4.2.5   Infinite Detail and the Problem of Copying

In 1989, Jack Carroll of IBM's T J Watson Research Center [Carroll 89] identified two critical problems for HCI. He called them 'Infinite Detail' and 'Emulation'.

Infinite detail is the problem that the finest details of a user interface can have enormous effects on its acceptability to users. For instance, the arrangement of items on menus can crucially effect search times, error rates, the learning and use of functionality, and even the way in which a system is understood; the naming of buttons or commands can have major effects on the types of mistakes a user might make; and a few extra or missing pixels in an icon design can radically alter its comprehensibility.

The fundamental problem here is that the complexity of the interactions between users, tasks, technologies and environments is immense. What is more, the effects of these interactions seem to be inherently non-linear. Thus we have very little chance of predicting what the effect of small changes might be. It is quite possible that interface usability varies chaotically (in the mathematical sense of the word [Gleick 87]) with small changes in the design parameters.

There is really no way around the infinite detail problem. By its very nature it seems to preclude the possibility of there ever being developed for HCI such rigorous theory that, for a given usability requirement, the precise interface that would satisfy it could be specified. Instead, we must use approximate techniques and we must rely heavily on prototyping. Prototyping avoids the problem of prediction by allowing the developer to iterate towards an optimum design by a process which proposes plausible solutions and then refines them in the light of evaluation (e.g., [Connell and Shafer 89]).

The problem of emulation is very similar. It is that, even when we find a user interface design that works well, we don't know exactly why it works so well. This means that if we copy it for some other purpose, it might fail — especially if we change it in any way, however slightly. Again, it is the infinite detail problem and the lack of theory which underlie the emulation problem. We cannot say just what it is about the design that works so well, and if we change any of the parameters — of the design, the users, the tasks, or the environment — we risk it not working at all. It appears that the sensitivity of an interface to perturbation is proportional to its degree of refinement for its purpose.

4.2.6 Procurers

The people who purchase computer systems are usually not those who use them. They are also not normally HCI specialists. This means that they are to some degree ignorant of the users, their tasks, and even the environments in which they work. They are also to some extent ignorant of the difficulties of good HCI design and of the kinds of solutions to the problems it presents.

To many procurers, a GUI simply makes interfaces more attractive and more expensive. They do not see the potential savings in user errors or the benefits in productivity and morale. To them, the extra effort in initial analysis (such as task analysis and conceptual modelling) simply makes projects longer and more costly. They are not aware of the vital nature of such analysis if usable, safe interfaces are to be built.

It is telling that a survey by Spikes Cavell [Spikes Cavell 92] on behalf of Groupe Bull showed that 32% of an IT department's time is spent on enhancing existing systems. Much of this so-called 'enhancement' is, in fact, putting right problems with the original design that users have complained about since the system was installed. It is also well known that many such problems are simply tolerated throughout the life of the system because of the prohibitive cost of late changes [Boehm 81]. As noted in Chapter 5, the specification of interface requirements helps to ensure that the user is considered at an early stage of development.

4.2.7 Developers

IT departments, consulting companies and software houses too are ignorant of the problems of HCI design. They use techniques and tools (such as the structured analysis and design methods and their accompanying CASE tools) which virtually ignore users and have almost no concern for user interface design. It is interesting and instructive to note that software development organisations almost never possess HCI skills — although they may well possess expertise in databases, networks, and other relevant specialisms. Some IT managers actually believe that the widespread use of graphical user interfaces effectively solves the HCI design problem. Nothing could be further from the truth — and the belief, and the attitude towards usability that it indicates, is perhaps a source of potential hazard.

4.2.8 Technology

User interface technology is a rapidly moving target. GUIs were first developed in the early 1970s, but their introduction into general use is now accelerating rapidly. It is a safe bet that in a few more years text-only user interfaces will no longer be developed.

Other interface technologies which too have been available for some time are also enjoying greater popularity. Speech input and output technology has reached the stage where it is now widely available in off-the-shelf products such as personal computers and toys. Radical new interaction styles such as virtual reality [Rheingold 91] will soon be moving out of the amusement arcades and into the workplace. Yet the practitioner's understanding of interface technology is lagging several years behind the leading edge of its deployment.

4.3   APPROACHES TO USER INTERFACE DESIGN

4.3.1   Structured Analysis and Design Methods

It is often said of structured analysis and design methods (SADMs) that they are system- or data-oriented and that they neglect the end-user. Unfortunately, an inspection of the leading SADMs reinforces that view.

One problem seems to be that the SADMs confuse the end-user requirement with the business or system requirement. Another, more serious, problem is that the SADMs provide little or nothing in the way of a *method* for capturing user requirements. If we look at the major SADMs in use today (e.g., SSADM, JSD and Yourdon), we find that the requirements capture stage is uniformly weak in the area of understanding user — as opposed to system — requirements. SSADM, for instance SSADM Version 3 [Longworth and Nicholls 86] and SSADM Version 4 [CCTA 90]), generates a document called the 'Problems/Requirements List' (in Version 4, it is the 'Requirements Catalogue'). This is in no sense-task oriented, and it is difficult to see how an end-user could make sensible comment on the abstracted, system issues it typically contains. In fact, most SADMs have very little to support the acquisition and recording of requirements as opposed to the analysis of them.

The CORE requirements specification technique [System Designers 85] is an approach which attempts to fill this gap. It not only provides a structure for collecting requirements and notations for writing them down, it also supports the concept of 'viewpoints'. A user 'viewpoint' is a typical perspective to be taken in a requirements capture exercise, and at least one attempt has been made to specialise CORE for user requirements capture [Gardner 86].

It is important that the user perspective on the system is taken into account throughout analysis and is reflected back to the user when he or she is required to make any judgement about its adequacy or completeness. Michael Jackson's views go some of the way along this path. In his discussion of the 'entity action step' [Jackson 83] he says: 'the input must be extensive and rich. We cannot afford to miss valuable clues to the user's view of reality' (p. 67). However, he goes on to say, 'we are not concerned with the specific techniques of interviewing, of gathering of evidence, but rather with the use the developer should make of the evidence gathered' (p. 68) Yet, the way in which the analyst communicates with the user is a powerful determinant of the quality and content of the analysis, and analysts who do not employ adequate techniques for requirements capture cannot hope to produce sound results.

Yourdon's is perhaps the method which is most sensitive to these problems. In one exposition [Yourdon 89], careful distinction is made between different classes of user. This in itself is not uncommon, but Yourdon goes on to discuss in detail the issues involved in working with end-users ('operational level users'). In particular, he mentions that the analyst must be allowed to communicate directly with such users, that the analyst must be able to talk in the users' own terms, and that analysts must be sensitive to the scope of a particular user's view of the system. However, although a detailed understanding of users is an important element of discovering their requirements, Yourdon's method still omits the detailed task modelling which is the most essential step.

4.3.2   Prototyping

Prototyping is one of the most useful techniques at the human factors analyst's disposal. Despite this, it is not a mainstream software development technique. Only recently has prototyping become acceptable to many developers, and only now is it beginning to be embodied in popular methodologies (e.g., SSADM Version 4). Take your favourite system development books off the shelf and look for the word 'prototype' in the indexes and it is almost certain that you will not find it. Although many have long argued that the popular models of system development are inadequate (e.g., [Swartout and Balzer 82, Hartson and Hix 89]), prototyping has not seemed acceptable as a solution.

This is partly because prototyping is seen to be costly and unmanageable. Its costliness is probably a justifiable complaint. However, considering costs is meaningless without also considering benefits, and it seems clear that effort spent early in a project to ensure that the requirements specification is a good and accurate one is amply repaid in later stages — especially during so-called 'maintenance' and 'enhancement'. Barry Boehm's famous relationship between life cycle phase and the relative cost of fixing an error suggest that an error is 'typically 100 times more expensive to correct in the maintenance phase on large projects than in the requirements phase' ([Boehm 81], p. 40).

The argument for prototyping at the analysis stage is that it offers significantly greater confidence that the decisions made in the analysis are correct than do other techniques. I strongly agree with Jackson [Jackson 83] who says, 'If a decision is error-prone, it should be subjected to the earliest possible confirmation or refutation' (p. 370).

Perhaps the most complete defence of prototyping as a development style is to be found in [Connell and Shafer 89]. They argue that, 'The benefit of prototyping to the analysis phase is the development of requirements specifications that are more meaningful to the users and will serve as a more accurate baseline for the following development phases' (p. 74). From this, most of the justification for the whole approach can be derived.

Of course, without adequate tools, developing prototypes can be a prohibitively expensive process. One very large project with which I am familiar produced a 'rapid' prototype in BASIC in six months. Within the context of a multimillion pound project, this was considered cost-effective. However, it is easy to see that for smaller projects, such an amount of effort would soon become excessive.

The purpose of prototyping requirements is to provide a simulation of the required system that a user can work with in order to assess whether the functionality, dialogues and task structures which it embodies are what are required in the final system. Building a sophisticated prototype from scratch in low-level languages like C, BASIC and Cobol is always going to be costly. What the analyst needs are tools supporting languages at a much higher level of abstraction and which are related specifically (or are tailorable) to the application domain [Windsor 92]. Such tools are very thin on the ground today. Where they are best developed is in the area of commercial data processing applications where the 4GLs (4th generation languages) fill this role. Unfortunately, while 4GLs do allow more rapid development of prototypes, they are often difficult and unwieldy in use and are limited in the range of interaction styles and functionality which they can support. However, the prototyping facilities of some of the leading 4GLs are undergoing improvements all the time.

For small-scale projects, or for small explorations within larger projects, tools such as Visual Basic or HyperCard can be useful. Although these embody rather low-level programming languages, the facilities they offer for the rapid construction of animated screens make them very effective if the appearance of the screen is the primary focus of the prototype. A recent trend towards rapid application development (RAD) has brought prototyping to a wider audience. RAD is, essentially, a type of incremental prototyping where 4GLs are used to create the core of a business application, in cooperation with user departments, in a very short time. Additional functionality is added later. While RAD does not, typically, involve human factors experts, it has proven very popular with users because of the increased level of their involvement in the development process.

The essential value of prototyping human-computer interactions is that it *validates* our understanding. We have discussed at length just why it is so very difficult to understand the human factors requirements for a system. Prototyping gives us an end-user evaluation of a concrete example of the manifestation of our current understanding. We can test directly whether our beliefs about the requirements for computer support are valid.

Prototyping also helps to *elaborate* the user interface requirement. By asking a user to do a job using a prototype for an extended period and under a variety of conditions, omissions in the requirement may be revealed. A particular instance, which is not at all atypical, occurred during the development of an oceanic air traffic control (ATC) user interface for the UK Civil Aviation Authority (this is described in [Storrs and Windsor 92]).

The incident involved clearance delivery officers (CDOs). These users have the job of speaking to aircraft as they approach the controlled airspace. The aircraft requests clearance to enter the airspace at a particular time, place, height and speed, and thereafter to fly a particular path. The CDO checks this against the flight plan which has been filed by the aircraft's operating company and passes the request, along with any changes, to the planning officer who attempts to fit the aircraft into the current traffic pattern. When this has been done, the planning officer passes the clearance to the CDO who then communicates it to the aircraft.

In our analysis of the CDO job, we read the appropriate procedure manuals and interviewed about ten CDOs as well as their managers. We observed CDOs at work on many occasions for many hours, discussing the work with them and taking careful notes. We produced analyses of the CDO tasks and discussed them with users to check that they were complete and accurate. We then designed a CDO prototype and produced the prototype components for it. We showed many users the designs on paper and then arranged a set of user trials, first with the components separately, and then together in a partially operative prototype. We then modified the design in the light of the considerable feedback we had amassed and produced a new and fully operational prototype CDO workstation.

This prototype worked well and the users were very happy with it. However, we noticed that, occasionally, during user trials with simulated air traffic, a user would jot down an aircraft call-sign on a piece of paper. When we asked why, we were told that, when they were too busy to deal immediately with a calling aircraft, they put it on a queue of pending calls and came back to it when they were free. They maintained the queue on paper. The requirement to support this task — indeed, the existence of the task itself — had been missed entirely by our team during observation and interviews, and by all the users who had inspected our task analyses, design documents and earlier prototypes. It was only by running a fully functional prototype with a realistic simulation that this had emerged at all. Without this prototyping step, we would happily have gone on to a full design without realising we had made a fairly significant omission in the requirements.

Prototyping user interfaces also helps to *resolve conflicts* between requirements. In the oceanic air traffic control project just cited, there was a serious conflict in the requirements for how many electronic flight strips (tabular displays of the proposed flight profile of an aircraft) should be displayed on the screen. Because of the amount of information on a flight strip, there was a physical limit to how many could be displayed simultaneously on even the large screens that were to be installed. Planning controllers wanted to see strips for all of the flights under their control, but this amounted to about twice the screen's capacity. All of the strips could be displayed if the information in each strip was reduced, but the controllers also wanted to see all of the information on each strip. To resolve this problem, we built a total of eleven working strip-boards with variations in the size of strip and access mechanisms for strip retrieval. Eventually, we reached a position of compromise on the requirements that all our trial users found acceptable. The important point here is that the use of prototypes helped to make the constraints and trade-offs clear to users and helped them to experience, and thus evaluate, the effects of particular design options as they attempted to do their jobs.

Finally, prototyping the user interface is a powerful means of *clarifying* vague requirements. Another example from the same ATC project was the planning of controllers' work from a queue of messages. These messages consist mostly of details of aircraft requesting clearances and are sent to the planners by the CDOs. Taking messages on a first-come-first-served basis can be inefficient, as giving a clearance to one flight may block the airspace for several later ones which would otherwise have clear paths. So the planners need to see several requests ahead so that they can make tactical decisions. But how many is 'several'? By varying the lengths of queue that were visible in our prototypes, we rapidly discovered the minimum queue length which would be adequate for this purpose, and we thus turned a vague requirement into a relatively precise one.

4.3.3 User-centred Development and the Importance of System Image

Donald Norman discusses the concept of a 'system image' [Norman 86]. This is the image of a computer system which we form as a result of using it. On the basis of the system image, users form mental models of the system which they then use to interpret what the system tells them and to help them understand how to control it. Each piece of software presents its own image to its users, and the software designer can consciously design this image to make it more helpful. It is possibly one of the most important aspects of the user-interface designer's job to make sure that the system image is the right one. This means that the system image should be one which helps the user to form appropriate models for interpretation and control of the system. With the formation of appropriate models, then the user's actions with or through the system will be less likely to be in error, and the user's beliefs and actions that arise as a result of the system will also be less likely to be in error.

Creating an appropriate system image for a piece of software to present to the user is the job of the user-interface designer. Because instilling an inappropriate model in the user's mind, or failing to enable the user to form a coherent model, can lead to error and thus potentially unsafe operation, there are two guidelines that should be followed when designing interfaces for safety-critical systems:

- Reflect the user's model;
- Be consistent.

For safety-critical systems, it is not wise to take the risk of inventing a novel system image. It is much safer to design the system to reflect the model that the user already has of his or her tasks. To achieve this, we must somehow understand how users construe their jobs and the domains in which the jobs are performed, so that we can reflect these same concepts back to the users through the system images of the software we design. HCI practitioners have, during the past decade or so, developed an approach known as 'user-centred development' which enables the designer to do just this.

The essence of the method is to focus on the actual or potential end-users of the systems to be developed and to use their understanding of the jobs to be supported to guide us towards a detailed design that will help them to do those jobs. To achieve this, we must begin with an understanding of who the users will be, what the purposes are of the tasks they perform, and under what conditions they will be doing their jobs. We then move directly to design and, by a rapidly iterating cycle of prototyping and evaluation, we simultaneously tease out the detailed requirements and evolve a design which will satisfy them.

There are two crucial elements in this. One is to analyse the tasks that the users will perform. There are many techniques for this (e.g., see [Diaper 89]) but the essential result we are looking for is a structured description of the activities of users that also motivates and sequences those activities. The other element is to build a model of the users' view of the domain in which they perform their tasks. This so-called 'conceptual model' should describe all the objects in the users' world, the relationships between them, the states they can be in, and the actions that can be performed on them by the users.

The task analysis and the conceptual model, taken together, provide us with a complete view of the user's job. If we base our system image on the image that the user already has of the domain, the actions that can and should be performed in it, and the consequential changes of state resulting from those actions, then we will reduce the time that the user takes to form a mental model of the system and we will ensure (as far as is possible) that the model formed is appropriate. In fact, the user will find the system 'intuitive' — the concepts it contains will be familiar and the results of actions will be predictable. Such immediate familiarity will have a positive effect on the safe operation of the system.

4.3.4 Usability Metrics

There is a notion in user interface design of 'usability metrics' (see [Shackel 86]). These are, simply, performance requirements, stated prior to design, which define what we expect users to be able to accomplish with the interfaces we provide for them. The usability metrics first specify the type of user, the task they are performing, the training they should have received, and the conditions under which they operate. They then state the metric to be used to gauge performance (normally, how quickly the task should be achieved and with how many errors). It is a way of defining in operational terms otherwise vague statements of requirements.

It is easy to see the attraction of the approach. If we can state precisely, in advance, the level of performance our users are expected to achieve, we may also be able to determine how safe we expect our system to be. However, the approach has serious problems. One is that current human performance is almost never known in sufficient detail for anyone to be able to do more than guess at what the required performance should be. (Nevertheless, 'at least as good as the present system' is a phrase that is frequently found in invitations to tender, statements of work, and even safety cases.)

Another problem is that, as has already been mentioned, there is not yet in HCI either sufficient theory or a sufficient accumulation of experience for us to be able to state the design process that would guarantee particular performance levels. By contrast, in the much older engineering disciplines, sufficient theory and experience does exist such that, by adding very wide margins of error (typically 100% or more), performance requirements can be guaranteed. One day, HCI may reach this stage of maturity. At present, there is still debate in the field of HCI about how to talk about the discipline [Storrs 89, Storrs 94] or even whether HCI is a discipline at all [Dowell and Long 89].

A further problem with usability metrics is that hundreds of detailed performance requirements demand lots of expensive acceptance testing, which suppliers prefer not to propose and procurers prefer not to pay for.

4.3.5   Formal Methods

One of the greatest hopes of recent years is that the production of safer software might be aided through the use of formal methods. User-interface developers have not been immune from this optimism, and the past decade has seen some important work on formalising everyday notions (e.g., [Payne and Green 86]) such as 'consistency' (interestingly, also referred to as 'safety' in [Harrison and Thimbleby 85]), and the use of formal languages to describe interactive systems (e.g., [Alexander 87, Dix 88]).

The benefit of using a mathematical description of requirements or software designs is that, as the development proceeds, the description is *in principle* verifiable at each stage. In reality, the difficulty of carrying out logical proofs for any but the smallest such description and for any but the simplest of properties is prohibitively expensive. Automated support for theorem proving is still relatively unhelpful. In addition to this practical problem, there are two more fundamental difficulties.

One is that notions such as consistency, which are central to the development of effective user interfaces, may be inherently unformalisable. ([Reisner 90] provides an interesting insight as to why this might be so and also how human error can arise from one kind of inconsistency.)

The other is that, while formality may help with verification, it does not help at all with validation — that is, confirming that what we have described corresponds to our intention, or, indeed, corresponds to the real world in any way whatsoever! In fact, because formal languages require specialists to interpret them, it is almost impossible for the originators of a requirement directly to validate it against a formal design at all.

The formalisation of concepts such as 'what you see is what you get' (WYSIWYG), interface separability, and the 'reachability' of states within interfaces, can be very helpful as a way of clarifying these often vague notions. However, as a practical user-interface development technique, formal specification is a long way from the useful tool that many expected it would be.

4.4 WHERE ARE THE HAZARDS IN USER INTERFACES?

From what has already been said, we can begin to see the many areas in user-interface analysis and design that can lead to hazards. We can also see where we need to direct our effort in order to reduce these hazards. The essential difficulties are in providing interfaces which will allow the accurate assessment of present and future system states and will control the safety-critical system to achieve desired states. We need to enable the interpretation of state information under both normal and abnormal conditions. This seems like an obvious statement, but the pitfalls are legion.

In the Airbus disaster described in Section 4.1 above, the operators misinterpreted the state of the system. Sometimes, however, the state may be uninterpretable. The problem of 'cascading' alarms when problems begin to arise is well known. As various parts of the system go into an abnormal state, each triggers its alarms. It is extraordinarily difficult to extract the significant information from these signals, to prioritise it, and to diagnose the underlying problem. User interfaces which simply dump all this information in the users' laps and leave them to get on with it are just adding to the problem. In safety-critical systems, this is a recipe for disaster, particularly when action — and it must be the correct action — is required speedily, such as in an aircraft in flight.

Other interface problems can be more subtle but just as dangerous. In the Airbus case, the state was accurately represented but falsely interpreted. At one point in the Three Mile Island incident, the system state was accurately interpreted from the information available, but the interface was not representing the true state of the system. In this case (as also mentioned in Chapter 2), the operators' consoles had lights which were taken to indicate whether particular valves were open or closed. In fact, the lights indicated the states of the switches that controlled the valves, not the states of the valves themselves. The disaster was caused when a valve stuck open while the corresponding light indicated that the switch was closed.

Operators are well known to have particular problems. They cannot maintain a high state of vigilance indefinitely — in fact, not for very long at all. They are easily confused by inconsistencies — not just the mode-related inconsistencies described earlier, but any kind of inconsistency in the presentation of information or in the control actions needed by an interface. They cannot form or maintain a useful model of a complex system's state simply by watching its indicators — if they do not maintain continued, active involvement with the system, they need a period of time during which to learn the current state before they can take any safe action in an emergency. Operators tend to explore their systems — the Chernobyl disaster can be thought of as an extreme case of such exploration leading to hazards.

All these and many other operator traits are well documented. They should not be ignored by analysts and designers. They cannot be fully compensated for with training programmes or procedural checks. They must be taken into account as being fundamental characteristics of the overall system. All too frequently, they are not. System certifiers should look for these well-known and common design faults, and their assessment should be included in the safety case where this is required (see Chapter 12).

4.5   PUTTING USER INTERFACES IN CONTEXT

In a striking study of London Underground control rooms, Heath and Luff [Heath and Luff 91] analysed hours of video tape in minute detail in order to understand the social dynamics of the control activity. The study was carried out as part of a larger project to replace older operations rooms and to amalgamate different functions. They found some surprising interactions taking place between controllers. For example, one controller, in charge of maintaining and updating the timetable and rescheduling trains dynamically as problems arose, was observed to be talking aloud to himself as he made changes. Another controller, responsible, among other things, for making public announcements about services, appeared to be able to monitor his colleague's mumbled monologue and telephone calls while he worked so that he could interrupt his own activities and make appropriate announcements. This highly effective style of working was not part of a documented procedure or of the official training for either job. Yet it was clear that if the two operators were separated (as was proposed in the new operations room design) the efficiency of their activities would suffer.

This is an example of how the environment in which an operator works (the arrangement of the furniture, the ambient noise, light and heat, and the proximity of colleagues and other resources) can have a major impact on the way a job is performed. The social and organisational environment in which people operate can also have a significant effect on their performance, determining to a great extent the modes of use of equipment. Many of the latent errors that occur are attributable to the practices that arise due to these social and physical environmental causes.

Safety is a property of systems taken as a whole, and the design of safe user interfaces must take into consideration the nature of the entire system. The user interface is the only means by which people are able to monitor or control a safety-critical system. As such, its designers must be aware of the impact it will have on the way work is organised and on the impact that the organisation of work will have on the use of the interface.

It is particularly important that the safety case (see Chapter 12) made for a system gives due weight to the usability of the system under all its modes of operation. In many respects, human operators exhibit many of the problems of software (difficult to test, unmanageably complex, full of hidden latent problems, unpredictably sensitive to unexpected combinations of events) when it comes to their propensity to fail in safety-critical systems. It is sometimes said that because software does not fail in the same way that hardware does (see Chapter 1 for a discussion of this), assigning a probability to its failure does not make sense. The safety case for software is therefore normally based on the use of appropriate development methods and quality assurance procedures. For user-interface software the human and the software problems are compounded, to the extent that most of the existing methods for software development are inadequate to guarantee an acceptably safe result. The safety case for a system involving human-computer interaction should show that user-centred, task-oriented and design-led development techniques have been applied in the analysis, design and implementation of the user-interface software. Without this assurance, we would be taking unnecessary and unwarranted risks.

4.6   PEOPLE ARE NOT LIKE OTHER SYSTEM COMPONENTS

Finally, it must be said once more that people are not much like other system components. Further, the innate variability of people and the differences between them have already been mentioned and they must be allowed for in order to ensure safety. To some extent, they can be ameliorated by careful recruitment and training. However, there are many forces acting to increase variability. People's ability to work varies throughout the day. So-called circadian rhythms, possibly due to underlying variations in body temperature, affect many cognitive skills. Attention rises and falls; ageing and learning affect speed and ability; interactions with other people affect all these variables so as to change them unpredictably.

Learning and experience can have a positive or a negative outcome. If the system image that is presented to the user through the interface leads to an inappropriate model of the system, then learning can have seriously detrimental effects on behaviour. Similarly, if the operator is misinformed by colleagues or through his or her own inferences, or if a system change is made which is not properly communicated, a faulty understanding of the system can arise that may lay dormant for long periods before it is revealed, perhaps to disastrous effect, during a system failure.

Different groups of individuals develop different cultures. For example, different groups of shift-workers, with low levels of interaction between groups, can develop quite radically different ways of working. These differences can be very difficult to eradicate, as each group believes its own way of working is the best — or only — way. Some of these modes of behaviour may be associated with successful operation in some circumstances but may not be appropriate to all circumstances (a deep understanding is missing). They may even be irrelevant to success or failure but are always entered into alongside the relevant behaviours (they are the so-called 'superstitious behaviours'). Other cultural differences come from past experience. Consider the way a North American will, through habit, attempt to turn a light off by flipping the switch down while the British try to flip the switch up. Deeply ingrained experience like this is very hard to overcome, with any amount of training — especially in high-stress situations, when 'unlearned' habits may recur spontaneously.

Unlike valves and electric motors, people have feelings. Unlike programmable electronic devices, they also have free will. The attitudes of an operator to a system and its user interface can crucially affect that operator's motivation. It can affect the amount of effort the operator puts into learning or understanding the interface. It can affect how much trust the operator has in what the interface is communicating to him or her. It can affect the length of time an operator will persevere with the interface to achieve or regain control in the event of difficulties.

People react to problems in safety-critical systems with varying degrees of anxiety or fear. A well-designed interface that keeps the user well informed, that displays information in manageable ways, that indicates the amount of uncertainty in the information in ways that can be understood, and which allows control actions to be taken in a forgiving environment, can help to reduce fear-induced stress. Stress can drastically alter an operator's capacity for rational thought. Decisions taken under stress are found to be poorer because they are typically taken on the basis of less information. Even perception suffers under stress. A kind of peripheral blindness or 'tunnel vision' can arise so that visual signals are missed and auditory communications may be mis-heard or missed entirely. User interfaces that take account of these factors can mitigate their effects (e.g., by not relying on peripheral visual signals during emergency situations, and by eliminating or reducing other stressors such as loud noises, flashing lights and high temperatures).

Free will is always the joker in the human factors pack. Whatever the training, procedures, or user interface, the operator may decide — for reasons an outsider can only guess at — to do something unorthodox or hazardous. For this reason, user interfaces must be designed defensively so that operators cannot — even for what seem to them to be good reasons — do something which is dangerous. A good example of this is the way that many models of car require the ignition key to open the petrol cap. This ensures that the engine cannot be left running while the petrol cap is open. Almost every petrol pump in every garage displays signs telling the driver to switch off the engine before filling the tank, but the car designers have taken the sensible precaution of not relying on the driver always to follow the instructions. Most drivers are unaware that they are being protected in this way, feeling only, perhaps, that the car's user interface is something less than totally convenient. Indeed, strict adherence to the principle of ease of use may occasionally be in conflict with the need to ensure safety. The good user-interface designer will, through a thorough analysis, be able to identify these occasions and to make an informed decision about the trade-off.

4.7   CONCLUDING REMARKS

In this Chapter, I have tried to illustrate the difficulty, and the sources of the difficulty, of designing a user interface to support the safe operation of a system. To achieve a good user interface requires special techniques: techniques which focus on the end-users and the tasks they will have to perform and which carefully validate the user requirements. Such techniques (e.g., task analysis, conceptual modelling and rapid prototyping) already exist but are almost never used in safety-critical systems development. It is certain that the adoption of these techniques would improve the safety of user interfaces.

It is difficult for some people to accept that the normal engineering approaches will not work when there are human operators in the system. These approaches have had such widespread success in the electromechanical domains for which they are appropriate that it is a pity they will not do for developing socio-technical systems. But people are not simple machines and they cannot be treated in simplistic ways. Over the past couple of decades, the discipline of human-computer interaction has developed ways of coping with the inherent difficulties of user-interface development. It is by no means a perfected set of techniques, but it embodies a way of understanding and avoiding many of the causes of human error and it is certainly the best set of techniques on offer.

REFERENCES

[Alexander 87]   Alexander H: *Formally-Based Techniques for Dialogue Design.* In Diaper D and Winder R (eds) *People and Computers III.* Cambridge University Press, 1987, 201-213

[Boehm 81]   Boehm B W: *Software Engineering Ergonomics.* Prentice-Hall, New Jersey, 1981

[Carroll 89]   Carroll J M: *Infinite Detail and Emulation in an Ontologically Minimized HCI.* IBM Research Report RC 15324 (#67108) 10/12/89 IBM Research Division, T J Watson Research Center, Yorktown Heights, New York 10598, 1989

[CCTA 90]   CCTA:*SSADM Version 4 Reference Manual.* NCC, Blackwell, 1990

[Connell and Shafer 89]   Connell J L and Shafer L: *Structured Rapid Prototyping: An evolutionary approach to software development.* Prentice-Hall, New Jersey, 1989

[Diaper 89]   Diaper D (ed): *Task Analysis for Human-Computer Interaction*. Ellis Horwood, Chichester, 1989

[Dix 88]   Dix A: *Abstract, Generic Models of Interactive Systems*. In Jones D M and Winder R (eds) *People and Computers IV*. Cambridge University Press, 1988, 63-77

[Dowell and Long 89]   Dowell J and Long J: *Towards a Conception for an Engineering Discipline of Human Factors*. Ergonomics, 32, 1513-1535, 1989

[Gardner 86]   Gardner A: *A Naval Engineering Standard for Human Factors in the Design of Military Computer-Based Systems*. HUSAT, 1986

[Gleick 87]   Gleick J: *Chaos*. Viking Press, New York, 1987

[Harrison and Thimbleby 85]   Harrison M D and Thimbleby H W: *Formalising Guidelines for the Design of Interactive Systems*. In Johnson P and Cook S (eds)*People and Computers: Designing the Interface*. Cambridge University Press, 161-171, 1985

[Hartson and Hix 89]   Hartson H R and Hix D: *Toward Empirically Derived Methodologies and Tools for Human-Computer Interface Development*. International Journal of Man-Machine Studies, 31, 477-494, 1989

[Heath and Luff 91]   Heath C C and Luff P: *Collaborative Activity and Technological Design: Task Co-ordination in London Underground Control Rooms*. Proc. E-CSCW '91, 65-80, 1991

[Jackson 83]   Jackson M A: *System Development*. Prentice-Hall International, New Jersey, 1983

[Longworth and Nicholls 86]   Longworth G and Nicholls D:*SSADM Manual*. NCC Publications, London, 1986

[Norman 86]   Norman D A: *Cognitive Engineering*. In Norman D A and Draper S W (eds) *User-Centred System Design: New Perspectives on Human-Computer Interaction*. Lawrence Erlbaum Associates, 31-65, 1986

[Payne and Green 86]   Payne S J and Green T R G: *Task Action Grammars: A Model of the Mental Representation of Task Languages*. Human-Computer Interaction, 2, 93-133, 1986

[Reason 90]   Reason J: *Human Error*. Cambridge University Press, 1990, 2

[Reisner 90]   Reisner P: *What is Inconsistency?* In Diaper D, Gilmore D, Cockton G and Shackel B (eds.) *Human-Computer Interaction: Interact '90*. North-Holland, 175-181, 1990

[Rheingold 91]   Rheingold H: *Virtual Reality*. Secker and Warburg, 1991

[Shackel 86]   Shackel B: *Ergonomics in design for Usability*. In Harrison M D and Monk A F (eds): *People and Computers: Designing for Usability*, Cambridge University Press, 44-64, 1986

[Spikes Cavell 92]   Spikes Cavell and Co. Ltd:*IT Development Environments: The International Perspective*. October 1992

[Storrs 89]   Storrs G: *A Conceptual Model of Human-Computer Interaction?* Behaviour and Information Technology, 8(5), 323-334, 1989

[Storrs 94]   Storrs G:*A Conceptualisation of Multi-party Interaction.* Interacting with Computers, 6(2), 173-189, 1994

[Storrs and Windsor 92]   Storrs G and Windsor P: *Rapid Prototyping for User Requirements Capture.* Proc. Air Traffic Information Systems '92. Aviation Technology Communications Ltd, London, April 1992

[Swartout and Balzer 82]   Swartout W and Balzer R: *On the Inevitable Intertwining of Specification and Implementation.* Communications of the ACM, 25(7), 438-445, 1982

[System Designers 85]   System Designers plc. *CORE — The Method*. Issue 1.0, November, 1985

[Windsor 92] Windsor P:*SIRIUS: An Object-Oriented Framework for Prototyping User Interfaces*. In Gray P and Took R (eds):*Building Interactive Systems: Architectures and Tools*, Springer-Verlag, London, 200-242, 1992

[Yourdon 89]   Yourdon E: *Modern Structured Analysis*. Prentice-Hall International, New Jersey, 1989

5

Specification of safety-critical systems

5.1 INTRODUCTION

Every system to be developed has requirements which must be documented in a specification. For systems associated with safety, the specification is a crucial necessity. Principle features of a specification are that it is a document and that it provides a description of the system to be developed. However, a specification may fulfil many roles, and the developers of systems need to have a clear view of any given specification's intended purpose. In this chapter, the various types of specification are discussed, together with their roles in supporting the safety analysis of a system. Where there is extensive human interaction with the system, it is recommended that the human roles and tasks are defined within the specification. The objective of this is to ensure that when safety is assessed, the interactions of human tasks with machine processes are clearly understood.

A specification forms the focal point for negotiation between the various parties who have an interest in the system to be developed or procured. It allows the procurer to identify to the supplier what is needed and then provides, in the future, a basis for assessing what has been supplied. Thus, prior to procurement it is used to identify and resolve differences between what is being offered and what is needed. It also provides the criteria against which the safety assessor can inspect the design of the system at any stage to ensure that safety issues have been properly addressed. Thus, for all the parties who have an interest in procurement, the specification provides a description of what is to be, what is to be done, and what should be produced. Without the specification it is impossible for the various parties to prove that the system does or does not meet its requirements, or even that it does what it is supposed to do.

In this chapter specifications of a total system are considered. Where a system includes significant human activity to meet its overall goals, the total system specification should consider the human as an integral component of the system rather than as being separate from it.

5.2   THE NATURE OF SPECIFICATIONS

The object of a specification is to provide a description of what is required. As the design and implementation of a system proceed through the life cycle, so a specification is required for the definition of what is to be produced at each stage. Thus, the format, context and role of the specification changes from stage to stage. At each stage, however, the specification must cover four aspects of the system:

- What the system is to do — its functions;
- How well it is to carry out those functions — its performance;
- The context within which it is to operate — its constraints;
- The features against which it will be judged — its acceptance criteria.

The role of the specification is always to provide a clear description which forms the basis of an understanding of what is to be constructed. As a result, there are a number of different specifications which are required during the design and implementation of a system. How these various specifications are developed and maintained depends on the life cycle model being used as the basis of the system development project. In a linear life cycle model they are carried out in sequence; in an interactive life cycle model they may not all be developed at the same time in the first place, but they are employed and refined in parallel. The principal specifications which need to be produced are:

- Statement of Need;
- Functional Requirements Specification;
- Functional Specification;
- Architectural Specification;
- Implementation Specification.

5.2.1 Statement of Need

This is a high-level description of the goals that must be met by the total system. It covers a full description of those goals and should identify all those persons who have an interest in the functionality of the system — not just users, but also managers, maintainers of the system, and all who expect to derive strategic advantage from it. These are the so-called stakeholders in the system. This specification should also define the procurement or development constraints on the system, including cost, time for development, interfacing requirements, and, importantly, the safety policy which is to be applied in its operation.

5.2.2 Functional Requirements Specification

This is a more detailed description of the system under consideration from the perspective of the client organisation. It includes a full description of all principal system functions and their mappings onto the organisational goals which are to be met by the system. The principal functions are decomposed to a level of unitary function, i.e., to the level of functional components that can be developed and operated in a relatively autonomous fashion. The functional requirements specification should also include a definition of all performance requirements that the system needs to meet.

Also contained in the functional requirements specification should be an analysis of the safety of the system. This should include the identification of hazards that may arise due to the operation or existence of the system. The hazard identification process needs to determine if there are processes or materials which can cause accidents resulting in death, injury or environmental incidents. If such hazards are identified, then procedures for the design and analysis of relevant parts of the system must be defined.

Lastly, the basis for acceptance should be defined. At this stage it is not usually possible to define specific acceptance criteria.

5.2.3 Functional Specification

This is an interpretation of the functional requirements specification from a design perspective and is concerned with the logical partitioning of the functions. It decomposes functions to a level where they can effectively be allocated to design teams.

It should also detail all the interfaces between the system and the external world and include definitions of responsibility for the management of those parts of the functional requirements specifications which relate to them. In cases of the system interfacing to other equipment, someone responsible for that equipment needs to 'own' the specification for the interface and be responsible for controlling any changes to it.

The functional specification should also include numeric values of the performance of both the human and the machine components of the system, as well as for the overall system itself in carrying out its various functions.

Where safety-critical aspects are identified, detailed design and evaluation procedures should be defined for their development. A preliminary hazard analysis needs to be carried out at this stage, and this will require a statement of tolerable risk for the system (see Chapter 1 for a discussion of tolerable risk). There should also be a statement of the protection policy that is to be adopted for the system, and this should cover issues such as the use of hard-wired protection systems, shut-down procedures, and manning-level requirements. The preliminary hazard analysis should seek to make the designers aware of the type and scope of hazards to which the system gives rise or against which the system must provide protection. The functional specification should also include a definition of the acceptance criteria for the system and the approach to be taken for acceptance testing.

5.2.4 Architectural Specification

This is a mapping of the logical system functions onto one or more specific system architectures. This specification needs to be developed in parallel with the functional specification, since it is only in the context of a specific allocation of functions to processes and physical architectures that performance and safety aspects can be evaluated. It should include the formal allocation of functions between the human and the machine components of the system

During the development of the architectural specification, the consequences of failure of design features can be analysed and estimates made of the likelihood and consequences of various hazards. The architectural specification should include numeric values for the system's performance and its acceptance criteria, as well as descriptions of the acceptance test procedures.

5.2.5 Implementation Specification

This is a detailed description of the chosen architecture specification with its mapping to the functional specification. It represents a guide to the design and implementation teams and also to the integration and test teams in the development and production of the desired system. It is an evolving description in that it is used to document all changes to the design which occur as a result of design decisions made during the development of the system. It is this description of the system which will be used in any formal analysis of the system for safe operation. It will necessarily include the trade-off between various performance requirements and their implications for safety targets.

5.3   THE LIFE CYCLE CONTEXT

There are many models of the development life cycle which set out to ensure that a system meets the requirements laid down for it. In the early days of software development, the basic approach was to write a specification and then develop code. The code was then evaluated against the specification and, where differences occurred, the code was fixed (or the specification amended). This model has come to be known as 'code and fix'.

The code-and-fix approach has many flaws, not least of which is that fixes may leave fragments of code which are not normally exercised. This renders it difficult to trace the operational flow through the system and to prove that code does what it is supposed to do and nothing else.

Dissatisfaction with this approach led to the proposal of various life cycle models designed to lead to well-constructed systems which can be analysed and maintained effectively. A widely used model is the 'V' Model which defines a series of stages, each of which needs to be complete prior to the commencement of the next. In effect, the product of each stage is a specification for the next stage of work. At the end of each stage (see Figure 5.1), there is a process of verification to check that the product of that stage is a faithful translation of its specification (the product of the former stage).

This model has significant advantages for use in safety-critical systems

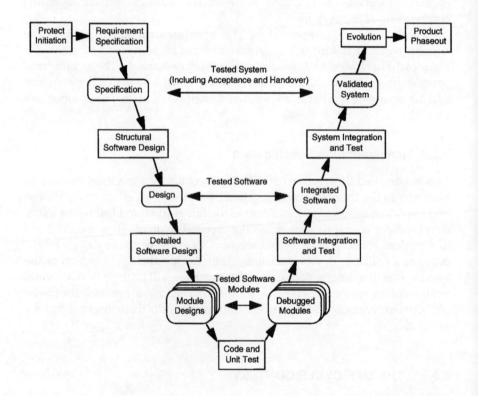

Figure 5.1 *The V Model of system development*

development in that the specific stages and validation steps can be completed by different teams. This independence of testing from development provides confidence that systematic errors are unlikely to occur due to misunderstandings within the design team. However, such a life cycle model imposes significant overheads on the design process in terms both of the time required for development and the rigidity of the design solution. This means that for large and complex systems the rigour of the process is often reduced to allow the procurement goals, such as cost and timescale, to be met.

It is not being suggested that safety should be sacrificed to meet procurement goals. What is at issue is that in many systems the safety-critical aspects form only a small part of the total system and that it is only these parts that need to be developed in the most rigorous manner.

A major issue with any new system is that its functionality is inevitably different from that of any system which it replaces. New systems are procured to meet more stringent performance goals, to comply with new regulations, and to overcome deficiencies in the existing system. The new features of the new system may include or necessitate radical changes in working practices and different partitioning of functions between humans and machines.

The result of changes to the manner in which human users interact with the machine component of the new system is that it is usually difficult to assess the full consequences of the changes until the system has been brought into use. It must be recognised that the safety of the system depends on there being no flaws in it and on its being used in a safe manner by those who operate it. Therefore, its design must include a description of how it is to be operated. The problem of describing the use of the system can be overcome by constructing prototypes which allow various options for user-machine interaction to be evaluated prior to full development of the system. However, the use of prototypes does not fit comfortably with the V life cycle model since the process of prototyping has to be completed at the model's first stage. A more appropriate life cycle model — the Spiral Model (see Figure 5.2) — was proposed by Boehm [Boehm 88].

In Boehm's spiral model there is an explicit place for the use of prototypes as part of the process of understanding various aspects of the system specification and reducing risk in the overall programme. When we apply this model to the development of safety-critical systems, the process of risk analysis covers safety risk as well as technical and procurement risk. The centre of the spiral represents the initial conception of the system, and the specification at this stage will be the statement of need. As each cycle of the spiral is completed, so the various forms of specification are produced, as defined in Section 5.2. Associated with each cycle of the spiral, the safety analysis will have different objectives — starting from identification of what hazards need to be considered and moving through to providing assurance that adequate protection has been provided against all identified hazards.

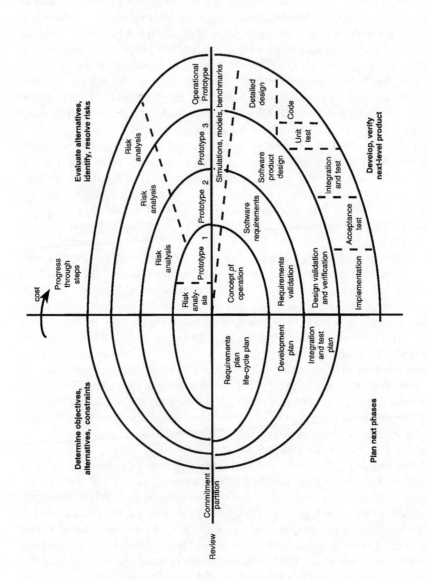

Figure 5.2 The spiral life cycle model

5.4 RISK ANALYSIS IN THE SPECIFICATION PROCESS

Whichever view is taken of the development life cycle, there is a need to carry out a safety risk analysis. A safety risk is defined as a function of the probability that an undesirable event will occur and the outcome or consequence of that event. Therefore, the safety analysis of the system must include procedures for identifying the hazards, quantifying the probabilities of their materialising, and determining their consequences. To meet this need, a number of procedures and analytical techniques have been developed. Collectively, these are referred to as 'predictive hazard evaluation'. The basic steps in the evaluation procedure are shown in Figure 5.3 (after [CCPS 85], and briefly described below.

5.4.1 System Description

A description of the system to be evaluated must first be prepared in a suitable form. The description must capture all the system's key aspects and functions. It must include all interfaces to external processes. Where time-related aspects are critical, it must include timing diagrams in which absolute times are defined. Where timing is in terms of sequencing, state transition diagrams, showing the state descriptions and the transitions from one state to another, can be used.

The system description must be in a form which is readily understandable by those who are to inspect it, so the choice of design representation should be made with the need for safety analysis in mind. The most natural, and perhaps appropriate, description is that which is, or is to be, used by the designers of the system. Where the design representation does not include the human components of the system and their activities, a suitable task analysis description is required.

Where the design representation is not understandable to all those involved in its inspection, or where, for some reason, it is deemed to be incomplete with respect to key aspects of the system, an alternative or additional representation will be required. However, the danger here is that any changes which have been made to the original design representation, due to specification changes or the discovery of design errors, may not be transferred to the alternative representation. Then the system analysed would not be that which is being, or has been, developed.

5.4.2 Hazard Identification

This is the process in which the system description is subjected to critical analysis to identify the hazards associated with it. Hazards are usually those which are introduced by the system, particularly in the event of a failure. They may also, however, be existing hazards which the system under development may be required to obviate.

The process of hazard identification requires an extremely detailed

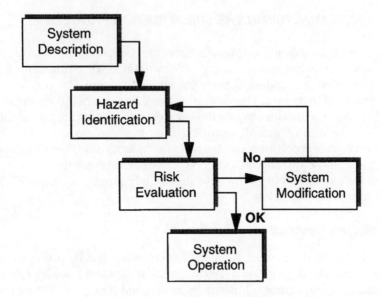

Figure 5.3 *The risk analysis process*

review, with the viewpoints of designers, users and other key stakeholders in the system being considered. A number of structured methods have been devised to support the process, two of the most widely used being hazard and operability studies (HAZOP) and failure modes and effects analysis (FMEA), both of which are briefly described in Chapter 1.

5.4.3 Risk Evaluation

In the process of risk evaluation, the identified hazards are analysed for their likelihood and consequence. Each hazard must be considered in terms of ultimate (i.e., system) rather than merely local consequences, for safety is a system issue. Thus, the risk evaluation must be carried out on the total plant rather than only on, say, the computer system controlling the plant.

If the consequence of a given hazard is determined to be relevant, its possible causes are then explored. Frequently the immediate trigger of the hazard is not the ultimate cause, and causal chains need to be followed to determine the sources of failure which could cause it to occur. In examining the causes of a hazard, the various protection systems which may already be included in the system to eliminate or mitigate the hazard are assessed for their adequacy.

Risk evaluation may be qualitative or quantitative. When it is qualitative, the causes and consequences are not attributed numeric values but only relative estimates, such as 'high', 'medium' and 'low' (see Chapter 1 and also [Redmill 97]. In some cases this is adequate as the basis of design decisions.

The current trend in safety-critical systems, however, is to carry out a quantitative evaluation. When this is done, standard methods, such as fault tree analysis (which is defined below), may be used. Chapter 3 offers a review of some of the methods that are used to provide a quantification of the human contribution to system failure.

An important part of risk evaluation is the determination of the tolerability of risk (see the discussion of the 'ALARP' principle in Chapter 1). In this, the risk (a function of the likelihood and consequence of the hazard) is compared with the predefined level of tolerable risk. Where the assessed level of risk is less (i.e., the analysis shows the system to be safer), the risk is deemed tolerable. Where the risk is assessed to be greater, it must be reduced if the system is to be brought into service. As risk reduction is often costly, a number of considerations need to be made and these are discussed in Section 1.4 of Chapter 1.

5.4.4 System Modification

It the risk associated with any hazard is found not to be tolerable, the possible means of eliminating or mitigating the hazard are considered. Such means may be built into the system (in software or hardware) or they may be external (such as re-siting the system or enclosing it by a concrete wall). They may include the introduction of additional protection measures, the setting up of special operational procedures, or, in the worst case, a major redesign of the system. Where significant modifications are made, the risk assessment procedure must be repeated.

5.4.5 System Operation

It is likely that modification will occur during the operational life of a system. Whenever modifications are made to a safety-critical system, they should be assessed for their effects on system safety, and the evaluation should take the course described in Sections 5.4.1 to 5.4.4 above.

The system description used for the safety assessment needs to include the modifications and be up to date, otherwise the assessed system will not be a true representation of the system as it exists. Each modification is then considered with respect to its interactions with the rest of the system and with the outside world.

When no significant changes have occurred, the safety inspection may be limited, but care should be observed in deciding on this, for even those modifications considered harmless at the time of making them may in fact have unexpected interactions with other components or with future modifications.

5.5   FAULT TREE ANALYSIS FOR SUPPORTING THE ANALYSIS OF LIKELIHOOD IN RISK EVALUATION

In order to produce a risk evaluation, there has to be an estimation of the likelihood of the occurrence of an event. There are two parts to this process. The first is to examine component failures and how they form a chain of events to cause a significant incident, and the second is to determine the overall likelihood of each event in the chain. To support this evaluation there need to be databases of known component failures and the occurrences of incidents. The analysis can then be supported by historic evidence from which component failure rates and the frequency of given types of incidents can be derived.

But for new systems there are no historic databases, so we must build up a picture of component failures and how they lead to an overall system failure, and this picture must include, or must lead to a derivation of, the likelihood of the system failure. This is largely done by extrapolating from known failures and combining the information in a realistic manner. A key analytical method used in this analysis is based on a tree structure and is known as 'fault tree analysis' [Vesely et al 81].

In this technique, a tree of faults and causative events is built up, with the top events being the hazardous incidents, such as explosions, which were identified in the hazard identification process (see Section 5.4.2 above).

Starting with each top event in turn, the high-level faults which could lead to them are identified, and these are made to form the next level of the tree. Each of these is then broken down to form successive levels, thus forming a chain of causation until the initiating events are identified. These, which are usually component failures, form the bottom level.

Faults are combined by logical AND and OR statements which are placed at the points of branching in the tree. Two faults at a given level are ANDed if they both have to occur for the fault above to occur. They are ORed if either fault would result in the occurrence of the fault above. In well-designed safety systems, the interlocks should all provide AND statements, for this ensures that a hazardous incident cannot be triggered by a single event but always requires a combination of events.

Probabilities of occurrence (usually based on historic data) are attached to each event on the tree, and from these the likelihood of hazardous events can be calculated. Branches identified as having negligible probabilities may be ignored, but this needs to be done with care.

5.6   HAZARD ANALYSIS OF HUMAN-CENTRED SYSTEMS

5.6.1   An Integrated Approach

One problem that we face in considering computer-based systems is the way in which we understand and model the roles and tasks of the human components — usually the operators. In many cases the approach taken is to treat the operator as a completely separate entity. The system is then designed with external sources and sinks, collectively labelled 'HCI' (human-computer interaction). This is treated as a separate 'system', the responsibility for its design being the HCI specialist. In complex systems, there may well be a separate activity to analyse the operators' tasks, using some form of task analysis aimed at defining appropriate manning levels and training requirements. It should also address human hazard issues in detail.

Problems may arise if the representation of human activity is not integrated with the representation of the machine design. Without a common representation of the two viewpoints, machine and human, it is difficult to imagine the consequences which a change in one representation may have on the other.

One approach to the problem of the integration of representations is the SUSI™ methodology which adopts a common notation for representing the human tasks and system processes [Chudleigh 93, Clare 94]. The notation to be used is not fixed within the methodology but may be one of a number employed for the description of software-based systems and which represent processes and the information exchange between them (for example, Hatley and Pirbhai [Hatley 88]). Importantly, however, the chosen notation must have real-time extensions. This is because the issues of time and sequence are key attributes of human tasks. However, an approach based on such an information model is valid only as long as the human activity is primarily information-intensive — for example, where the human processes involve:

- Decision taking (including problem solving);
- Information transfer;
- Classification and sorting.

This type of representation is less effective when the human processes have significant motor skills or introspective reasoning components, since these are not easily represented in terms of processes and information and control flows. However, the vast majority of human work falls into the information-intensive class, especially where automation has been used to replace the manual aspects of tasks.

The common representation of human tasks and system functions must cover all relevant aspects. The growth of structured design methods in software has led to a variety of methods with a number of conventions for producing an explicit representation of the system. Key among these methods are various data flow, process flow and state transition representations, such as CORE [Mobbs 85, Mobbs 86] and Yourdon [Yourdon 79]. These descriptions provide powerful means of representing the software design and are more or less suitable for differing categories of software systems. A key feature of such design methods is that they are used to represent the specification as well as the design, at least at the functional requirements level.

When a data flow and process model is adopted for the human components, we can generate an integrated representation of the whole

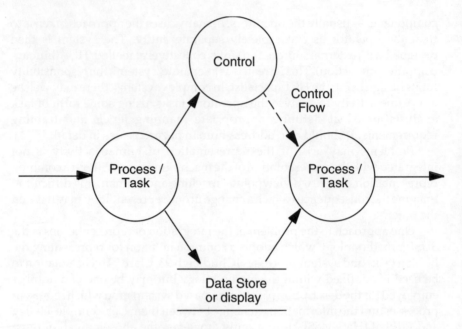

Figure 5.4 *Process and flow representation*

system. Using this integrated representation we can explore the consequences of failure in a consistent manner across the entire system, rather than within only one or other part of it. Figure 5.4 illustrates the conventions used in this type of representation. The key components are:

- A circle, which represents a machine process or human task;
- A dotted circle, which represents a control, machine processor or human task;
- A solid line, which represents a data or information flow;
- A dashed line, which represents a control flow (stop, start, etc.);
- A pair of parallel lines, which represent a data store or visual display.

Part of the convention is to represent visual displays as data stores, since data may be written to a screen, but there has to be an explicit human process to read the data. Auditory signals usually act as control flows and are represented accordingly.

5.6.2   Developing the System Representation

In order to develop a representation of the system's functional model, it is necessary to consider both the human tasks and the machine processes in the context of a total system. It is likely that a description of the functions associated with the machine processes will already exist, and in some cases so will task descriptions. If task descriptions do not exist, they need to be prepared.

The approach to developing an integrated system representation set out here is based on a staged process which starts with an identification of the system goals and proceeds to a functional decomposition of the system. The nature of the process is such that there need to be several iterations in developing the system descriptions. The principle steps are:

- Identify the goals of the system;
- Define the constraints on the system;
- Define the boundary of the system;
- Decompose the functions until human tasks and machine processes can be separately identified.

The methodology is illustrated by considering a worked example applied to a medical imaging system.

5.6.3 An Example: A Medical System

The role of specification is illustrated in the evaluation of a medical imaging system. The system in question was the result of a research programme which sought to provide automation in the inspection of cervical smear specimens. The research programme had addressed the key areas of image processing and specimen preparation. A system concept had been developed. As part of a review of the system, the question was asked, 'Is the system concept safe when deployed in a laboratory environment?' 'Safe' is defined as no wrong diagnostic decisions arising from error.

The system concept was reviewed in the context of a typical laboratory where it could be used, and it was revealed that it did not include key aspects which should be in a functional requirements specification or functional specification. A critical issue was the recovery from errors in the forms received from clinics. It was decided that prior to a safety review, a functional specification would have to be developed. This was done by a small team which included the head of a test laboratory. This meant that the requirements of an operational laboratory were now included alongside the key technical issues of image processing and specimen preparation.

The following sections describe how the functional specification was developed. Lastly, a formal safety review was carried out using a HAZOP study of the specification. The process was carried out in accordance with the SUSI methodology as described above.

This system is illustrated in Figure 5.5 and briefly described here to provide a context for understanding the methodology.

(a)   System description

Specimens are submitted by clinics and doctors' surgeries, together with forms containing details of the patients. Specimens will have been taken either as part of a routine testing programme, because of previous doubtful tests, or because of clear clinical indications.

On receipt of a patient sample, the data from the relevant form are entered into the laboratory computer system. Form details are matched with previous records for the patient, if any exist.

To each sample bottle is attached a unique bar-coded label. Each bottle is submitted to the robot slide preparation system (RSPS) which applies a layer of the specimen contained in the bottle to each of two slides.

At this stage a form is printed, together with patient detail labels for attachment to the slides, the original sample bottle, and the original patient form.

The next stage is a checking and sorting process, the objective of which is to cross-check that the original matching of the sample bottle and form were correct, that slides have been correctly produced, and that details on the original form have been correctly entered. Slides are then assigned for processing.

The slide scanning system (SSS) scans the slides and, when abnormal or unclassifiable objects are found, it enters the most significant of these into an image data file. The images of objects which are doubtful or unclassifiable are sent for interactive review. Interactive review is conducted by a specialist reviewer who classifies the sample on the basis of an inspection of images displayed at a work station. Work stations for both medical and technical reviews consist of a display screen for the images from the SSS and a microscope for looking at the slides, stained for visual inspection.

The administrative process reconciles patient data and diagnoses in issuing reports to clinics. Finally, patient slide pairs are placed in a long-term archive. The overall process includes a number of quality control checks and error checks to ensure that errors are minimised.

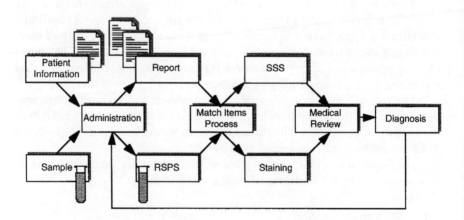

Figure 5.5 Schematic of the cervical screening system

(b) Development of the data flow description

(i) Identify goals

The principal goal of the cervical screening system is the inspection of specimens for abnormalities. This goal is to be met with a high level of certainty in the correct identification of both abnormal and normal cases. The principal stakeholders in the system and their requirements are:

- Patients, who require a reliable system which does not produce false reports;
- Doctors, who require the provision of reports which can be interpreted easily for patients;
- Management, who require the production of quality control and audit data;
- Staff, who require feedback on individual performance.

(ii) Constraints

The system is to be deployed in existing laboratories and it will have to be integrated with existing patient record-keeping systems. The existing practice is to retain all forms and specimen bottles submitted to the laboratory for a period of 3 months as part of a quality control and audit procedure. All inspections of slide specimens are stored for 20 years, or for 5 years after a patient's death. The equipment will be operated by existing grades of staff with appropriate training.

(iii) Boundary definition

The boundary is defined as the interface between the total system and the external world with which it interacts. Principal boundaries are:

- Clinics and surgeries;
- Hospital management systems;
- Archive store for retained slides.

Based on the analysis of the boundaries, a context for the system can be defined as shown in Figure 5.6.

(iv) Decomposition of functions

Starting from the context shown in Figure 5.6, the top-level decomposition is developed (see Figure 5.7). From the description of the system we can identify two principal functions: those of preparation and screening, and review. The preparation and screening function is principally concerned with the analysis of material in order to present the medical review process with the best data on which to make a diagnosis. The review process then

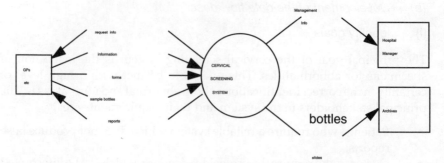

Figure 5.6 *The context of the system*

takes that data and develops a diagnosis from it.

At this level, two additional functions can also be identified. First, the specimens and forms from the medical clinics and General Practitioners' surgeries need to be input to the computer system. The form details then need to be compared against existing records to find matches with patient histories. Where inconsistencies are found, further information will be sought from the sources of the samples. The second function identified is that of accessing existing records from the laboratory computer system. This function is separated at this level since it relates to an existing system which also provides functionality for other processes outside the scope of the cervical screening system.

Each process is then further decomposed until there is sufficient detail from which to conduct a hazard analysis. Figure 5.8 shows the decomposition of the preparation and scanning processes.

A key process at the lower level is 'match items'. In the design of the system it was recognised that there had to be a double check on the initial

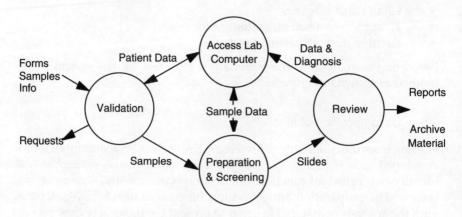

Figure 5.7 *Top-level decomposition*

pairing of patient forms and patient sample bottles. The forms and labels on bottles are usually hand-written and there are often inconsistencies in names, e.g., Maggie and Margaret may both refer to the same person, or between several samples with similar patient names in a batch (e.g., Smith). In the manual system there are a number of checking procedures to trap any errors that occur at the first matching of patient forms to samples. In the automated system there are fewer opportunities to cross-check for errors, so an explicit process was defined. It was also recognised that there were other quality control and sorting processes that could be carried out at this stage.

(c)   Hazard identification

The description of the whole system was subjected to a hazard and operability study (HAZOP) — for a description of which see [MOD 96]. The study team consisted of a leader, two designers (one of whom acted as recorder and who was also a human factors specialist), a doctor, and a senior medical technician. The latter two were both members of a laboratory where the screening of cervical smears was carried out without automated inspection support. The objective of the HAZOP was to identify potential hazards and hazardous situations in the preliminary design of the automated screening system which would need to be addressed in subsequent design stages. Examples of the activities in which errors could lead to hazards included:

- Cross-checking the original matching of a patient sample and its documentation;
- Visual quality assessment of prepared slides;
- Attachment to each slide of a label bearing the patient's name and date of birth;
- Assignment of samples for priority processing where there are clinical indications.

The execution of this process would be supported by a screen to display status information and the use of bar-code reader pens to cross-check bar-coded items against patient details.

(d)   The benefits of early safety analysis for the medical system

The process of considering the safety issues associated with the system quickly highlighted shortcomings in how the system would be deployed in a real environment. The process of considering the functional requirements specification and functional specification from the viewpoint of an operational laboratory identified a number of functions that would be required for operational systems. The safety review of the functional specification also identified a number of aspects which would require design attention. The early identification of these factors meant that they could be addressed during the design of an operational system. Thus later and more costly redesign activities could be avoided.

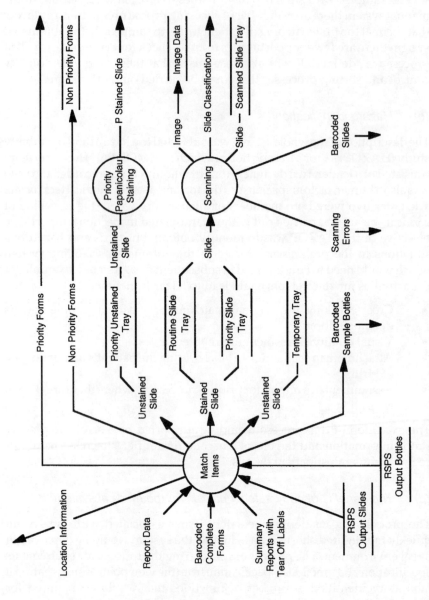

Figure 5.8 The preparation and scanning process

5.6.4　The Specification's Role in Analysis of the Medical System

The investigation of the Cervical Screening System was carried out as part of a project to understand how such systems could be designed in an intrinsically safe manner. Considerable work had been undertaken to solve the problem of automatic recognition of healthy cells. In order to achieve the recognition performance, the need for the precise application of a layer of the specimen had been identified. Because these technical issues had driven the research programme, the description of requirements was largely framed in terms of the desired technical performance. Since the goal of the project was to ensure that the system, if fully developed, would be safe, the first stage was to review the safety of the system concept as it existed.

The review of safety was conducted as a series of hazard identification studies from various viewpoints. The review from the usability viewpoint initially encountered a significant problem in that no specification existed which corresponded to a complete statement of need or functional requirements specification.

A cursory review from the usability viewpoint in the context of a working laboratory showed that significant areas of the functional requirements specification had not been addressed. These related in particular to the need to resolve errors which occurred outside the system boundary, such as when different names were given on the specimen and the form.

As a result of the initial analysis, a function requirements specification was developed which encompassed the total system, i.e., not just the screening systems and slide preparation, but the whole process from specimen receipt to report delivery. Developing this functional requirements specification immediately focused the attention on areas of operation of the system which had not previously received design attention.

The development of the functional requirements specification had additional outcomes which would also be key to the deployment of the system. One of the most significant of these was the issue of competence levels of staff associated with the various human functions that would be required. It became clear to the laboratory manager that the deployment of such a system would require a different type of employee in terms of skills and experience. This in itself would lead to the creation of new grades and changes to the organisational structure. Thus the safety review highlighted both machine and human aspects.

This study was only concerned with a system at an early stage in its life cycle. It was at the concept development, or early prototype, stage. However, the development of the functional requirements specification for the total system placed the programme on a sound footing for the development of a safe system. If the project had continued without this all-encompassing view of functions contained in the integrated specification, there is little doubt that any developed system would have been difficult to use effectively. In addition, that difficulty of use would have led to an increase in errors which would have resulted in unsafe diagnosis of patients' conditions.

ICLUSIONS

/-critical system there is a need to carry out a safety analysis. e process of safety analysis is the ability to describe the system. ~~~~~ription is significantly improved if the process of specification in the context of the development life cycle is clearly understood.

As system development proceeds, the specification goes through a number of stages, from the initial statement of needs to a description of the implementation of the system under development. The process of safety analysis and the associated management of risk can be significantly enhanced if the various specifications are seen as living documents which are continuously refined throughout the life of the system. By maintaining the various specifications, it is possible to review the safety issues from disparate viewpoints at any stage of the system's life cycle.

Where an iterative life cycle model is adopted, such as the spiral model, the safety analysis can be initiated at an early stage. The advantage of this is that it can guide the system design so as to minimise the need for modifications arising from subsequent safety analyses. The later the need for a modification occurs in the development of the system, the greater the cost of making it.

When the system includes a significant amount of human interaction, it is recommended that the overall system description includes the human as a component. By including the human component, overall functionality can more easily be understood. The early consideration of human tasks means that the role of the operator can be explicitly defined, and this significantly aids the process of understanding the relationship between machine functions and human tasks.

In order to support the overall description of the system, appropriate tools are needed which fully represent all key aspects with respect to safety. Where the processes are information-intensive, a process and data flow representation can be used. The use of system representation methods such as Yourdon [Yourdon 79] allow an integrated human and machine description to be developed. The SUSI methodology has been developed by defining an integrated approach to system description, hazard identification and risk evaluation, and this is particularly appropriate for representing the total safety-critical system.

REFERENCES

[Boehm 88]   Boehm B W and Papaccio P N: *Understanding and Controlling Software Costs*. IEEE Trans on Software Engineering, October 1988

[CCPS 85]   American Institute of Chemical Engineers: *Guidelines for Hazard Evaluation Procedures*. Center for Chemical Process Safety, New York, 1985

[Chudleigh 93]   Chudleigh M and Clare J: *The Benefits of SUSI: Safety Analysis of User System Interaction.* In Górski (ed): Proceedings of the 12th International Conference on Computer Safety, Reliability and Security. October 1993 (SAFECOMP '93) p. 123-132

[Clare 94] Clare J, Chudleigh M and Catmur J: *SUSI: a Methodology for Safety Analysis of Control Rooms.* In Murthy et al (eds): Computers in Railways IV-Volume 2—Railway Operations (Railcomp 94), p. 155-160, September 1994

[Hatley 88] Hatley D J and Pirbhai I A: *Strategies for Real-Time System Specification.* Dorset House, 1988

[Mobbs 85] Mobbs A J: *CORE and its Applicability to Naval Command System Requirements.* Report XCC3/TN37/85, MOD(PE), ARE Portsdown, April 1985

[Mobbs 86] Mobbs A J: *STARTS Debrief on CORE.* NCC Publications, 1986

[MOD 96] Interim Defence Standard 00-58: HAZOP Studies on Systems Containing Programmable Electronics. Ministry of Defence, 1996

[Redmill 97] Redmill F: *Practical Risk Management.* Chapter 8 in Redmill F and Dale C (eds): *Life Cycle Management for Dependability.* Springer-Verlag, London, 1997

[Vesely et al 81] Vesely W E, Goldberg F M, Roberts N H and Haasl D F: *Fault Tree Handbook.* Report Number NUREG-0492, US Nuclear Regulatory Commission, Washington, 1981

[Yourdon 79] Yourdon E N: Classics in Software Engineering. Yourdon Press, 1979

6

Interface design for safety-critical systems

6.1   INTRODUCTION

The design of interfaces has always been a focus of the human factors discipline. In early human factors this was characterised by the so-called 'knobs and dials' ergonomics. It is now focused on the boundary between people and the software and hardware they use for systems control. The term 'human-computer interface' (HCI) is used to describe the boundary where man interacts with the software and hardware of a system. Exact definitions vary and this boundary can be seen as physical or conceptual, or both. What is agreed are the principals which contribute to an effective human-machine interface and that these principles are manifest through the design of controls and displays. All systems have some reliance on human operators and the ergonomics discipline views man as an integral part of the systems loop (Figure 6.1).

   This chapter aims to address the issues to be considered in the design of interfaces to support the operation of such complex systems with the emphasis on the formatting and design of VDU- (visual display unit-) based displays as the most commonly used media. It is intended to address the issues in interface design that arise in large-scale complex systems based in a high hazard and safety-critical context. This chapter will not address the issues that arise in the design of displays which aim to provide intelligent decision support or knowledge-based systems, but see Chapter 4 for an introcuction. This chapter builds on this and focuses on the issues that need to be addressed in the design and evaluation of safety-critical interfaces. The chapter is structured into five sections:

- This section examines the characteristics of the different elements that comprise an interface, the user, hardware and system software;
- Section 6.2 looks at the process of display design;
- Section 6.3 provides an overview of the structuring and operational aspects to be considered in display design;
- Section 6.4 introduces the issues which are important in representing information on display formats;
- Section 6.5 provides an overview of interface design issues in abnormal and emergency operating conditions.

The information that is presented to the operator by the system and is input into the system by the operator is the key to effective human-machine interaction, and the role of the human-machine interface is to facilitate this interaction. The interface itself is a property of both hardware and software, whilst the operator will come to the interface with certain skills, training, and knowledge to allow him to use the information and operate the system, the way in which the interface is designed will govern the ways in which this interaction can occur. Human beings are very adaptable and so will be able

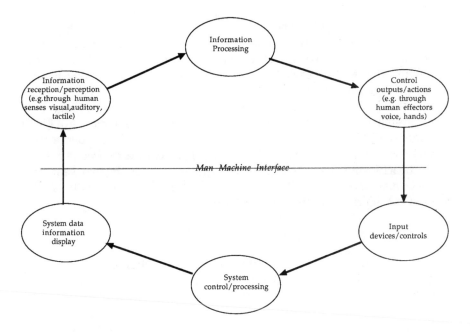

Figure 6.1 *Diagramatic representation of the transfer of information across the man-machine interface*

to operate all but the most poorly designed interfaces to some degree, often developing informal ways of overcoming problems. There are numerous examples of this in texts such as [Norman 88] and most people can identify common items such as video recorders, washing machines, clock radios, etc. which have features apparently designed without consideration of the way they will be used in the real world. Whilst these are not safety-critical systems, the human factors principles in preventing error and optimising use are the same. A poorly designed interface may induce or at least increase the likelihood of a human error being committed. Where the tasks of the operator are safety-critical, the consequences of such an error may be catastrophic.

Many of the accidents cited in other chapters of this book, such as the nuclear incident at Three Mile Island, had the design of the man-machine interface given as an underlying contributory cause of human error. Countless examples of errors induced by even simple but safety-critical interfaces are reported in the human factors literature (for example, [Kletz 94]). Simple, real life, examples include the overpressurisation of a system because the display was given in bar units instead of the lb/square inch familiar to the operator and used in other system displays, or the misdiagnosing of a failure by an operator because an indicator (when lit) showed that the control signal to operate a valve had been sent rather than indicating that the valve was now open and the operator's model of the system was that a light indicated that the valve was open.

The complexity of technological systems has continued to grow over the last two decades and the demands for the safety and efficiency of these systems has continued to increase. Increasingly the overall reliability of such systems is limited by the reliability of people who operate them. As technology has improved and the reliability of the hardware components has increased, so the human operators have become one of the more unreliable components of a system. It is therefore imperative that the interfaces, which provide the means of interaction between the system and the people who operate them, allow the operation of the system to be as efficient and error free as possible.

In achieving design to support error-free operation the two elements of the human and the technical system with its hardware and software both need consideration for their characteristics and limitations.

6.1.1 User Characteristics

The design of any interface centres on the population of users who will interact with it. In a safety-critical system context such users are generally well-trained individuals who understand the functioning of the system and its limitations.

In large systems, the complexity of information that an operator requires to perform the task is such that often the visual sensory channel is the only one with sufficient discrimination to be effective. Auditory channels are more readily overloaded, for example, as occurred in the Three Mile Island crisis, where over 60 different alarms and warnings were activated and

Table 6.1 *Human characteristics which impact on interface design*

The human perceptual system	The size of text and information, perception and discrimination of colours
Anthropometric and biomechanical characteristics	The layout of the workplace, console design, reach distances
Human cognitive performance capabilities (allocation of function)	At a cognitive level includes issues such as the limitations of human information processing, vigilance and the way in which people make decisions, at a physical level the design of lighting, and the acoustic and thermal environments - e.g., for hearing alarms
The individual in the organisational context	An indirect but important influence on interface operation and design. Operation as a team or individual, means of communication available, training and procedural support (see Chapter 7), levels of operating performance, resources and social interactions
Individual skills, abilities and knowledge	The ability to use full range of facilities on the interface, understanding of abnormal situations, ability to manipulate controls and use the information effectively

sensory discrimination was insufficient to deal with the vast amount of information. Speech recognition and speech synthesis technology is not yet sufficiently advanced to have the width of application that the range of current visual displays have. So visual displays are still the primary media for the communication of complex information to the operators of complex and large-scale systems.

To achieve an effective interface design the information system should be structured to facilitate the interchange, i.e., transmission and reception, of information. In addition, the interface should be able to support normal, abnormal (see Chapter 8 and Section 6.5) and emergency operation. Alongside the primary system operators, the design should consider other system users such as maintenance personnel, supervisors, managers, and systems software personnel.

People have a set of characteristics which need to be considered in the design of any interface which impact on their performance and use of the system. They bring to the system an adaptation and flexibility that cannot yet be mirrored by automation but at the same time each individual has a different range of knowledge, skill, motivation and other factors which are commonly labelled individual differences. These are documented extensively in the human factors and psychological literature but can be summarised (along with examples of the elements of interface design they may impact upon) in Table 6.1.

The range of factors which influence how an individual will perform are known and performance-shaping factors (PSFs) and performance-influencing factors (PIFs). These may be individual, social, organisational, technological and environmental. They impact on how effectively an individual will perform a task and the likelihood of an error being committed (see also Chapter 2).

6.1.2 System Characteristics

If a system design is to fully incorporate human factors, then ergonomists will need to be involved from the outset of the design process. The human factors aspects of the design will then be integrated and carried out in parallel with the hardware and software design. However, just as people have limitations on their performance capabilities so will the system. These need to be considered at the outset by human factors engineers to ensure that expectations of system functioning are realistic and pragmatic. The human-machine interface ideal may be influenced by a range of factors which constrain the technologies available for use, such as:

- The availability of hardware and software;
- The requirements of the procurement process;
- Limitations imposed by the structural design of the control room, building or room in which the equipment will be located;
- The need to interface with existing equipment;
- Environmental constraints — such as the need for ruggedised displays or restrictions on lighting, etc.;
- Industry standards, guidelines and regulations;
- Maintenance issues such as access, the availability of spare parts, etc.;
- Emergency operational requirements.

A helpful discussion of the constraints is given in [EPRI 84].

At the commencement of the design process key factors influencing the nature of the system will be decided. These are:

- What is the purpose of the system;
- Where is the system intended to be used;
- When will the system be used;
- Who will use the system.

For safety-critical systems the question must also be asked how safe is the system — and how can we assure that it is designed to and retains an acceptable level of safe operation?

(a) Hardware design

Whilst this chapter focuses on the design of the information exchange between the system and its users, the human factors input into the design of the consoles and hardware in which the interface will be mounted is no less important in assuring system safety and shaping human performance. The reader is referred to [Grandjean 80, McCormick 83, EPRI 84]. In summary the issues which should be addressed are:

- User anthropometrics;
- User biomechanics;
- Layout of the individual and group workspace;
- The physical environment (visual, acoustic, thermal and vibration);
- Job and task design;
- User support;
- Display and control hardware.

The number of VDUs for a given control room application can be calculated based on a realistic worst case operational scenario. This is based on two unrelated failures occurring concurrently with a VDU screen dedicated to alarm handling in the situation. This gives a *minimum* of three VDUs and includes the possibility that a display failing may constitute one of the failures.

(b) VDU vs panel-based displays

In safety-critical systems VDU displays are probably now the most common interface technology employed, replacing or being used in conjunction with panel-based displays to give 'hybrid' interfaces — panel-based displays being wall or floor mounted panels offering information displayed in the format of dials, LED indicators, etc. The use of VDU-based displays offers many advantages including increased flexibility in the type and format of data display; the possibility for the same data to be displayed in a variety of formats if required; the easy manipulation of data on line and the ready updating of displays. In addition, VDUs are generally cheaper to install than panel displays, may be more easily maintained and replaced if required and (an important factor in environments such as offshore, where space is at a premium) they take up less operational space than more conventional panel displays. VDU displays are most commonly found in the format of CRT (cathode ray tube) displays.

However, the use of VDUs introduces a range of issues in the design of displays particular to this type of display media. The operator uses the interface to perform the required control tasks, but in addition has the subsidiary tasks of navigating between the alternative display formats in order to access the display page with the required information, and then locating the required information on the display. Care must be taken in design to ensure that these additional tasks do not significantly increase

Table 6.2 *VDU- and panel-based displays: a comparison*

	CRT	**Panel**
Advantages	• Offer flexibility in data format • Same information can be given in several different forms • Greater capacity to display data • Data can be manipulated on line • Cheaper to install and replace than panel displays • Take up less operational space • Maintenance is often easier and display formats more readily modified and changed	• Plant information is continually on display and does not have to be searched for • Several operators can easily work side by side • Data access is immediate - this may decrease response times to alarms for example. • The operator has a readily available overview of plant status
Drawbacks	• Only limited data can be displayed at any one time • The operator has an additional task in accessing the data • System response time and the display structure may delay response time in the event of an incident • If the CRT or power supply fails a back up system is needed	• Panel displays have a fixed format and so may not ideally support all task operations • They are custom made and so may be expensive to install and difficult to modify • The panels often require a large operational area

operator workload and that such systems are in line with operator capabilities and limitations (for example, they do not require the operator to carry significant data in memory from one display to another).

Under some operating scenarios panel-based displays offer several advantages over VDU-based formats. For example, in an emergency context all information is available and does not have to be retrieved from the system. Controls can be located appropriately for the display of the parameter they control and can be designed to facilitate the particular type of control input required, in contrast to the operation of VDU-based displays where input devices must allow a range of different control inputs to be made. Table 6.2 summarises the benefits and drawbacks of the two media.

(c)   Controls

Hardware devices that allow a user to make an information input into the system are collectively labelled 'controls'. However, as this term is commonly applied to mechanical devices, the term 'input devices' is now in more common usage to include those devices which incorporate some software in their system input. There has been a body of research to look at different input devices, for example [Carey 85] and the DTI's *A Guide to Usability* [OU / DTI 90] provide a useful summary of the applications, strengths and weaknesses of the different options. Devices in use include track balls, joysticks, tablets, keyboards, data glove, lightpens and touchscreens.

The important factors in the selection of input devices are consideration of the tasks the devices are to support, the interaction between devices, suitability for the environment (e.g., requirements for ruggedisation) and the control-display relationship. For complex tasks the devices cannot be selected in isolation, either from each other or from the displays for which they provide the means of interaction. As with the design of displays the characteristics of the user population need consideration, in particular to ensure display control compatibility (see, for example, [McCormick 83]). Issues which are commonly cited as a basis for the selection of devices are error rates, reaction time and accuracy requirements.

6.2   THE DISPLAY DESIGN PROCESS

Modern technologies provide highly flexible computerised systems with a range of CRT displays, providing potentially vast amounts of data about the system and its functioning. Sensors and systems data abound and it is common to provide all the data available for display to the operator in one form or another. In most instances the operator does not need all this information to perform the task and an information overload can hinder task performance. A systematic approach to display design is needed to ensure that the final product will provide the information to complete the task, in a format that supports the user and in the context of a display structure that is transparent and easy to use.

This section looks at the process of display design in a series of simplified steps outlined in Table 6.3. These are discussed step by step in the following sections. The indiscriminate display of information should be replaced by careful consideration of the information needed to perform the functions allocation to system users. To ensure that the interface design meets the requirements of the end-user and is acceptable in a work environment, the user interface has to be based on user analysis in terms of user characteristics (see Section 6.1.2), tasks and the user conceptual model. The analysis of the information and control flows to support user task completion are based on the task analysis. Users of safety-critical systems, as already noted, generally have a high degree of expertise, and are selected and trained according to

Figure 6.3 *Outline stages in the design of VDU-based display formats*

Display design stage	Elements for consideration in design
Functional analysis	• Allocation of function between people and machines • Identification of required system functionality (normal and emergency operations)
Task analysis	• Identification of tasks to be performed by all personnel who will operate system (normal and emergency scenarios)
Information analysis	• Identification of information requirements of operators for each task scenario • Identification of information to be displayed on VDU-based formats
Display formats	• Decision on range of formats to be used • Outline of customised formats • Decisions on the use of colour, coding techniques, etc.
Allocation of information to formats	• Allocation of information to different formats, based on task requirements • Design of customised formats (e.g., plant-specific) • Redundancy requirements
Structuring of formats and display navigation	• Form of navigation • Control specification • Control display integration • Integration with other displayed information
User trials and evaluation	• Feedback on usability to allow design verification and iteration

defined criteria. This allows interface requirements and the associated operator support (see Chapter 7) to be well defined. As Avouris [Avouris 93] points out there is an increasing use of KBS (knowledge-based system) tools to support operator cognitive tasks such as planning, diagnosis and prediction. Systems can be put in place to automatically shut down systems or to prevent manual overrides or user intervention for defined periods of time. This allows the operators to have time to identify causes of failure and to respond appropriately.

6.2.1 Functional Analysis

In designing a system there will be a division between the functions the system will perform and those to be carried out by a human. At a high level there will be functions that are contributed by both. The way in which functions are assigned to be completed by either the system or the human operator is known as allocation of function. It includes decisions about the way in which tasks will be designed, how many people will be required to operate the system and what functions will be automated. There has been a trend (which is perhaps now less prevalent) towards the automation of function in preference to some functions being performed by a human operator. The human factors perspective is that of examining the functions which are performed well by people and those which machines can complete with superior reliability, and taking into account the overall job design, entering into a dialogue with the technical system designers as to the best allocation of functions for that system. The impact of automation is perhaps best summarised in a paper by Bainbridge on the 'ironies of automation' [Bainbridge 87], the key points of which are discussed in Chapter 2.

6.2.2 Task Analysis and Definition of Information Needs

Once the functions are transparent, the next stage is to analyse the tasks to be carried out. Task analysis is a basic human factors method providing documentation of the user element of the system. If a systems approach is taken to design, and the user is viewed as an integral part of the system, then the task analysis provides a parallel to the system descriptions, documenting the 'human factors' elements of system operation. There is a wide selection of potential methods which can be applied in the context of analysing tasks to determine information needs for display design purposes. Methods can be selected based on the type of analysis required, e.g., temporal task information (e.g., timeline analysis [Kirwan and Ainsworth 92]), a breakdown of the task into its detailed components (e.g., Hierarchical Task Analysis), analysis of the information exchange between people and the system (e.g., job process charts [Tainsh 85]), etc.

One commonly used method of task analysis, Hierarchical Task Analysis or HTA, is described in Chapter 7, whilst a review of task analysis methods is given in [Kirwan 92].

In modern technological systems many of the simpler operating tasks are carried out automatically by the control system (in particular the tasks that involve the execution of a procedure or monitoring for a change in system state). This has had a variety of impacts on the design of the HCI in a process context. The tasks of the operator focus increasingly on the more cognitive-based tasks, which include problem solving and decision making. A simple classification of the task types that an operator may be required to complete in the operation of a process is given on the previous page in Table 6.4. Tasks such as problem solving may manifest themselves in terms of planning and supervisory activities, whilst tasks such as determining a valve state may be perceptual in nature.

Task analysis provides a breakdown of the elements of physical and

Table 6.4 *Process operator task types (after [Astley 91])*

Procedural	Following a predetermined sequence of events
Monitoring	Sampling information to determine the correct states and variables of importance. These include the identification of deviations and the manipulation of the information to determine changes in system states
Decision making	Choosing between alternative responses on the basis of available information
Problem solving and fault diagnosis	Process of resolving uncertainty about system states
Fault detection	Perception of a fault having occurred or being imminent
Motor tasks (i.e., control actions)	Any operator action on system state or configuration
Prediction	Judgement of likely future system states
Communication	Accurate transmission of information without any processing of the information from transmission to reception by the individual or group who will use it

cognitive activity, for each of these elements the information needed to be exchanged with the system in order to achieve the task goal can be specified. The basis for the design of the information and its content and layout in the displays can be drawn from the task analysis.

Changes in systems have led to changes in the emphasis on the types of tasks carried out by operators and consequently the information they require to support these tasks and the way in which it is displayed. This change has impacted on the information-processing demands placed on people in the workplace. Tasks no longer tend to follow predicted sequential patterns, but rely on operators to draw upon repertoires of skills and their system knowledge to carry out tasks. Using Reason's skill-, rule-, and knowledge-based hierarchy [Reason 90] (see Chapter 2) as a reference, the shift has been away from skill- and rule-based operation towards knowledge-based operation. Bainbridge [Bainbridge 92] points out the main types of skill which can be acquired to make information processing in such contexts more efficient as it is easier (in the sense that the cognitive workload is less) for people to operate at a skill based level. Bainbridge illustrates that display design can be used to support and enhance such skill-based operation, for example by optimising the use of operator's pattern recognition skills in the layout of information on displays.

6.2.3 Analysis of Information Requirements

The second stage is the analysis of information requirements based on the task analysis. There are methods of task analysis which provide a framework for such analysis, for example [Tainsh 85, Astley 91, Piso 81]. In a safety-critical systems context the safety-critical tasks may first be identified to assure the appropriate attention is given to the information that will be safety-critical. It is important to identify what this information may be based on tasks and human functions as this may not coincide with an engineering view of safety-critical information. If the displays are to be VDU-based there will be additional tasks, and the associated displayed information, involved in the operation of the display system itself. In effect there are three basic kinds of information to be displayed to the system user:

- Information from the system on its operational status (current, historical, predicted);
- Information on display options and navigating around the display system itself;
- Information on control inputs (options available and feedback).

The analysis should identify the information required to perform each task element.

This analysis will identify what the content of displays need to be in broad terms, how the same item of information may be used in different tasks and what different items of information will be required to complete a single

Table 6.5 Questioning procedure for information analysis (source: [Wilson 95])

What information is to be displayed?	Why is it necessary?	What else could be displayed?	What should be displayed?
Where is the information to be held?	Why there?	Where else could it be displayed?	Where else should it be displayed?
When is the information to be accessed or communicated?	Why then?	When else could it be communicated?	When should it be communicated?
Who is to hold or use the information?	Why them?	Who else could hold or use it?	Who else should hold or use it?
How is the information to be presented?	Why that way?	How else could it be presented?	How should it be presented?

task. The output of this stage should therefore be:

- An analysis of the tasks to complete the functions allocated to operators;
- What information is needed to achieve them;
- What the content of that information needs to be;
- How the same information is used for different task elements;
- What information needs to be located together on a display(s) to allow it to be used for a particular task.

Whilst control systems can offer data from a vast array of sampled data points, it is a human factors problem to determine what is to be displayed, how it should be displayed and the means of accessing the information. A structured questioning process is proposed [Wilson 95] which can be used to determine information requirements (Table 6.5) for structuring a display system.

6.2.4 Allocation of Information to Display Formats

Having analysed operator tasks and identified the information that will be required from the system by the operator (and to be communicated to the system for control purposes) the next stage is to structure the information into 'pages' that can be displayed on the system. There are no systematic and readily usable human factors techniques for achieving this, it is highly reliant on the skill and experience of the human factors specialist. Format decisions will be based on the types of tasks the display is to be used for and the type of data for display, which include:

- Parameter values;
- Trends;
- Historical data;
- Alarms;
- Control values;
- Status;
- Navigational information;
- Help information;
- Feedback on control inputs.

Formatting displays for VDUs is therefore much more complex than in the situation where there is one system control to one display. The information has to be managed and matched to users' abilities and task requirements.

6.2.5 Structuring the Display System Pages and Interactions

The required information content and number and type of formats can be set using the task analysis information as a framework, taking into account available data points and the limitations of the control system software. Where information must necessarily be spread over several VDU 'pages', decisions have to be made about the division of such information into pages.

A means of accessing and moving between these pages must b
and carefully designed to ensure that the interaction does not add s
to the task load of the operator. The system design must be ap
ensure that displays are structured and formatted in a consistent anu
coherent manner, whilst ensuring that information is presented in a usable
and accessible way. Options and approaches to the structuring of displays
are discussed further in Section 6.3, whilst display formats and the design of
display content are considered in Section 6.4.

6.2.6 User Trials and Evaluation

Once the initial design of the interface is complete there is a need to verify the
user-system interaction and evaluate usability. Sweeney, Maguire and Shackel
[Sweeney 93] offer a framework which classifies usability evaluations in
terms of three dimensions: approach to evaluation, type of evaluation and
time of evaluation in the product life cycle. Whilst Table 6.3 lists user trials
and evaluation last, there is a place for evaluation throughout the display
design process, in particular in terms of the involvement of users wherever
possible. In some safety-critical systems designs the design team may
include the control room operators. An approach which offers the practical
input of users into design as well as in the evaluation process. Approaches
to evaluation can be user, theory or expert based utilising types of evaluation
which can be diagnostic, summative or metrication. These usability indicators
can be summarised as follows:

- *User based:* these evaluations are based on factors such as performance,
  non verbal behaviour, attitude, motivation, stress and cognition.
  Measures include task times, percentage completed, error rates,
  questionnaire and survey responses and ratings.
- *Theory based:* Based on idealised performance and predictions of
  usage. Measures include prediction of task performance times, learning
  times and likely ease of understanding.
- *Expert based:* This involves expert evaluation based on opinion, rating
  of usability properties and consideration of conformance with
  standards, guidelines and design criteria.

In safety-critical systems such evaluation can highlight problems in normal
and abnormal operating conditions where task performance can have a
significant impact on maintaining system safety and may mean the difference
between the operator being able to recover a system fault or a system
shutdown. Factors such as the time taken to perform safety-critical tasks and
likely error rates can provide an indication of the task performance and
reliability that can be expected from the user and important feedback into the
design process

6.3   STRUCTURING THE DISPLAY SYSTEM

In a practical sense the questions most often asked (in the experience of the author) during the design, evaluation and modification of VDU-based process displays are:

- How many VDU screens do I need?
- How should these be arranged in the console hardware?
- How should my VDU displays relate to other displays (e.g., panels)?
- What redundancy and diversity should be designed into the system?
- Is my design and the provision of displays adequate for an emergency situation? What core functions need to be provided as a back-up and what form should this take if my main control position fails?
- What data should be displayed and in what format?
- How should the information be divided between pages?

These issues illustrate the need to consider both representational (the way information is displayed) and operational (the 'mechanics' of user-system interactions) aspects of system design. This section is concerned primarily with the operational aspects of the system, its structure and associated interactions, while Section 6.4 addresses the formatting of display content.

6.3.1 Structuring the Display System Pages and Interactions

There are three principal elements to the structuring of displays: the division of information into display 'pages', the relationship between pages and the macro structure of the information with its associated navigational mechanisms. In a safety-critical systems context the user must be able to locate information accurately, read it on the screen, interpret it unambiguously and respond appropriately. The design of the system must support this.

The key principles which apply to the design of the operational elements of the system apply equally to the formatting of the display content and can be summarised as follows.

(a)   *Maintain consistency throughout the display system*

Consistency ensures that users know what to look for, where and the format it will take. It assists learning and reduces memory load. It applies equally to the rules which govern navigation around the system as to abbreviations, the use of colour, symbols and other representations.

(b)   *Do not overload working memory*

Users should not be expected to hold detailed information in memory, for example whilst switching between display pages, the often quoted 7 ± 2 bits of information provides the limitation of human information processing capacity. This, however, should not be all constraining, as there are many examples where information can be presented in a way that aids remembering.

(c)   *Ensure display control compatibility*

The mapping between displays and controls should match user stereotypes and expectations.

(d)   *Provide appropriate feedback*

Effective feedback to indicate to a user the response of a system to an input is essential. For frequent or minor operations this could be very simple, becoming more substantive for inputs with increasing complexity.

The feedback should be appropriate, timely and offer an indication of any further options or actions required, or if the group of actions has reached completion.

(e)   *Enable efficient information assimilation*

Design displays and use formats that allow users to readily understand and use the information they provide. For example, the use of object-oriented displays (e.g., [Wickens 86]) to allow deviations to be readily detected may support fault detection more readily than alphanumeric lists of parameters (see Table 6.6).

(f)   *Allow reversal and recovery*

Design the system to allow actions (intended or unintended) to be reversed and for errors to be recovered (e.g., through diagnostics and help).

(g)   *Anticipate and engineer for errors*

People will make errors however well designed the system. The system should anticipate errors and prevent the user from making some of the more serious errors. User support should include effective information on erroneous actions (see above).

(h)   *Locus of control*

It is important for the operator to feel he has control over the system and that it will respond to his actions. Sneiderman notes: 'Surprising system actions, tedious sequence of data entries, incapacity of difficulty in obtaining necessary information and the inability to produce the action they want all build anxiety and dissatisfaction' [Shneiderman 86].

This is particularly the case in situations where users are largely experts, as for safety-critical systems contexts.

(i)   *Design for transparency and clarity*

Whilst design will be based on an understanding of the user's model of the system and its functioning, a system whose functioning is transparent to the user will ensure that such user models are formed based on accurate information and reinforced by the system operation. (Transparency may not always be appropriate but is generally helpful where users are expert rather than naive, in particular in situations where fault diagnosis may be required.)

6.3.2 Dividing Information into Pages

The tasks of the operator are threefold:

- Operating the system;
- Locating and manipulating information presented on VDU formats;
- Navigating between display formats.

The process of designing displays must take into account these three elements and examine not only how personnel will operate the system and what information will be displayed (e.g., the information content of the system), but how that information will be divided into display 'pages' and how those pages will be structured and accessed. The required information content and number and type of formats can be set using the task analysis information as a framework, taking into account available data points and the limitations of the control system software. Where information must necessarily be spread over several VDU 'pages', decisions have to be made about the division of such information into pages. There are several rules of thumb that can be applied to such decisions, taken in the context of the tasks to be carried out, safety-critical tasks and the means of navigation used to move between display formats.

- All variables affecting a particular process state, should, as far as possible, be displayed together (detailed information can be included on displays at lower levels of the hierarchy).
- Information should be grouped and partitioned in ways that are meaningful to the operators of the system.
- Operators should not have to carry information in memory from one display to another.
- The structure of the displays should be transparent.
- A simple functional relationship (e.g.,. opening a valve shown on one display format increases a pressure variable shown on another format) can be divided across two display pages. If the opening of the valve was to have a multiple effect then all variables should be shown on one page.
- All effects of control actions should be simultaneously observable.

Display type	Description	Suitability for tasks
Table 6.6 *Display format options (source: [Wlson 95])*		
Mimic/schematic	A functional or geographical representation of the plant using symbols to represent plant items such as vessels, valves, etc., connected by lines to indicate pipework and flows	Usable for a wide range of tasks. Allows functional relationships to be easily established and so can assist in problem solving. Overlays can be used to avoid clutter resulting from the display of a wide range of variables
Loop	Loop displays mimic the more traditional faceplate displays of three-term controllers	Analogue displays can show the relative values and relationship between measured value, setpoint and output. However, the displays are oriented towards the tasks of an engineer rather than an operator. In a simplified format they can provide an easy comparative reference, but if group loop displays are used they cannot be normalised and so comparison is difficult. Useful for monitoring and fault detection tasks. May be used to support problem solving
Deviation displays	In its simplest format the display comprises a horizontal line which represents a setpoint with a small vertical line indicating the magnitude of the deviation from it. The deviation display can indicate deviations in both positive and negative directions	Used principally for monitoring and fault-detection tasks
Graphical/trend	The main use of trend/ graphical displays is for displaying trends on one or more plant variables. VDU resolution imposes a	Trend information may be used for monitoring historical trends, but is likely to be used as a diagnostic or predictive

Table 6.6 continued on next page

displays	limitation on the scale detail that may be used and data points. A maximum of 4-6 charts should be placed on the screen at any one time	The displays are useful for fault detection and under some circumstances fault diagnosis, where the fault conditions lead to identifiable patterns being formed. They allow monitoring of key variables at a glance on one display. However, there remains some debate concerning their effectiveness
Object-oriented polar-coordinate displays	Object-oriented displays aim to integrate information from multiple channels to enhance operator performance in recognition of total system state. They take the format of a geometric figure with variables indicated as points at each of the angles of the object. Variables are normalised so that the normal system state is shown as the geometric figure in its correct form. Deviations are detected by changes in shape	
Alphanumeric displays	Whilst most formats have alphanumeric labels and information included, some formats are solely alphanumeric. Design recommendations for such displays are numerous	Such displays provide a high degree of detailed information and so may be used to support problem-solving or decision-making tasks. Where the alphanumerics are coded (e.g., using colour), the displays may be used for fault detection or for monitoring progress through a sequential list
Sequence	The control of a process will often take the form of a predetermined sequence of operations. Increasingly such sequences are automated. However, it is important for the operator to be able to monitor progress through the sequence. Sequence displays may take the form of textual lists, flow charts or networks	The displays provide a means of monitoring progress. They may allow anticipation of future system states (prediction and decision making), the detection of faults and potentially fault diagnosis

6.3.3 Navigation

In a VDU-based system the need to display a large amount of data on a limited number of screens means that there are likely to be a large number of potential 'screens' of information for display. Some will have a fixed content, other screens may allow the user to display the information of his choice, for example in the format of overlays or windows. The effective use of such displays is dependent on the way the display pages are conceptually structured and accessed by the user. It is increasingly common to have representative end-users involved in design. In some instances there will be options built into the display system for users to format and structure their own displays within a framework. The 'Windows' system commonly in use on many personal computers provides an example of such an option in which users can configure displays and develop their own icons. Most systems are hierarchically structured and the main issues to be considered are:

- The method of selecting a page or window of information for display (e.g., menu, keyword or command, index, direct interaction);
- The naming and grouping of items within the page, window or menu;
- The number of levels in the hierarchy of pages or windows;
- The amount of data or information that can be presented on one display.

Specific guidance on each can be found by reference to the human factors literature, e.g, [Shneiderman 86, OU/DTI 90].

Full advantage should be taken of available technology and the use of devices to maximise access to data without the need for a full or permanent display. Examples might include the use of windows and overlays, dynamic simulation of 'what if' scenarios, rapid recall of historical data and so on. Ideally a design team should include users, control system/software and human factors specialists — this allows the formats to be designed for user tasks without restricting the displays to what is familiar and known to users. If users are not involved in the design process then the ways in which users use information and manipulate displays can not always be fully understood as the designer's view of the world will not coincide with that of the user.

Once a preliminary set of formats has been designed these should be verified by conducting user trials with representatives of the group who will use the system. These highlight any issues and problems and allows provision for operator support, such as operating procedures and training, to be evaluated alongside the interface. It is likely that several iterations to incorporate feedback from such user trials will be made in the design cycle. It is important in a safety-critical context for user trials to consider emergency and abnormal as well as 'everyday' scenarios.

6.4   REPRESENTATION OF DISPLAYED INFORMATION

As the complexity of operator tasks has increased so has the associated complexity of the VDU interface. There is a wide variety of human factors research which has focused on the formatting of VDU displays in detail, for example the use of coding to enhance information presentation for VDU-based formats (as summarised in Table 6.6). However, there are no easy to apply methodologies available which will assist the designer in translating an examination of the user's tasks or information requirements into display design recommendations.

In practice the task analysis forms a framework within which the information can be related to the type of task it supports. Information that is common to several tasks can be considered and decisions made about how it will be displayed, in some situations task-based formats will provide the best assistance to the operator in completion of his task. In other situations functionally based displays such as mimic/schematic formats will assist the operator in developing appropriate operational strategies.

6.4.1 Format Options

Recommendations can be made for the display formats best suited to different task types, although research has largely focused on the display formats that are currently widely used rather than on the use of new technologies for information display. Formats commonly encountered in a process context and the tasks they are commonly used to support are given in Table 6.6. If information is used only to perform a singular task then the choice of display type can almost be a one-to-one mapping, choosing the most appropriate format for the task type. However, generally, situations are more complex and expert judgement must be applied by the designer to consider what format(s) will best support the range of tasks to be carried out, making appropriate trade-offs and assuring safe operation.

Decisions on trade-offs will take into account not only the needs of the operators of the system, but of the full range of users which may include management, maintenance personnel and supervisors. In particular it is also important to note the changes in needs and information support between normal and emergency operation. Under emergency conditions there are several factors which mean that existing information may not be presented in the most useful format. For example, information requirements under emergency conditions may need to support the tasks of assuring the safety of the plant and people via safe shutdown, whilst normal operation also focuses on safety but requires details of production parameters too.

Table 6.6 gives commonly used process display formats but in considering the types of formats best suited to the tasks, plant specific and customised formats may also be considered. For example, a graphical 3D, colour-coded presentation of a turbine in a power station may be the most effective to present temperature information. In situations where 2D and 3D mimic displays have both been in use, users have expressed preference for 3D displays citing the ability to be able to pinpoint an item of equipment exactly when leading an operator to a fault and being able to understand system functioning more accurately. However, this is dependent on whether the displays are functional (i.e., show the functional flows of the system) or topographical (i.e., mimic the geographical layout of the plant [Vermeulen 87]. In selecting appropriate formats, Hartson and Hix consider the importance of system transparency, i.e., the level and format of detail that the user is given to allow them to understand the underlying system [Hartson 89]. The level of transparency that is appropriate to a safety-critical systems context will vary and will be dependent on factors such as the type of fault diagnosis the operator is expected to carry out and the underlying complexity of the system itself.

6.4.2 Coding

Within VDU-based design coding is used extensively as a means of representing information in formats that match people's perceptual capabilities. By coding information, for example by the use of colour, displays can be made to appear less dense and more information presented on screen. Coding therefore can be used to group related information, hightlight information and improve readability.

Decisions about the use of coding are an integral part of developing display formats. The codes that will be used and applied should be fixed to ensure consistency in their use throughout the system. There is a variety of research providing guidelines for coding, this is summarised in Table 6.7 [Wilson 95] for the main types of code but may also include the design of symbols and other graphical representations.

Table 6.7 *Coding guidelines (source [Wilson 95])*

Coding type	Description of code	Application of code
Size	Use of different sizes of same display element to convey changes in magnitude, etc.	Useful way of representing magnitude for level, temperature, etc. When it is the only code only three different sizes are recommended in the code for easy discrimination Non-linearity of CRTs can make comparative judgements difficult if variations in magnitude are small
Shape	Use of different (usually geometric) shapes to	A relatively large number can be used and be readily distinguishable (up to

	represent categories of items displayed, e.g., different classes of marine vessel on a display of a sea channel	10) To be effective, good resolution (e.g., to ensure a hexagon is not mistaken for a circle) is required and effective contrast, especially if colour is used It is a useful means of coding for search, counting and comparative type tasks
Alpha-numeric	Use of letters and numbers to form codes to (uniquely) identify elements within a system, often used for labelling purposes, e.g., to label valves with a common code on plant and on the display	Avoid confusability between similar letters, numbers, e.g., X and K, S and 5 They should form a natural category and stand out from other items on the display
Brightness	The use of different luminance levels to highlight certain items of information or to indicate degrees of magnitude, e.g., temperature	No more than two levels should be used to be distinguishable and effective
Spatial	Use of the location on the display to give a meaning to an item of information, e.g., to identify it as a menu item	Should be used consistently throughout the system to indicate: title pages, information fields, alarms, active and static display areas
Colour	Use of different displayed colours to provide differentiation between items on a display or to impart inherent meaning (colour should only be used as a redundant code) Specification and the selection of palette colours can be difficult	Colour has a variety of uses which are task dependent: To identify and classify information As a formatting aid To collate related information across different displays Reduction of clutter Visual display structure Aid to visual search To connotate danger, etc. To highlight items or status change
Flash	Used to attract attention to items of information on the display	Up to four blink rates are distinguishable, but less is preferable Rate should be between 1 and 4 Hz It is useful for redundant coding It should be used sparingly Operators should be able to cancel the blinking

6.4.3 Colour

Colour is a particular form of coding which requires careful use in design and so merits a particular mention here. Users almost always prefer colour displays and believe it enhances task performance, even if there is evidence to show that monochrome displays offer equal or better task performance. Colour may be used in one of three ways in display design:

- As part of a graphical representation of the real world (e.g., virtual reality);
- For abstract representations, such as procedures and flow-charts;
- For schematic representations, such as mimic displays.

However, colour should be used with care and following human factors guidelines as used incorrectly it can 'play tricks' on the perceptual system. For example, the use of small yellow characters on a grey background means that the characters will appear black to the eye. The use of a saturated red will make items appear to move forward on a display, whilst blue makes them recede (see for example [Hanson 79]), other perceptual 'tricks' include chromatic induction and assimilation or spread effect. Colour should generally only be used as a redundant code and follow existing population stereotypes (e.g., meaning of red for stop, etc.). The aims of colour coding may include [Hopkin 80]:

- The reduction of clutter;
- The collation of related information across different displays;
- Imposition of visual structuring on the display;
- The portrayal of a unique feature in a unique way;
- Facilitation of visual search;
- Highlighting information;
- Improvement of the ability to remember information;
- An indication of a (pending) change of state;
- Reduction of misreading (e.g., for columns of data);
- Addition of connotations of danger, caution or safety.

6.5   ABNORMAL AND EMERGENCY OPERATION

Although Chapter 8 considers the issues related to abnormal operating conditions in detail, the chapter concludes by emphasising some of the issues which are particularly pertinent to interface design in safety-critical systems.

There will be a range of system operating conditions, both planned and unplanned (e.g., planned maintenance) for which operators will use the control room-based man-machine interface as a locus of control (for example, as opposed to the manual operation of valves on a process plant). The operators of the system in such an instance may not always be those who operate the system on a day-to-day basis, for example an annual shutdown may involve a special operational team. In any emergency scenario there will usually be additional personnel in the control room involved in operation, unless a risk assessment has shown this to be undesirable.

Individual systems will vary in the optimum way of dealing with the presentation of information in such situations and the degree to which the operator is permitted to interact with the system (for example, safety systems may prevent human intervention for a defined period as the system is brought automatically into a safe status). In the design process consideration should be given to the following:

- Are the displays used for 'normal' operation adequate to support the range of emergency operational scenarios that may be encountered?
- Are any 'special' displays required to give information that is not required under day-to-day operation?
- What additional support may operators require (e.g., procedures, training in understanding and using displays)?
- How will operator workload change, and how will this be managed?
- What redundancy of information presentation and control inputs are required (in some cases this may be a remote control panel allowing the safe shutdown of the plant; in others, a console replicating many of the critical control functions and information display)?
- Will the information be required in other geographical locations (e.g., an emergency control room)?
- If the operator is to be prevented from active control for a period, then how is the operator informed of system status, in particular if he is required to resume control after a period?

Under emergency operation, arguments for using existing formats focus on the need for familiarity, for example, the fact that the operator will have a greater automation of his skill in using displays that are in a familiar format. This means that he can locate information quickly and move between formats easily; assuming of course that the displays in questions are used and accessed on a regular basis. Under emergency conditions it is likely that an operator will be dealing with many unknowns and this will make demands on his cognitive processing leaving less spare capacity to deal with unfamiliar displays and to search for the required information. However, it can equally be argued that certain information needs to be located together on one, or a series of, displays and to be rapidly accessible when required. What is appropriate will be determined by analysis of each situation. In the consideration of emergency and abnormal scenarios the display system itself needs to be addressed alongside the control system. For example, the impact on operation of one VDU undergoing maintenance, or a failure which could range from the loss of an electron gun on one VDU to a total power failure. Design of an interface must consider the means of control under such situations, the familiarity of the operator with these interfaces and their use in training and emergency exercises.

6.6   CONCLUSION

This is a simple overview of a very practical approach to designing display formats and ensuring that they will function in practice. There are many issues relating to the detailed layout of information on displays, and these have been well researched. Guidelines can be found for the design of VDU-based displays and standards and legislation govern their use in the workplace (for example, ISO 1993, and the Display Screen Equipment Regulations 1994). There are still some issues of dissension, for example is a pump which is displayed as a solid red symbol on a mimic format, on or off? However, as human factors are increasingly incorporated into design a more user-centred approach to design is emerging which should both enhance productivity, by supporting the user in his task performance, and increase safety by minimising the potential for error.

REFERENCES

[Astley 91]   Astley J A: Task *Analysis for Industrial Process Control*. University of Aston. Unpublished Ph.D. Thesis, 1991

[Avouris 93]   Avouris N M, Van Leidekerke M H, Lekkas G P and Hall L E: *User Interface Design for Co-operating Agents in Industrial Process Supervision and Control Applcations*. International Journal of Man Machine Studies, 38(5), 747-890, 1993

[Bainbridge 87]   Bainbridge E A: *The Ironies of Automation*. Automatica, 1987

[Bainbridge 92]   Bainbridge E A: *Multi-plexed VDT Display Systems: A Framework for Good Practice*. In Wier G and Alty J (eds): HCI and Complex Systems. London, Academic Press, 189-210, 1992

[Carey 85]   Carey M: *The Selection of Input Devices for Process Control*. Report No CR2773 CON, Warren Spring Laboratory, Stevenage, 1985

[EPRI 84]   Electric Power Research Institute: *Computer Generated Display System Guidelines*. Report No EPRI NP-3701 Vols 1 & 2, EPRI, California, 1984

[Grandjean 80]   Grandjean E and Vigliani E (eds):*Ergonomic Aspects of Visual Display Terminals*. Taylor and Francis, London, 1980

[Hanson 79]   Hanson D C (ed): *Special Issue on the Applciation of Colour to Displays*. Proceedings of the Society for Information Display, Vol. 20, No.1, p. 1-42, 1979

[Hartson 89]   Hartson H R and Hix D:*Human Computer Interface Development: Concepts and Systems for its Management*. ACM Computing Surveys, 21, p. 5-92, 1989

[Hopkin 80]   Hopkin V D: *Human Factors Research, and Some Emerging Principals, for the Use of Colour in Displays*. Displays 1980

[Kletz 94]   Keltz T: *What Went Wrong? Case Histories of Process Plant Disasters.* Gulf Publishing Company, Houston, 1994

[Kirwan 92]   Kirwan B and Ainsworth L K: *A Guide to Task Analysis.* Taylor and Francis, London, 1992

[McCormick 83]   McCormick E J and Sanders M:*Human Factors in Engineering and Design.* McGraw Hill, New York, 1983

[Norman 88]   Norman D:*The Psychology of Everyday Things.* Basic Books Inc., New York, 1988

[OU/DTI 90]   Open University and the Department of Trade and Industry: *A Guide to Usability.* Open University, Milton Keynes, 1990

[Piso 81]   Piso E: *Task Analysis for Process Control Tasks: The Method of Annett et al Applied.* Journal of Occupational Psychology, 54, p. 247-254, 1981

[Reason 90]   Reason J: *Human Error.* Cambridge University Press, 1990

[Shneiderman 86]   Shneiderman B: *Designing the User Interface.* Addison Wesley, 1986

[Sweeney 93]   Sweeney M, Maguire M and Shakel B: *Evaluating User Computer Interaction: A Framework.* International Journal of Man Machine Studies, Vol. 38, 689-711, 1993

[Tainsh 85]   Tainsh M: *Job Process Charts and Man-computer Interaction Within Naval Command Systems.* Ergonomics, 25, 555-565, 1985

[Vermeulen 87]   Vermeulen J: *Effects of Topographically Presented Process Schemes on Operator Performance.* Human Factors, 29, 383-395, 1987

[Wickens 86]   Wickens C D: *The Object Display: Principles and a Review of Experimental Findings.* Champaign IL, Dept of Psychology, University of Illinois, Tech, Rep, No. CPL 86-6, 1986

[Wilson 95]   Wilson J R and Rajan J A: *Human Machine Interfaces for Systems Control.* In Wilson J R and Corlett E N: Evaluation of Human Work, Taylor and Francis, London, 1995

7

Training and operator support

7.1 INTRODUCTION

In the management and design of industrial and commercial operations, a prime motivation is to ensure that tasks are carried out to a satisfactory standard of safety and productivity. Training and operator support both serve this end as they are concerned with ensuring operator competence. Training refers to activities designed to change the operator's skills, typically through practices such as on-the-job instruction, classroom teaching, and simulation training. Common forms of operator support—operator manuals, procedural guides, on-line help facilities in computer systems, and even well designed and well labelled interfaces — do not set out to change skills but to assist their execution. However, they can also help the operator to learn.

Many companies still separate the development and administration of training from the development of methods to support operators, with different people in different departments assuming responsibility for training, for operator manuals, and for interface design. This can lead to inconsistency, confusion, and an improper balance between training and support.

This chapter sets out the factors to be considered when designing and developing measures to assure competence through operator training and support. The following themes will be developed:

(i) Understanding the task that the operator may be required to carry out, especially in the area of safety-related tasks;

(ii)   Understanding the processes of skills acquisition, including factors governing learning processes;
(iii)   Understanding the role of task context in making design choices;
(iv)   Ensuring safe and competent operation through support;
(v)   Ensuring safe and competent operation through training;
(vi)   Assessing performance.

7.2   UNDERSTANDING THE TASK

7.2.1   Types of Safety Task

Safety issues include dealing with emergency situations such as escape and self-evacuation, supervision and coordination of evacuation, using emergency equipment, supervising plant during emergencies, and the coordination of and liaison with emergency and relief services. They also include handling normal situations in a safe and effective manner, such as carrying out normal operations to minimise current and future risk, and carrying out maintenance. These are discussed as follows.

(a)   Escape and evacuation

Operators must know and understand the conditions under which they are expected to quit their workplace or evacuate the installation entirely. They should be able to recognise the alarms signalling evacuation and identify the physical conditions which indicate that evacuation is necessary and the situations in which they themselves are permitted to attempt to deal with the problem. They should be able to locate safety equipment, including clothing, breathing apparatus and fire extinguishers.

(b)   Supervision and coordination of evacuation

Duties include the responsibility for coordinating the evacuation of others, for example, by sounding alarms, advising on the accessibility or restriction of various escape routes, scheduling evacuation and advising when posts may be safely evacuated. This entails monitoring current status, maintaining communications where possible, and planning.

(c)   Co-ordination of and liaison with emergency and relief services

This may involve calling for assistance, taking steps to assist with access, providing information to emergency services concerning current plant status, damage reports and missing personnel. It entails understanding personnel safety systems, the geography of the installation, and the likely extent of damage in different areas of the installation.

(d)   Using emergency equipment

This involves an understanding of the proper use of emergency equipment, for example, putting on safety clothing and breathing apparatus correctly and using fire extinguishers and evacuation vehicles. It also entails carrying out routine inspections of such equipment, testing it and taking steps to make good any deficiencies.

(e)   Supervising the plant during emergencies

It may be necessary for operators to maintain supervision of a system during emergencies with a view to containing an accident. For example, an operator may need to control the venting of gases to reduce excess pressures or to manipulate manifolds to permit flows through some parts of the plant while isolating others. Moreover, the operator may need to maintain the availability of system emergency services.

(f)   Normal operation to minimise current risk

It is important that normal routines are carried out so as to preserve and promote plant safety. Management, engineering and planning operations should be carried out in accordance with approved procedures, and essential constraints should be properly observed. Automatic safety features should be designed to minimise the chances of operators straying beyond permitted operating procedures.

(g)   Normal operation to minimise future risk

In hazardous circumstances people become stressed, and there is a danger that actions taken to deal with current problems could fail to leave the system in a state where future problems may be dealt with efficiently. Staff are most likely to act within safe operating boundaries if they are able to follow set procedures and can rely on things being where they should be and systems being in anticipated states.

(h)   Permit-to-work systems

Proper adherence to permit-to-work systems is essential to ensure that maintenance staff are able to carry out their tasks with maximum safety. It is also necessary to ensure that proper records are kept of the location of all personnel during normal circumstances such that they can be located during emergency situations.

(i)   Maintaining safety systems

It is necessary routinely to inspect all equipment that might need to be relied upon during emergencies so as to ensure that the system has been left in a safe state — for example, checking the availability of breathing apparatus and ensuring the functioning of alarms, standby equipment and services.

(j)   Maintenance tasks

Maintenance tasks almost invariably entail risks and are often undertaken in hazardous environments. Furthermore, working in difficult or restricted positions can be physically stressful and can create a spatial disorientation and a distortion of normal tactile feedback which may disrupt normal working skills. When a technician has acquired a high standard of skill, it is often assumed that this will transfer to all situations, but this is not always the case if working conditions vary. Mismatches may not result in mistakes but could slow work down. This could be crucial if the maintenance technician is being exposed to risk.

7.2.2   Task Analysis

The term 'task analysis' refers to a number of methods of examining work in suitable detail, either to identify areas of potential human error or to establish essential task information to facilitate design. Task analysis offers several benefits in the design of training and support. It prompts an examination of the context in which work is undertaken. This means that the resultant training and support are directed towards the functional requirements of the task in question.

Task analysis also helps the designer to understand where training and support should be developed, and it specifies how people do their jobs. Task analysis of some form has a role to play in the design and development of all systems in which human beings are employed — Diaper discusses its importance with regard to human-computer interaction [Diaper 89] and Kirwan and Ainsworth deal with its application to complex systems such as process control [Kirwan and Ainsworth 92]. They also describe a number of different methods of task analysis and show how they are applied to a variety of different human factors applications. Task analysis is a prerequisite to each of the major stages of human factors design, including:

- The allocation of functions between human beings and equipment;
- Information requirements specification;
- Job design;
- Interface design;
- Documentation design;
- Training design and the maintenance of skills.

A popular and effective strategy for carrying out task analysis is Hierarchical Task Analysis or HTA [Annett et al 71, Duncan 74, Shepherd 89]. In HTA a task is first expressed in terms of a goal that the operator is required to attain. If, from this general description, the analyst is unable to suggest how the task is to be carried out to a satisfactory standard, then the goal is redescribed as

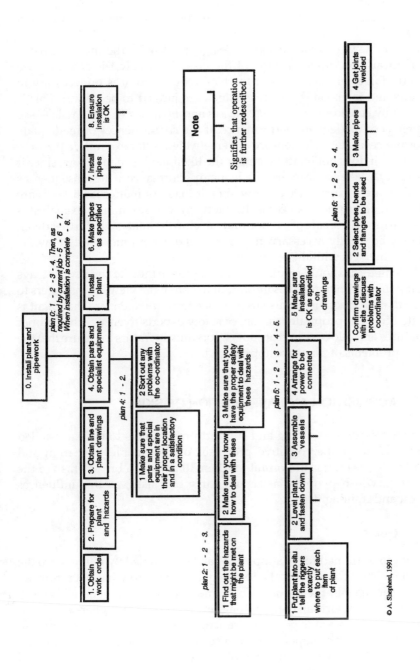

Note

Signifies that operation is further redescribed

0. Install plant and pipework

plan 0: 1 - 2 - 3 - 4. Then, as required by current job - 5 - 6 - 7. When installation is complete - 8.

1. Obtain work order

2. Prepare for plant and hazards

3. Obtain line and plant drawings

4. Obtain parts and specialist equipment

5. Install plant

6. Make pipes as specified

7. Install pipes

8. Ensure installation is OK

plan 2: 1 - 2 - 3.

1 Find out the hazards that might be met on the plant

2 Make sure you know how to deal with these

3 Make sure that you have the proper safety equipment to deal with these hazards

plan 4: 1 - 2.

1 Make sure that parts and special equipment are in their proper location and in a satisfactory condition

2 Sort out any problems with the co-ordinator

plan 5: 1 - 2 - 3 - 4 - 5.

1 Put plant into situ - tell the riggers exactly where to put each item of plant

2 Level plant and fasten down

3 Assemble vessels

4 Arrange for power to be connected

5 Make sure installation is OK as specified on drawings

plan 6: 1 - 2 - 3 - 4.

1 Confirm drawings with site - discuss problems with coordinator

2 Select pipes, bends and flanges to be used

3 Make pipes

4 Get joints welded

© A. Shepherd, 1991

Figure 7.1 An illustration of hierarchical task analysis, showing a maintenance task

a set of component sub-goals along with a plan which governs when each sub-goal should be carried out. Each sub-goal is then addressed in order to decide whether a solution is forthcoming or whether further redescription is warranted. The result is a hierarchy of sub-goals and plans, as illustrated in Figure 7.1.

As a framework for analysis and design, HTA helps the analyst to focus systematically on different parts of the task. The analyst is able to cease redescription at various parts of the hierarchy where it is unnecessary to continue and focus on those areas which are judged to be critical. This is achieved by considering the *consequences* and *likelihood* (i.e., the risk — see Chapter 1) of inadequate performance. If the consequences of inadequate performance of a task element are extremely high — for example, a threat to life — then it does not matter whether the likelihood is high or low, since it is necessary to take action to ensure adequate performance. If the consequences of inadequate performance are low, then it may not matter if mistakes are made, so redescription can cease. In safety-critical situations, the analyst is advised to err on the side of caution and develop the analysis to further detail than may be strictly necessary, in order to understand more fully how the operator interacts with the environment.

Task analysis should result in the stipulation of the safe and proper ways of doing things. Therefore it is important to obtain the views of managers to ensure that productivity goals are met, of engineers to ensure that plant is operated in an appropriate fashion, of safety experts to ensure that proper procedures are followed, and of supervisors and operators to ensure that the routine problems of plant are appreciated.

7.3   ACQUISITION AND APPLICATION OF SKILL

It is unnecessary for training and support practitioners to have a detailed understanding of the cognitive processes of the people they train and support. However, it is important for them to possess a broad grasp of the manner in which people master and carry out tasks, as this influences support and training design.

7.3.1   Goals

All work to be trained for or supported can be discussed by reference to the *goals* to which it contributes. The term 'goals' refers to what operators are trying to achieve in their work. Goals and sub-goals are generally set by operating personnel. Managers and engineers indicate the higher-level goals in the task analysis, while people more acquainted with the details of how to operate plant, such as supervisors and senior operators, indicate goals at a greater level of detail to ensure that plant is operated safely, efficiently and with maximum convenience.

7.3.2   Input, Action and Feedback

Skill can be described by reference to the main flows of information and control between the operator and the system under control. These are illustrated in Figure 7.2. In this model the operator, seeking to attain a goal, monitors the system environment in order to identify system states associated with the goal. The monitored information is used to determine whether the system deviates from the required goal state and, hence, whether action is necessary. It is then used, along with the operator's knowledge of system states and understanding of the workings of the system, to select an *action* to move the system towards the goal state. During the action, the operator monitors *action-feedback* to ensure that what he or she intended to do is done. Following the action, the operator monitors *system-feedback* to establish whether the action was, indeed, appropriate in moving towards the target goal state, or whether further action is necessary. Skilled performance is where the operator is able to monitor system states and choose appropriate actions in an efficient manner, often without heavy reliance on feedback, to regulate actions.

When we develop skills through training, our aim is to ensure that the operator knows things such as: which action cues to monitor, how long to wait for suitable system feedback, which actions are most appropriate to different sets of circumstances, how to plan, and what system information is most helpful in carrying out this planning.

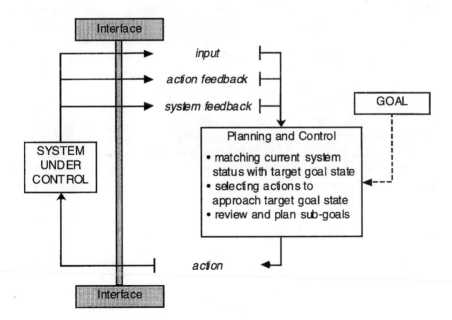

Figure 7.2 *The input-action feedback model, showing the information flows between the system under control and the operator doing planning and control to meet the set goals*

7.3.3   Knowledge, Skill and Practice

Knowledge is an essential component of competence, but for it to provide an effective basis for *skill*, the trainee must be given the opportunity to *practise* the task so as to adapt the knowledge. Teaching knowledge without the opportunity to practise is of limited value [Stammers 87]. Knowledge supports the acquisition of skills by guiding actions, helping to make choices, suggesting analogies, indicating the relevance of appropriate feedback, and categorising information and action. The practical implication of this is that the knowledge to be taught should be considered at the same time as the practice that should be provided.

It is common for training centres to organise knowledge teaching around formal courses, to be followed by practical experience on a later occasion. This separation has administrative benefits, as classroom teaching can be scheduled for several people at once. However, it is likely that trainees will not understand how the knowledge previously acquired relates to subsequent practice. A far more satisfactory approach is for trainers to identify practice opportunities in conjunction with the knowledge required. It may be less convenient to administer, but with techniques such as computer-based training, where knowledge and practice can be presented on an individual basis, even these administrative and practical difficulties can be overcome.

7.3.4   Extrinsic Information and Feedback

When trainees develop skills, they must do so using information that will be present in the real task situation. During training, however, we are at liberty to provide additional information that is not present in the real task. This is called 'extrinsic information' [Annett 69]. We may give the trainee hints concerning which action to choose. We may provide feedback, to confirm that an action was correct (e.g., 'well done') or to provide information concerning the modification of a response (e.g., 'you stopped too soon'). This information will not occur in real tasks, and will be useful only providing that the trainee uses it to focus on intrinsic cues that can be used on a subsequent occasion without recourse to extrinsic information.

As training progresses, extrinsic information must be removed to ensure that the trainee learns to cope with only those features that will be present during real operations.

7.3.5   Phases in Learning

A long-established notion is that learning develops through a series of phases [e.g., Fitts 62, Anderson 87]: an initial 'cognitive' phase where the learner generates actions based on task knowledge; an 'associative' phase where correct response patterns are laid down by practice with appropriate training support; and an 'automation' phase where practice continues in order to consolidate response patterns. This progression follows the transition from conscious response, entailing reasoning about the system, to practise, entailing consolidation of automated responses.

These phases map onto Rasmussen's 'skill-rule-knowledge' classification [Rasmussen 80] (see Chapter 2 for a more detailed description). The cognitive phase relies on the learner carrying out the training task via the application of knowledge, as in Rasmussen's 'knowledge-based' form of responding. When the mode of response becomes more predictable and the operator has learned to associate response options with situational patterns, performance is 'rule-based'. When response patterns are consolidated, prompting automatic response, performance is 'skill-based'. These phases imply a framework for organising training which is discussed later.

7.3.6   Individual Differences

People differ in the knowledge they possess at the start of training, the manner in which this knowledge is organised, the ways in which they like and are able to learn, the speed with which they can assimilate new material, and their confidence in dealing with new situations. Individual differences in learning mean that training methods should be flexible and adaptive to the rates of learning of individuals. Some people may be generally slower than others. Some may be slow in some subjects but able to catch up in others. The main aim of industrial training is to ensure that all personnel attain a satisfactory level of competence. If one trainee takes more time to learn than a colleague, this may be of no consequence, provided both are fully reliable when their training is complete.

7.4   MAKING TRAINING AND SUPPORT DESIGN CHOICES

Before designing support or training, we must first decide which parts of the task need to be supported and which parts should be trained for.

7.4.1   Identifying Support and Training Options

Attainment of a goal or sub-goal depends on knowing both when and how it should be attained. If a goal is to be attained by unaided performance, the operator must be trained to know both when and how to carry out the tasks which result in the goal. On the other hand, if a goal is to be carried out via supported performance, then the support (job aid, colleague, alarm, etc.) will indicate when the goal should be attained. Gaining the knowledge of how to attain the goal may be achieved by specifying what to do, in greater detail, within the context of the job aid or by relying on the operator's skill. Ultimately, any support solution must depend upon the operator knowing how to do something; all that support can do is instruct an operator what to do in greater or lesser detail.

The different assignment of support and training can be illustrated by reference to the analysis of the simple task of cooling a reactor vessel, shown in Figure 7.3. Carrying out this task requires the operator to switch strategies in accordance with the values of parameters such as level and temperature. An operator performing unaided would need either to remember a number of procedures or to understand certain operating principles in order to generate an appropriate strategy. If the consequences of error are not severe or there is a frequent requirement for performing the task, there may be benefit in training the operator to perform unaided. However, if the consequences of error are severe and the task infrequent, then job aids are justified, or even demanded.

We can often provide job aids to support performance in a number of ways. Figure 7.4 shows the simplest variant, where the job aid supports the

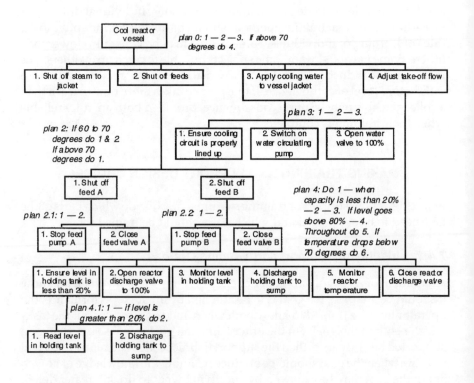

Figure 7.3 *Analysis of a simple cooling procedure*

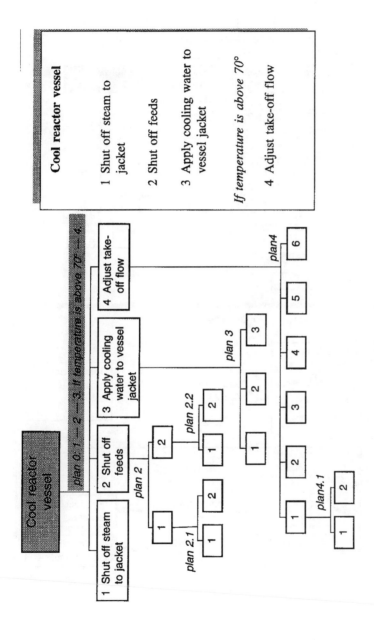

Figure 7.4 *A hierarchical task analysis, partitioned into aspects for support and training. Elements in bold boxes represent the parts of the task to be dealt with by the simple procedural guide (on the right). The remainder of the task must be trained for.*

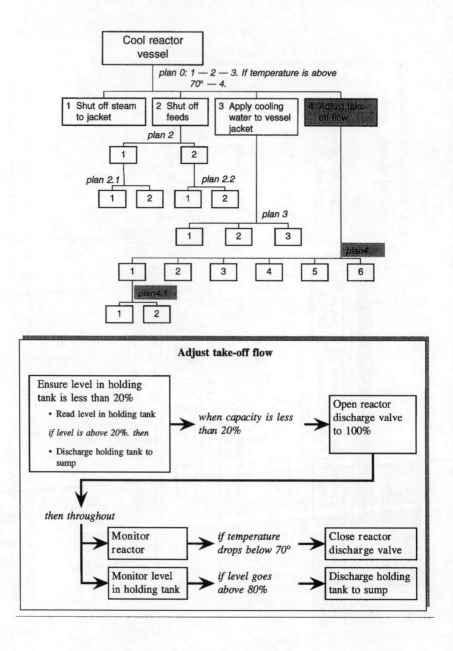

Figure 7.5 *Another task analysis, where the job aid provides support to the operator at the lower levels of the task*

top line only. This job aid supports plan 0, requiring the operator to be trained to carry out the tasks to which the job aid refers. While the job aid refers to 'shut off steam', 'shut off feeds', etc., this is only to specify *when* to perform the tasks and not *how* to do them. Therefore, the operator must be trained in the constituent sub-skills.

Figure 7.5 shows another way of supporting the same task. This leaves the higher levels of the analysis to be trained, requiring a job aid to be prepared for sub-goal 4 'adjust take-off flow'.

It would be easy to provide a comprehensive operating manual combining support at the higher level (as in Figure 7.4) with support at the lower levels (as in Figure 7.5). In this way, an operator referring to the manual would use the higher-level support as both a general guide and an index to more detailed support.

7.4.2   The Role of Context in Design Decisions

Judging whether a part of a task should be treated by training or support or both and, indeed, what type of training or support should be provided, depends very much on the context in which the task is carried out. Contextual factors influence the consequences of making mistakes and the frequency of the task, and thus the opportunity to practise and maintain skill. Several factors need to be considered, some environmental and some technical:

- Task difficulty;
- Representation of information or resources used in task performance;
- Task feedback;
- Predictability of events;
- Controllability of events;
- Frequency of events;
- Severity of consequences of error;
- Recoverability;
- Stressors;
- Access to help;
- Environmental and movement constraints;
- Costs of training support;
- Legal, industrial and cultural compliance.

(a)   Task difficulty

The more difficult a task is the more time a trainee will need for the acquisition of skills.

(b)   Representation of information
or resources used in task performance

The representation of information or resources, such as controls, dials, and escape hatches, can significantly affect the manner in which they are used within a task. Depending upon their design, such resources may remind the operator of what could or should be done next. The extent to which an object prompts the operator in its use has been called 'affordance' by various authors (see [Lansdale and Ormerod 94]). When affordances are poor, the operator must be more highly trained or explicitly supported because the display itself will not prompt the selection of appropriate responses. The representation of information via the human-machine interface and the relationship between information display and operator support is detailed in Chapter 6.

(c)   Task feedback

When an action is carried out, the operator monitors feedback from the system under control to confirm that the action either is correct or needs to be regulated in some way. In some systems such feedback is readily available, while in others it is unavailable or delayed and of little use. Tasks with adequate feedback are easier to master on the job.

(d)   Predictability of events

If the set of events triggering a particular task can be anticipated, the instructor and trainee can prepare for learning and free themselves from other activities. The instructor may then help the trainee to attend to important features, demonstrate suitable responses, monitor the trainee's learning, and provide support as necessary. If the onset of the task is unpredictable, then no such preparation is possible and instruction is less effective.

(e)   Controllability of events

Instructors may have some control over events or discretion concerning when some tasks are carried out. For example, a cleaning or purging operation may be required at some time during a shift, but not at a precise moment. This means that an instructor would be able to schedule the task when it is most convenient to both trainee and instructor. In other tasks, the instructor may be able to influence circumstances directly and safely so as to enable more realistic practice.

(f)   Frequency of events

If events occur frequently, there is more opportunity to practise than where events occur rarely. Infrequent events limit opportunities for practice. Indeed, one of the ironies of automation is that, as tasks are increasingly automated, opportunities for operators to master and maintain skills are reduced [Bainbridge 87].

(g)   Severity of consequences of error

Where the consequences of error are serious, justification for off-job training is strengthened. If error consequences are unacceptable (for example, life-threatening) and other factors mitigate against the acquisition and retention of skill, then either expensive simulated training or operator support becomes justified.

(h)   Recoverability

Some tasks permit recovery from error and some tasks do not. If there is no easy recovery from error and the potential consequences are severe, then trainees cannot be permitted to carry out these tasks if there is a risk of their being unable to cope. Under such circumstances a training regime which includes simulation and careful demonstration is required. The processes for assessing error and recovery are discussed more fully in Chapter 3.

(i)   Stressors

Some real situations place stress on the operator and inhibit learning, thereby justifying simulation training. However, simulation should be carried out with care because the work stresses are still present. A written procedure may be warranted to overcome these problems. However, if the task requires the operator to make unique judgements, then even a decision aid may not suffice and some form of intensive skills training is necessary.

(j)   Access to help

Any factors which limit access to help in an operational context need to be taken into consideration. Working in hostile environments often creates conditions where access to help is limited.

(k)   Environmental and movement constraints

Environments in which there is insufficient light to see by, noise which may impair hearing, too little space in which to turn a page, too much inflammable material to permit an electronic aid or too much corrosive material to enable a job aid to be used, clearly place constraints on the adoption of support to facilitate carrying out the task to a satisfactory standard.

(l)   Costs of training support

Cost is a serious consideration in all organisations and it can influence the choice of training and support options for safety as well as for production activities. The use of a training simulator may be indicated as a solution, but excluded as an option if its cost is judged to be excessive.

(m)   Legal, industrial and cultural compliance

Often, things have to be done in one way rather than another because there is a law or company practice which decrees it. Sometimes it is necessary to demonstrate compliance with industry codes of practice or company standards, or, in the case of safety-critical systems, to document and implement a safety case (see Chapter 12). Such dogmatic prescription may not match the real needs of operators and may need to be challenged or compensated for.

While the above factors have been discussed separately, it is important to emphasise that their effect on design choices can only be considered by assessing all relevant factors in combination. For example, for an infrequent and difficult task where error consequences are high and recovery from error is unlikely, a job aid may be judged to be the most suitable option. However, this is not the case if access to help is limited in some way. Generally, all contextual information needs to be considered if a reasoned choice is to be made.

7.5   DEVELOPING SUPPORT AIDS FOR OPERATORS

Support comes in a variety of forms, including from colleagues and from artefacts such as documents and computer-based advisory systems. Generally, support from colleagues comes in the form of instruction or a 'local expert'. The focus in this chapter is on support that is provided by artefacts, such as written procedural guides and operating manuals designed and developed explicitly to assist the operator, rather than on the social aspects of support.

7.5.1   Types of Support

(a)   Reference sources vs job aids

Two different types of support are *reference sources* and *job aids*. Reference sources, usually in the form of manuals, consist of detail about a wide range of aspects of a job. They are not intended to be used in their entirety or as a matter of routine, but are made available for reference when required by the user or operator. Computer manuals, for example, are rarely meant to be read from cover to cover, but to be referred to as and when the user feels the need. If they are well written, constant reference to them becomes unnecessary as the user learns what to do. It is often obligatory for an industrial plant to provide an operating manual, in particular where there are safety-critical elements involved in operation. However, manuals are often bulky, as they must contain a great amount of information, and in many cases are never used.

Job aids, in contrast, are usually written to help people carry out a specific task reliably and safely, and are designed to be used as a matter of routine. They are useful in helping people to do essential things which are difficult to remember. As they are specific and limited in their subject matter, they are generally quite small and easy to handle.

Most job aids are *domain-dependent*, meaning that they are designed to support operators to carry out a specified set of activities in one particular context. It is possible to reduce domain dependency by providing job aids which provide support in the form of general heuristics. However, such aids are more likely to constitute a part of a training environment where the operator is being prompted to solve problems and to learn adaptable strategies.

(b)   Procedural guides

Procedural guides are lists of instructions setting out what an operator should do, step by step. This means that the operator must be competent to accomplish each of the steps. A part of competency is knowing when each step has been accomplished. This means that the operator must understand how to monitor feedback from the system and judge that the step has been accomplished, before moving on. The procedural guide might be written simply, stating only the actions to be taken and relying on the operator's skill and knowledge to judge when the step has been completed. Alternatively, the writer might choose to state the feedback for each step explicitly. This would be important if the task is carried out infrequently or in safety-critical situations (see also Chapter 8 on the use of procedures in abnormal situations).

In dynamic systems, where the operator's action affects the way a plant or item of equipment performs, a subsequent action is often cued by observing a change in the system's state rather than simply registering that a previous action has been completed successfully. It is important to state explicitly within the job aid the cues that must be monitored in this way. These cues should have been identified from the task analysis.

Using procedures is a complement rather than an alternative to training. Each step in the job aid prompts an operator to respond in a competent way. This implies that the operator must know how to do the things to which the job aid refers. For example, in the procedure in Figure 7.4, the instruction to 'adjust take-off flow' requires that the operator is competent in this action. Equally, the operator must know how to use the aid itself.

(c)   Checklists

Checklists are memory aids designed to remind an operator to do several things, but in no particular order. They may be used when making sure that several steps in the inspection of a device or system are completed, for example in a pre-flight maintenance or medical context. Although no fixed procedure needs to be remembered, operators must remember which items they have already dealt with and which remain. Close examination of how operators use a particular checklist may reveal that one sequence is preferred to all others, or is more efficient or convenient. If this is the case, then it is helpful to specify a *procedure* for making checks, to ensure that no check is overlooked. Often a checklist is needed as a record that certain things *were* done. Should a subsequent problem emerge, it may be important to know what was done, how it was done, when and by whom, and whether any difficulties were encountered. So, a checklist, whether presented as a fixed procedure or not, usually requires the operator to enter information of some kind on completion of each step. Checklists are common in the maintenance and preparation of large systems, dangerous systems and experimental systems, both to prompt conscientious action and to provide a record of what was done.

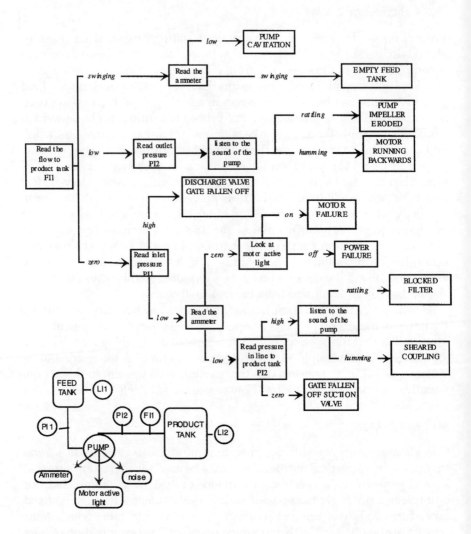

Figure 7.6 *A simple pump and tank configuration (bottom left) together with a fault-finding decision tree*

(d)   Decision aids

Decision aids, such as decision trees and decision charts, prompt operators to collect information and then aid them to interpret the information for the purpose of deciding what action should be carried out next. A perennial problem is that decision aids tend to restrict decision outcomes to a finite set of possibilities. In practice there is often a need for operators to identify problems which have not been anticipated—when job aids may be inadequate and the operator should be given decision-making training instead.

Decision trees are diagrams which take the operator through a series of questions, with the next question depending on the answers to previous questions. Thus, decision trees chart the flow of questioning until a solution to a problem is reached (see Figure 7.6).

Fault-finding charts record a set of symptoms and, for each, list a set of faults or solutions that should be considered if the symptoms in question occur. They do not take the operator systematically through a series of questions to arrive at a specified solution. Rather, they prompt thought, relying on the operator to exercise some skill. There are many other forms of chart which represent system functions for the purposes of diagnosis and decision making, and these are described by Kirwan and Ainsworth [Kirwan and Ainsworth 92].

(e)   Expert systems

Expert systems are computer programs which use knowledge derived from domain experts to provide advice to operators for the support of decision making. The operator, or the system, supplies information regarding current circumstances, and then the expert system generates advice consistent with the information provided and the stored knowledge.

(f)   Labels and notices

Labels and notices fall into the ambiguous territory between job aids and interface design. They may serve to prompt an operator on what to do next or what not to do. They may present a caution, in which case they need to be attention-gaining. They may define the functionality of a control and therefore serve to help an operator construct a suitable response to a situation which might otherwise be overlooked. To meet these ends, labels and notices must be designed and expressed carefully in accordance with the behaviour they are intended to invoke.

7.5.2   The Purpose and Scope of the Support Document

In developing any support document it is important to be clear about the exact purpose it is intended to serve. It may be impossible to satisfy all purposes within one document. Systems manuals may be required to support system managers and system engineers, rather than people carrying out operational tasks. Documents may be needed to provide evidence to various authorities and agencies that good practices are being adhered to. Procedural control may be required to ensure compliance with particular operating procedures where, for example, deviation from certain conditions may compromise safety. Documents (for example, plant manuals) may be used to enable easy point-of-need reference for the operator. Training manuals may be needed to guide the trainee in a step-by-step fashion through procedures to support learning. A job aid may be required to be used routinely when carrying out a task. It is rare for a single document to satisfy the full range of requirements.

Establishing the purpose of a support document also identifies the person or grade of person who will be using it. This is essential, as the design of the document must take account of the person's task requirements, their level of skill and knowledge, the extent to which they are engaged with other aspects of the task, their understanding of terminology, and the environment in which they work.

7.5.3   Setting out a Support Document

When the purpose of the support document has been established, care must be applied to its proper design. In setting out any support document we need to address four principal components:

- The means by which the user navigates through the document;
- The form of each instruction encountered;
- The clarity of presentation of instructions;
- The physical form of the document.

These are partially determined by the examination of the purpose of the document and by good human factors design principles. The following notes are necessarily brief, and the reader is referred to Hartley [Hartley 85] for a more extensive review of the key points in document preparation.

7.5.4   Navigation

Navigation is concerned with how the operator moves around the document, i.e., which instruction to move to next. In a procedural guide, navigation is concerned with cues to sequences. In a manual, it is concerned with effective indexing and the location of choices. In a flow chart, it is concerned with a clear route to different options according to different contingencies. In a decision table, it is concerned with the clarity with which the operator can read a table to determine which option to pursue. Navigational features for a variety of support methods are summarised below.

(a)   Navigational features in procedures

In a fixed procedure, the following common principles should be observed.

- Individual instructions should be numbered in order of execution;
- Instructions should be arranged to progress down the page or screen, or from left to right across the screen;
- It should be clearly indicated when an individual instruction starts and finishes by using a 'bullet point' or number and finishing with a full stop or semi-colon, or by containing the individual instruction in a box.

(b) Navigational features in manuals

The criterion of layout in a reference manual is that the operator should be able quickly to identify the item of interest. The following guidelines apply.

- Number each page;
- Provide a table of contents;
- Provide an index, making sure that the entries are expressed in terms which are relevant to the operator and relevant to the task in hand;
- Provide running headers on each page;
- Use page tags for different sections;
- Provide discernible topic headings and subheadings, using different typefaces and styles, indentations, and numbering, to ensure that the item sought can be rapidly located;
- Computer-based systems are ideal for providing screen-based help facilities.

(c) Navigational features in decision flow charts

Navigational features are especially important in decision flow charts, where the operator is required to trace different routes through the chart according to different answers to questions.

- Ensure that the alternative answers to questions are clear, exhaustive and unambiguous;
- Ensure that there is a route of progress for each logically possible alternative—a question which distinguishes between two states must have two routes, three states must lead to three routes;
- Join questions to instruction boxes by clear lines;
- Introduce arrows to ensure that the direction of travel along lines is clear;
- Place options on lines at the point where they exit from the decision box; place arrow-heads at the point of entry to the next instruction box.

(e) Navigational features in look-up tables

- Clearly label rows and columns;
- Provide a grid to enable cells to be located easily;

- For large tables provide shading on every 3rd or 5th row and column to aid the location of cells.

7.5.5   The Form of Instructions

Solving the navigational problems ensures that the operator gets to the required parts of the document quickly and reliably. However, the instruction must then be written clearly so that it can be interpreted promptly and without ambiguity. The following guidelines aim to create text which is most easy to understand.

- Use short, concise sentences;
- Use words which match the vocabulary of the user or operator;
- Include a glossary of terms;
- Express instructions in the active rather than the passive voice, e.g., 'start the pump' rather than 'the pump is to be started';
- Construct sentences in the positive rather than the negative as people generally understand positive concepts more easily;
- Use positive terms such as 'heavier than' in preference to negative ones such as 'lighter than';
- Avoid reducing the length of text by omitting key linking words such as: the, a, that, which, of, for;
- Mention instructions in the order in which they are to be carried out;
- In instructions, emphasise *when* things should be done (the condition), as well as *what* should be done (the action), and present them in the following order: condition—action;
- Make instructions explicit — when discussing amounts, be precise;
- Use punctuation appropriately and consistently;
- Avoid using sexist language.

7.5.6   Presentation

In addition to content and form, there are a number of features of presentation which need to be considered so as to ensure that the document serves its purpose most effectively.

- Excessive use of different text styles, such as italics, bold or underlined text, or different fonts, can make text look messy. Different styles should only be used to serve a purpose, to help the operator find his or her way around the text most efficiently.
- Excessive use of upper case (capital) letters is not recommended. When people read, they use the shapes of words as well as the individual letters.
- It is essential that consistency in the use of all text features is maintained throughout.
- Typefaces may vary in size. Size is useful as a basis for distinguishing headings from text and assisting navigation, but excessive and

inconsistent variations in size should be avoided.

- An important aspect of text size is the extent to which it is easy to read. This can be influenced by lighting conditions and the eyesight of the user or operator. Larger text may compensate for poorer lighting. Smaller text may be unavoidable in order to produce a smaller document, but it should be adopted with caution and conditions should be provided to ensure that it can be read.

It is also important to pay attention to the physical form of a job aid. This is of particular importance in hazardous environments. The support document must be easily located and convenient to handle and it must be designed to stand up to the environmental stresses to which it will be subjected (e.g., in a dirtproof or waterproof cover).

7.5.7   Prototyping the Documentation

When the information has been collected, it is important to develop a small part of it as a prototype of the final document to see how well it will work in the context for which it is intended. There is little point in spending too much effort in producing a document in an inappropriate form. Not only should the layout and type of content be considered, but so too should the manner in which it is to be stored and used in the workplace. It is advisable to enlist the cooperation of potential users and to seek their input into the design of the document.

7.5.8   Implementation

Manuals and job aids should be introduced with care to the staff expected to use them. This is necessary to ensure that they will be used in the manner intended and that operators develop confidence that the document will be useful to them.

Manuals should be demonstrated to potential users. Their layout should be explained, as should any formatting methods used to convey certain types of information. A simple walk-through demonstration should be used to show how the document can be used for different purposes. Users should be given the opportunity to practise using the document in different scenarios.

Job aids should be introduced during operational training. Operators then become used to using the job aid in the context of learning procedures. This ensures they can use the information effectively and that they have confidence in the help that it provides.

7.5.9   Maintenance of Documentation

Procedures must be established to make certain that documents are kept up to date in terms of technical changes and changes in customs and practices. Maintenance of documentation is a serious management issue. Unless a manual is routinely and conscientiously maintained, it will soon become out of date and lose credibility. Several provisions should be made.

- There should be a named person with responsibility for maintaining each document;
- The named person should have the skills and resources do this job effectively;
- Procedures should be established for the regular review of documents;
- There should be a standing instruction to all managers, engineers and supervisors to notify the person responsible for a document each time a relevant modification to plant or procedures is made;
- Document masters should be stored safely to enable rapid modification (e.g., on backed-up computer software).

7.5.10 Media for Support

Support may be presented through a range of media, including printed, auditory and electronic media, or a combination of these. Different contextual constraints will rule out some combinations and enable others.

The use of electronic media provides a number of facilities which allow flexible access to databases and enhance operator interaction with support devices. For example, some computer programs provide easy solutions for navigating through a library of job aids and so is ideal for the selection of the required document — such as present procedural guides, diagnostic aids, look-up tables and relevant system diagrams. Computer programs can respond flexibly to the operator's request for information. Information may be presented via a stationary visual display unit, a lap-top or palm-top computer, or a head-up display. The database may be contained within the device's memory or accessed via a communications link. The maintenance technician using a palm-top computer may need to load relevant support packages before leaving the maintenance centre, may carry the full database on a CD ROM, or may plug into a network at the equipment site.

7.6   ENSURING SAFE AND COMPETENT OPERATION THROUGH TRAINING

There are several common forms of training. All are satisfactory, provided that they are carried out well and applied in an appropriate context. They include:

- Knowledge teaching;
- On-job instruction;
- Simulator training;
- Part-task training;
- Team training;
- Refresher training.

7.6.1   Types of Training

(a)   Knowledge teaching

Teaching basic knowledge about a system's structure and function has long been regarded as an essential component of operator training. Generally this view has support because possession of some systems knowledge is necessary in the acquisition of skill (see Section 7.3.3). Knowledge training can help in teaching the names of locations and the parts of plant and equipment, and in justifying certain procedures and safety measures. It is also motivating if carried out well. Acquiring knowledge is also important if operators are required to develop reasoning skills to deal with situations which have not been anticipated.

There is a danger that this form of training is overemphasised for two reasons. First, it is relatively easy to generate content by presenting a simple account of how the system functions. This is bad practice because the knowledge content may not relate to the skill requirements of operators. Second, it is relatively easy to administer — all of the knowledge to be taught can be assembled and presented in a classroom session or in a computer-based learning package. This is bad practice if it separates the knowledge teaching from the skills that have to be practised to make use of the knowledge.

(b)   On-job instruction

A second common form of training is on-job instruction, where a trainee watches activities on a real plant and is introduced to certain tasks under the scrutiny and control of an experienced colleague or instructor. This approach provides the opportunity for the trainee to see how work should be paced, to observe the signals which must be responded to, and to gain a proper feel of the controls. But there are negative aspects which are particularly pertinent to hazardous industries. One problem is the risk to trainees. Another is that many crucial events occur infrequently, offering little or no opportunity for effective practice.

(c)   Simulator training

Simulator training is used widely where a company is able to afford a simulator or is otherwise obliged to use one. Simulators for training are often essential in safety-critical systems. A primary justification for cockpit simulators, for example, is that infrequent, possibly complex manoeuvres can be undertaken without risk to air-crew, passengers, the public and the aircraft, even if the trainee makes a mistake. A further argument is that safety-critical events are generally infrequent and opportunities to practise on the job are, consequently, rare. Training simulation enables critical events to be specially generated, so we encounter simulation training in most safety-critical domains, including power generation, military contexts, and medical situations such as the training of resuscitation techniques and anaesthesia.

Simulation training means providing a trainee with the opportunity to practise given aspects of operating skills safely within a training programme by the use of some kind of representation of the task to be learned. It is the *task* rather than the equipment that should be simulated, so it is necessary to analyse the task in order that precise simulation requirements can be established.

Simulators need not be close physical representations of control interfaces, although they often are. Simulation may be provided by using real equipment and manipulating events to enable the trainee to experience a range of operating conditions, but this is not feasible in all situations. More typically, simulators use their own technology and are often expensive. Patrick provides a useful review of simulation for training [Patrick 92].

(d)   Part-task training

A part-task trainer is a structured learning environment in which training can be given economically. It is used where the trainer identifies a particular part of a task to provide a focus for skills training. The 'part' may be a routine that needs special practice or a skill that needs emphasis before training in a more complex environment.

Often simulation training is expensive, inefficient in time, and only appropriate for practising complete tasks. The prior concentrating on difficult task components in special part-task exercises can make the utilisation of a full-scope simulator more efficient. In training an anaesthetist to monitor the well-being of a patient, for example, concentrating on the skill of problem detection will ensure that this aspect of the task will have been mastered to a reasonable degree before practising the whole task. Using a full-scope simulator to master basic skills of this kind is extremely wasteful of resources.

Part-task training has many benefits, including the opportunity to gain far more control over learning conditions than is often available in the simulation of complete tasks, and it is a relatively inexpensive way of gaining quick training benefits.

(e)   Team training

Many complex systems rely on the collaboration of a group of skilled individuals working as a team. Team functioning requires individual team members to perform their jobs effectively and the coordination of effort to meet the team's goals. It may even entail team members standing in for colleagues during periods of overload or disability. Team members must also monitor the performance of colleagues. This enables them to establish when information and support must be given or requested (see Section 9.3 of Chapter 9 for some interesting observations on functioning within teams and on team leadership). These various requirements place some interesting demands on team training, above and beyond the individual skills training of each member, and there is also a substantial supervisory training implication in any team function. Swezey and Salas provide a comprehensive review of all aspects of team training [Swezey and Salas 92].

(f)   Refresher training

Refresher training, sometimes in the form of on-job drills and sometimes in the form of simulated exercises, is a key component in maintaining operator skills in infrequent but crucial tasks. Refresher training is important in safety-critical systems as it is often necessary to ensure that operators are highly trained in skills they will rarely or never encounter during the normal course of operations.

7.6.2   The Processes of Training

Each of the forms of training described in Section 7.6.1 is acceptable in the right context, provided that it has been properly developed and that it contributes to a rational training programme. Advice on training design is often more dogmatic than is justified. Good training employs the different training options sensibly, to enable trainees to attain necessary levels of skills in an efficient manner. Understanding how these different training methods might be selected to complement each other within a training programme is best achieved by appreciating how the different phases of learning are accomplished. Training must provide conditions to enable the mastery of knowledge (see Section 7.3.3), followed by development through three phases of skill acquisition: the cognitive phase, the associative phase, and the automation phase (see Section 7.3.5). These four phases of learning translate into four stages of training:

- Stage 1: the acquisition of task knowledge;
- Stage 2: the application of knowledge to generate effective responses;
- Stage 3: the practice of skills to enable the trainee to discriminate information patterns and refine activity patterns;
- Stage 4: the practice of skills to consolidate them and master the timing, rhythm and accuracy required in the real task.

In Stage 1 we need to ensure that the trainee has the knowledge necessary to accomplish the task. A good training system will keep records of what people already know, or test them to see whether knowledge training is necessary. Training will be delivered either in group sessions or by individual instruction, for example using computer-based training.

Stage 2 deals with the cognitive phase and requires appropriate opportunity to be provided for practice, along with various practical training strategies to help the trainee accomplish the task and relate previously gained knowledge to it. The role of guidance and help, as well as feedback extrinsic to the task, is necessary to help the trainee accomplish this. The context for practice could be the real workplace, but this may be unsatisfactory for a number of reasons, including safety. Moreover, ideal conditions for practice are rarely available in the real workplace and some form of simulation is often warranted. However, simulators are often built without paying adequate attention to training facilities, either facilities integral to the simulator itself or those in the broader training context in which it is to be used.

In Stage 3 the trainee organises the skill in the associative phase. It is essential to ensure that an appropriate strategy is emerging that will ultimately meet the criteria set for the task. An effective training system monitors the trainee's developing performance in order to adjust training conditions according to progress. It is necessary to monitor progress both in terms of how successfully the trainee accomplishes the tasks set in training and the manner in which the trainee accomplishes them. In fault diagnosis training, for example, it is not sufficient merely to establish that the trainee can find familiar faults. It is also important to ensure that the trainee is carrying out the task using a strategy which enables some kind of analysis of the problem (i.e., a rule-based approach) rather than simply interpreting patterns. An analytical approach leads to the transfer of skill to non-routine situations, whereas pattern recognition does not.

Stage 4, the automation phase, requires that the trainee be given a realistic context in which to practise. Full task simulation comes into its own as the means of providing practice without risk. Alternatively, the trainee may be required to refine skills on the job with all of its attendant risk.

The training designer needs to organise the most effective means of accomplishing these various stages and to select training options from those available according to the contexts in which the task is carried out and training takes place. For example, the preparation of operators to diagnose faults may entail:

- Learning about plant layout (in a classroom, via computer-based training, via plant walk-through, or as a by-product of learning about plant start-up);
- Learning about plant functions by the above methods;
- Learning diagnostic rules and strategies (in a classroom or via computer-based training);
- Learning how to apply the diagnostic rules and strategies (via a part-task trainer or classroom exercises);
- Acquiring operational skills at diagnosis (through practice on a simulator or through instructor-supported work experience);
- Undertaking refresher training on a simulator according to a regular schedule.

Thus, training follows the route of imparting knowledge, then providing a context to enable the knowledge to be compiled into effective diagnostic procedures, and then refining skills in an operational or simulated situation. Training can be represented as a cycle of activity which continues until competence is reached (see Figure 7.7).

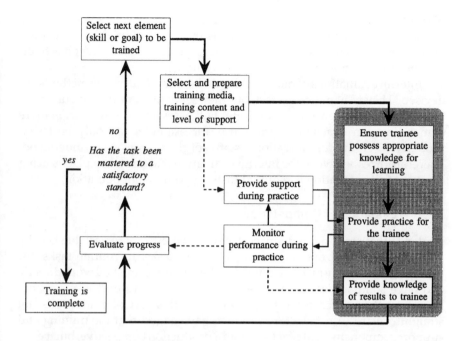

Figure 7.7 *The cycle of training activities*

7.7 ASSESSING PERFORMANCE

There is a need to assess operator training and support, both to evaluate whether the methods used are generally satisfactory and to determine whether individual operators have been trained to an adequate standard.

It is not easy to separate assessment of training from the assessment of support. To assess a job aid properly we need to establish that it serves its purpose when used by a competent person. Equally, to assess training thoroughly we need to establish that it results in competency in the real job situation, which means that the operator performs effectively with any support that might be available. Until all of the elements of training and support are in place, all we can do is make expert judgements that the design developed at a particular stage complies with best practice. This means that it is consistent with the principles used to carry out the design.

When a job aid is first developed, it can be checked to ensure that it conforms to the purpose for which it is intended, it complements training provision, it is set out clearly to aid navigation, and its individual instructions are expressed clearly. For proper quality control, this assessment should be undertaken by someone other than the original designer.

It should also be possible to check that the job aid can cope with the contingencies for which it was designed by carrying out a walk-through or talk-through of possible incidents as indicated by the task analysis on which it was based. We may review training in the same way. In some cases we might be able to test that certain skills-training modules achieve their purpose by evaluating the outcome with respect to the training objectives. However, this is not the same as a full evaluation.

Interim evaluations of methods should be carried out as the methods are devised. They may not be perfect, but they serve to identify obvious errors that are then inexpensive to eradicate. Full assessment, however, is more demanding. An evacuation procedure, for example, can only be tested properly when the full operational status of the workplace is understood. Therefore, irrespective of the interim evaluation that has taken place during design, a full evaluation is required to assure competent operation.

7.7.1   Evidence of Competence

Evidence that training and support have been effective is established by assessing how well operators carry out their tasks. In routine tasks, for example an assembly task, we can judge the quality of finished work directly and have some indication of whether training and support is successful. However, the inference we may draw from this sort of evidence is not straightforward. A satisfactory product may suggest that the training and support component is satisfactory from a production perspective, but it does not indicate whether work was carried out safely. Moreover, if performance is unsatisfactory, there may be several things that need to be put right which have nothing to do with training and support.

More complex tasks are even more difficult to evaluate. For example, a characteristic of fault diagnosis tasks is that they are required infrequently and irregularly. Evidence that the operator is competent may not extend beyond the specific situations that have arisen. This is often the case in safety-critical systems where considerable effort is usually made to reduce the frequency of the most critical events, so opportunities to assess performance on the job do not often present themselves. This infrequency is good news for the world at large, but operators will have little opportunity to prepare themselves to cope with problems when they do occur. In the assessment of training and support, therefore, simulation of some kind is nearly always required.

In evaluation, we need to establish that the operator has acquired the necessary skills and can apply them when required to do so in the operational situation. All we mean by people having skills is that they can carry out tasks to a specified standard. A person may be said to be skilled if they:

- Use an acceptable method;
- Minimise mistakes;
- Complete the activity in an acceptable time;
- Complete the activity with the pace and rhythm of someone with acknowledged expertise;
- Properly integrate the activity with other activities;
- Yield accurate results.

7.7.2   Factors Constraining Assessment

A major purpose of training assessment is to predict performance in the real situation. It is a reasonable assumption that a trainee unable to attain the relevant task measures in a training situation is unlikely to meet these criteria in the real situation. This may not matter too much if the company allows the trainee to make mistakes in order to gain experience, but it does matter if the trainee has to perform correctly every time. However, even if the trainee appears to perform satisfactorily, it is wrong to assume that this performance can be relied on to be repeated on the job, because other situational factors may be present. In understanding the relevance of assessment to performance in the real situation, we need to appreciate the following:

- First-shot success;
- Availability of real tasks for practice;
- Simulation and fidelity;
- Infrequency and forgetting;
- Self-reliance;
- Stress.

(a)   First-shot success

We may need to ensure that the operator is fully competent to deal with particular circumstances when they first occur. Landing an aircraft and evacuating a burning building are two examples. This is called 'first-shot performance', which contrasts with those situations where a lower initial standard is permitted because it is acceptable to learn from experience — for example, serving in a shop, where error consequences are not severe and recovery is usually possible. The term 'savings measure' is often used to describe the measurement of training when minimal or zero real-world experience is permissible in reaching an acceptable standard, given that a particular course of training has been followed. In training for safety-critical tasks, we are mainly concerned with first-shot performance.

(b)   Availability of real tasks for practice

Often, we are unable to train people to carry out the precise tasks they will encounter in the operational situation, either for reasons of cost in replicating the conditions for practice or simply because we are unable to anticipate the range of situations with which they will be required to cope. For example, it may be impossible to anticipate all faults which might occur in a system, but it is still hoped that operators will be able to locate unforeseen faults with a fair measure of accuracy.

(c)   Simulation and fidelity

In principle, the ideal context for assessing the benefits of training is the operational situation. Undertaking training assessment in circumstances in which the real task cannot be used, perhaps for reasons of safety, entails representing the task by some means. If the fidelity of the simulation is limited in any respect, the trainee's performance may differ from that required in the real task. Therefore, it is essential that the consequences of reduced fidelity are understood during assessment. The assumptions made in the design of the simulation can lead to infidelity, so they too should be clearly understood and explicitly stated.

(d)   Infrequency and forgetting

A well trained operator will perform tasks which occur regularly to a satisfactory standard on each occasion. For infrequently occurring tasks, there is a real risk of forgetting through lack of practice. There is, consequently, a danger of assuming that a skill judged to have been acquired through training will still be available in practice.

(e)   Self-reliance

An operator may be required to act autonomously in the real situation, independently of any human being or documented support. Alternatively, the operator may be allowed to rely upon support from a job aid or another member of the operating team when certain circumstances arise. These two different sets of circumstances should be acknowledged because the need for self-reliance places a higher competency requirement on the operator.

(f)   Stress

The real task may be carried out in conditions of personal danger to the operator, or in circumstances where the operator recognises that error will lead to an accident. Performance without stress is usually different to performance under conditions of stress. It is often not possible to simulate these *stressors* in a realistic way during training assessment.

7.8   CONCLUDING REMARKS

There are in principle a substantial number of tasks that need to be considered in conjunction with safety. They include not only those, such as evacuation, which relate directly to accidents, but also those concerned with normal operations. Normal operations must be carried out according to safe practice, both to maintain current and future safety and to ensure that plant and equipment are in a suitable state for coping with possible future emergencies. The temptation to distinguish between safety tasks and non-safety tasks, where the former concentrates on safety criteria while the latter concentrates on productivity, is misguided. A proper approach recognises that both safety and productivity are essential performance criteria. Safety comes to the fore in emergencies, but should not be omitted during apparently normal conditions.

In ensuring competent operation, training and operator support need to be considered hand in hand. They are not alternative solutions to performance problems but complementary solutions. The designer must make a decision to provide operators with support or not. Having made the decision, training to perform aided or unaided must be carried out. This principle should apply to 'normal' conditions as well as emergency conditions.

The role of context is emphasised in making training and support design choices. Context applies to both the real task and the training situations. One set of training and support options may apply to one context but not to another, even though the tasks are, on the face of it, similar. Careful attention to contexts optimises both the costs and benefits of providing training and support.

Training and support should not be developed *ad hoc*. If care is taken in the analysis of tasks, in properly determining which parts of tasks should be trained for and which supported, and in developing both in a coherent fashion, substantial improvements can be made to safe practices while not jeopardising productivity.

REFERENCES

[Anderson 87] Anderson J R: *Skill Acquisition: Compilation of Weak-Method Problem Solutions*. Psychological Review, 94 (2), 193-210, 1987

[Annett 69] Annett J: *Feedback and Human Behaviour*. Penguin Books, 1969

[Annett et al 71] Annett J, Duncan K D, Stammers R B and Gray M J: *Task Analysis*. HMSO, London, 1971

[Bainbridge 87] Bainbridge L: *Ironies of Automation*. In Rasmussen J, Duncan K D & Leplat J (eds):*New Technology and Human Error*. John Wiley & Sons, 1987

[Diaper 89] Diaper D: *Task Analysis for Human-Computer Interaction* (1st ed), John Wiley & Sons, 1989

[Duncan 74] Duncan D K: *Analytical Techniques in Training Design*. In Edwards E and Leeds F P (eds): *The Human Operator and Process Control*. Taylor and Francis, London, 1974

[Fitts 62] Fitts P M:*Factors in Complex Skill Training*. In Glaser R (ed):*Training Research and Education*. John Wiley & Sons, New York, 1962

[Hartley 85] Hartley J R: *Designing Instructional Text*. Kogan Page, London, 1985

[Lansdale and Ormerod 94] Lansdale M W and Ormerod T C: *Understanding Interfaces — A Handbook of Human-Computer Dialogue*. Academic Press, London, 1994

[Kirwan and Ainsworth 92] Kirwan B and Ainsworth L K: *The Task Analysis Guide*. Taylor and Francis, London, 1992

[Patrick 92]   Patrick J: *Training: Research and Practice*. Academic Press, London, 1992

[Rasmussen 80] Rasmussen J: *The Human as a Systems Component*. In Smith H and Green T (eds): *Human Interaction with Computers*. Academic Press, London, 1980, 67-96

[Shepherd 89] Shepherd A: *Analysis and Training in Information Technology Tasks*. In Diaper D (ed): *Task Analysis for Human-Computer Interaction*. Ellis Horwood, 1989, 15-55

[Stammers 87] Stammers R B: *Training and the Acquisition of Knowledge and Skill*. In Warr P B (ed): *Psychology at Work*. Penguin Books, 1987

[Swezey and Salas 92] Swezey R W and Salas E (eds): *Teams: Their Training and Performance*. Ablex Publishing Corporation, New Jersey, 1992

8

Design and support for abnormal situations

8.1 INTRODUCTION

When considering human factors in the design and evaluation of safety-critical systems, it is necessary to understand and take account of not only the demands associated with normal operations, but also those which derive from emergencies and abnormal situations. This chapter discusses the nature of emergencies and abnormal situations, from the perspective of the human user of the system. Particular emphasis is given to the manner in which the user makes decisions as well as the interface and system design issues which must be addressed if the system is to support its users.

8.1.1 Safety-critical System Functions

Safety-critical systems can support two different aspects of operations. First, they can assure safety during routine or normal operations. Second, they can be designed to assure safety during abnormal conditions, or emergency situations. Whereas it is possible for the same system to fulfil both of these functions, from the viewpoint of the user the functions are typically different. Systems which support normal operations and the prevention of emergencies frequently require different user intervention from those intended to support recovery from abnormal or emergency situations, or to mitigate their consequences. Although examples given are usually hardware interfaces such as control-room displays, all aspects of the interface between the user and the system should be considered, including organisational interfaces.

The functionality of a safety-critical system is determined by the nature of the process to be controlled and the range of potential operational configurations identified during the design process. The manner in which users interact with such a system should be specified according to the predictable capabilities and limitations of the users (see Chapter 5 for further information on specification). Although the user requirements will vary, as noted above, the manner in which the system and its interface support the user must always fulfil those requirements as they exist at any given time.

8.1.2 Abnormal Situations and User Requirements

Chapter 6 considered the manner in which the interface can be designed to support the user during normal operations, but the differing demands imposed on the user by abnormal operations and emergencies are likely to alter the user's interface design and system support requirements. This does not imply that the interface itself should be different, whether in terms of the human-machine interface hardware or the information presented via that hardware. Rather, it requires that the interface design recognises those additional demands which abnormal situations can impose.

Abnormal operations can be considered, at one level, as being any operation or activity with which the user is unfamiliar. However, such a definition would include training situations, infrequent system conditions such as certain maintenance states ('planned abnormalities'), and so forth. Of greater significance are operations associated with situations which constitute unplanned system state abnormalities (including emergencies). The system behaviour may differ from normal due to the nature of the abnormal event or emergency. A particular concern is that these differences may not immediately be apparent to the users. At such times it is essential that the interface supports the users and assists them in identifying appropriate actions.

8.1.3 User Recognition of Abnormality and Action Planning

In providing support to a user during abnormal or emergency operations, two particular issues become apparent. The first concerns the recognition of the abnormal situation, the second the identification of appropriate actions. Clearly, it is necessary for the interface to support the user in taking action, but this aspect of interface design has been discussed extensively in other chapters. Consequently, this chapter primarily considers the two issues which are specific to abnormal situations: recognising abnormality and identifying the appropriate actions.

It must be emphasised that recognition of abnormality does not imply diagnosis of the cause of the abnormality, although supporting such diagnosis should be one of the objectives of the interface (and is considered below as such). In the present context, recognition of abnormality concerns the awareness that the system state is different from that which the user expected, and hence that system behaviour may not be as originally predicted. Such recognition is essential if the user is to be able to identify appropriate recovery procedures. However, the interface, and indeed the system, must as far as possible also support the user during the period when the abnormality exists but has not yet been recognised by the user. It is during this period that users can have significantly adverse effects on the development of the situation, through inadvertently taking inappropriate actions.

Once the existence of an abnormal condition has been recognised, the second aspect of user support offered by the interface becomes important, that is, to assist the user to determine appropriate actions. It may not be necessary for the user to diagnose the situation, provided that the interface supports the user in taking actions which move the system towards a safe state. Alternatively, it may be necessary for the interface to ensure that the user does not inadvertently disable automatically actuated safety systems.

It is not possible to prevent a user from attempting to diagnose the cause of observed symptoms, and possibly not desirable. The interface design should aim to encourage correct identification of the cause of the event, and to minimise the likelihood of incorrect diagnosis, such as through the provision of appropriate feedback or methods for hypothesis testing. Thus, after recognition of abnormality, the two further aspects of support which also will be considered below are the development of appropriate actions and support for diagnosis of the underlying event. It is important that the interface supports an integrated approach to the three activities.

It should be emphasised that users tend always to attempt to explain and understand current events by reference to past experience. This leads to a tendency to adopt past practices and procedures for dealing with novel events, until such time as it becomes apparent that the event differs significantly from the past experience. As Duncan has noted, 'uncritical extension of heuristics and guesses, especially after prolonged practice in rather restricted contexts, is a common human failing' [Duncan 87].

The behaviour of the user and the nature of the required user support are discussed further in the following four sections. Section 8.2 considers in more detail the nature of users' behaviour in abnormal situations, both before and after recognition of abnormality, and hence the general nature of the support which they require. Section 8.3 reviews the nature of the system and interface support which could be offered to users, based on requirements which can be identified from predictable features of human behaviour. Section 8.4 provides a brief review of the manner in which training can better equip the user to respond to abnormal events, both in terms of recognising the abnormality and of reacting to it. Section 8.5 provides conclusions concerning the overall approach to assisting the user of a safety-critical system when confronted by an abnormal or emergency situation.

8.2   BEHAVIOUR IN ABNORMAL SITUATIONS

8.2.1   Characteristics of Abnormal Situations

Many complex processes, whether a nuclear power station, an air traffic control system, or an aircraft, operate within what can be described as a dynamic decision environment, which creates ill-structured problem and decision domains. These are characterised for the user by [Klein 89]:

- Multiple goals;
- Multiple problems;
- Multiple people (whether part of, or external to, the operating team);
- Feedback inadequacies;
- Decisions embedded in a decision-action-feedback process;
- Varying time pressures;
- Problems of role differentiation and specialisation.

Characteristics of the unfamiliar situation in such a context may be listed as:

- Increase in uncertainty;
- Change in goals and decision criteria;
- Change in patterns of trust;
- Increase in stress;
- Change in patterns of task requirements;
- Change in the appropriateness of mental models;
- Loss of understanding of the state of the plant or process.

The consequence of these characteristics of the situation is that three types of operator activity may be made more difficult:

- Recognition of an abnormal situation;
- Cognitive representation of the problem;
- Management of cognitive workload and cognitive demands.

In principle, an abnormal situation should not affect the overall plant objectives, which include safe operation. However, such a situation may change the immediate goal to that of the attainment of a safe state. This change in immediate goals may be associated with a change in the demands which are placed on a system. Some safety-critical systems may be protection systems (see Chapter 1), only being called on during certain abnormal plant states. Others may have a function during normal operations, but may fulfil a different function during abnormalities, or their importance to the attainment of a safe plant state may alter. Thus, for the user, the implications of the abnormal situation will be that the function of the system may change, and hence the manner in which the user interacts with the system to attain system goals may also change.

In this way, the abnormal situation may have one or more of the following effects on the role of the safety-critical system:

- Change its status from standby to active involvement in plant control;
- Change it from a system to be monitored for availability to a system to be monitored for performance;
- Change it from a 'low priority' system with respect to operator goals, to a 'high priority' system.

In order that a system can properly support the user during an emergency or other abnormal situation, the interface must be designed in recognition of the task characteristics and the situation characteristics, including both normal and envisaged abnormal operational scenarios. Clearly it is inappropriate to design an interface which is optimised only for predictable and routine operations.

It is in the nature of safety-critical systems that failures, including those of the system users, can be least tolerated during abnormal or emergency situations.

8.2.2   Characteristics of the User

In order to identify how a user can be supported in responding to an abnormal situation, and hence how the interaction between the user and the system can best be managed, it is important to understand the characteristics of the user. To do this it is necessary to gain an understanding of human information processing and decision making, and hence to understand the factors which can influence decision making during abnormal situations, and the failures which can ensue (see also Chapter 4 for a consideration of user attributes in the design of the system). From such an understanding it becomes possible to identify how the systems with which the user interacts can support decision making and control the opportunities for error by preventing inappropriate actions, minimising the consequences, and providing appropriate recovery mechanisms.

Of the studies which have proposed cognitive models of human behaviour [e.g., Reason 1987, Doerner 1987, Rasmussen 1986, Klein 1989], many have emphasised certain human characteristics associated with rationality and limited information processing which can adversely affect decision making. These adverse factors include:

- *Bounded rationality* — attentional limitations on cognitive activity and the use of 'satisficing' as a decision rule (satisficing is the acceptance of a plan which may be merely adequate rather than optimum);
- *Imperfect rationality* — the cognitive limitations that result from schemas or mental models which provide interpretive frames (where interpretive frames can be considered as the structures which shape the understanding of perceived situations);
- *Reluctant rationality* — the desire to avoid the 'mental workload' created during cognitive activity and hence the desire to reduce 'cognitive strain', such as by seeking familiar actions.

In the context of responses to abnormal events, these key features of rationality and limited information processing are critical to understanding how users may respond to the demands of an abnormal event and, hence, the requirements of system interfaces.

An important feature of human information processing is that all mental actions (such as evaluation, classification, assessment, problem solving, judgement, decision making) appear to take place within the context of a representational model. Such interpretive and representational models have been described in various ways in the literature — as frames, mental models, cognitive schemas, etc. They are learned; their use may be unconscious; they may differ among individuals; individuals may switch between models in a single situation. Moreover, they may be learned in different ways: formal training, trial and error, on-the-job training, subtle influences of task interfaces, and so on.

Rasmussen [e.g., Rasmussen 86] has developed a model founded on the notion of three levels of behaviour — skill-based, rule-based and knowledge-based — and this is described in Chapter 2. Representational models operate to define the context. At the skill-based and rule-based levels, they determine the appropriate behaviour. For knowledge-based behaviour they are the basis for the interpretation of data, the understanding of relationships, and the development of hypotheses.

People tend to operate at as low a level as is deemed consistent with an acceptable level of performance — there is a tendency for people to attempt to minimise cognitive 'effort' [Reason 90]. Even when behaviour at the knowledge-based level is required, this only occurs until such time as an appropriate set of rules can be determined, perhaps through matching characteristics of the situation to past experience. Once a sufficiently close match has been obtained, rule-based behaviour will occur.

A consequence of this is that people tend always to attempt to identify a match between the current situation and past experience. Furthermore, any similarities which do exist may be given undue emphasis, as a means of selecting an action plan to address the current situation. This in turn leads to two effects. First, there is a reluctance to recognise that the situation differs from that which has been previously experienced, and hence the abnormality of the situation may not be recognised. Second, even if the situation is recognised as being abnormal, users tend to respond according to previously successful actions, without necessarily recognising that they may not be appropriate in the present circumstances.

There is a set of basic error tendencies which can be determined from the cognitive modelling literature. These include the tendency to minimise 'cognitive strain', and to over-utilise stored knowledge structures, heuristics and short cuts in order to simplify complex informational problems. As Reason notes, humans are compulsive pattern matchers [Reason 88]. Cacciabue et al note that there appear to be two types of cognitive processing. The first is a controlled and conscious processing which is selective, resource limited, slow, laborious, serial, and supported by a set of task-oriented small programmes. The second is an automatic and unconscious processing which is apparently unlimited, fast, effortless, parallel, and based on two heuristics: 'match like with like' and 'select in favour of high frequency item patterns' [Cacciabue et al 89].

Reason provides an elaboration of the basic error tendencies, which include [Reason 87]:

- Change enhancing biases (the tendency to take decisions which magnify the changes occurring to the system);
- Resource limitation effects (the tendency to consider only those actions which are comparatively easy to execute);
- Schema properties (the properties of the user's cognitive model of the system will influence the user's expectations of system behaviour);
- Strategies and heuristics (the 'rules-of-thumb' and other methods for reaching decisions which the user would normally employ, and which may be effective in normal situations, may not account fully for system performance in abnormal situations).

Reason further describes a set of cognitive domains in which error may occur. These include:

- Sensory registration (the physical detection of information, such as an alarm horn triggering a response from the auditory system);
- Input selection (the allocation of 'attention' to the registered input - 'recognising' that the alarm horn is sounding);
- Short-term or working memory (the cognitive 'scratch-pad' which stores information for a few seconds while it is being processed);
- Long-term memory (the long-duration repository for all information to be used in decision making, such as past experience, knowledge of procedures, memory of system status);
- Recognition processes (e.g., the identification of similarities between the current and previous situations);
- Judgemental processes (e.g., the assessment of the relative merits of different options);
- Inferential processes (the identification of the probable development of an event on the basis of limited information);
- Action control (execution of the decision, e.g., operate a control, acquire further information).

The interactions between these two sets provide primary error groupings such as false sensations, attentional failures, memory lapses, inaccurate recall, misperceptions, errors of judgement, reasoning failures, and unintended actions (see Chapter 2 for a consideration of human error). Although it is not clear that this list of groupings is immediately helpful, as they tend to be descriptive and to address the surface forms of the errors, it does highlight the areas where the interface and system must support the user. For example, false sensations can be countered by training users to understand the situations in which they might arise and, therefore, which sensations to ignore. Attentional failures can arise as a consequence of information overload or underload and can be countered by appropriate task and interface design. Memory lapses can be mitigated by reducing demands on working and long-term memory, perhaps through the use of checklists and suitable interfaces.

The two categories which are most difficult to address, but which are perhaps most pertinent to abnormal or emergency situations, are judgemental errors and reasoning errors. It is these which are most prone to the three failure types — bounded rationality, imperfect rationality, and reluctant rationality.

The consequence of this for interface and system design is that the designer of the system must recognise the failures to which the user is prone, and must defend against those failures. Two methods for providing defences are to prevent inappropriate decisions, and to prevent the actions which might arise from them.

The optimum method for minimising the likelihood of an inappropriate decision is to ensure that relevant information is presented and that it is structured according to the user's needs and expectations. This is an essential requirement for any interface, but especially so for a safety-critical system whose overall performance must be assured during abnormal situations. Necessary information must be available to the user, presented in a manner which minimises the likelihood that the user will overemphasise certain aspects of the information or omit to take account of other aspects. In addition, the information presented through the interface must guide the user towards correct decisions.

The second method for defending against failures of decision making is for the system to prevent the user from executing inappropriate actions. This can be done through the use of interlocks and other means of limiting the options available to the user at a given time. While this may be appropriate for actions from which there can be no recovery, it may be difficult to identify in advance which responses are to be prevented, and also it calls into question the value of retaining the user within the control loop. If the user is retained in the loop, he is better placed to respond when necessary, rather than being required to intervene at a time when he is unclear of how the system is behaving. Further, it is the user's decision-making skills which justify his presence within the system. Any method for preventing him from executing those skills should be implemented with caution.

It is suggested that provision, or strengthening, of defences against human error is a classical engineering response to accidents, and that no system can be made 'foolproof' [Wagenaar et al 90]. However, this may be an incomplete analysis. Whereas it is probably correct that attempts to 'engineer out' opportunities for human error, perhaps through the use of increasing levels of automation, may be doomed to failure, it may be possible to provide 'cognitive' defences to enhance operator performance.

Interfaces must support the operator with respect to his current model of the system, and the sophistication, or otherwise, of his knowledge base. At the same time, the interfaces must recognise the limitations of basic human information processing, and hence must not impose unnecessary additional demands. For example, where possible the display formats provided for emergency operations should not differ in structure and philosophy from those used in normal situations, thereby avoiding unnecessary cognitive demands.

8.2.3   A Decision Model

The guidance provided in this chapter is based on a cognitive model which reflects the decision-making behaviour of users in abnormal or emergency situations.

Klein has proposed an empirical model, the recognition-primed decision (RPD) model [Klein 89], based on observations of real decision making in a 'natural decision environment', which models behaviour in decision-action-feedback loops. It suggests that analytical decision making is practised far less frequently than might be expected and that, typically, even expert decision-makers tend to utilise recognition-based decision making.

The model, presented in simplified form in Figure 8.1, represents the decision-making process as commencing when a proficient decision maker is confronted by a new situation. The first action is the evaluation of familiarity, and the judgement of the typicality of the situation, based on prior experiences (e.g., training and prior events). The judgement includes the selection of:

- Critical cues;
- Possible goals;
- Expectancies and causal relationships;
- Potential responses.

If a situation is recognised, an immediate response can be implemented. In more complex cases, the decision maker may need to evaluate multiple potential options. If problems are identified, he may choose to modify a standard response, or consider a new 'typical' response option.

The implications for system design are that the interface, and the information presented via that interface, must be designed such that they assist the user to recognise the atypicality of the situation. In addition, it must be designed to support the process of identifying response options, evaluating those options for potential problems and, if necessary, developing and assessing modifications to the response options. The critical aspects of this process are the recognition of atypicality and the identification of potential problems that may result from selected response options.

A Modified RPD model of a simple seven-stage process for responding to an abnormal situation has been developed and is presented in Figure 8.2.

When considering the four aspects of the judgement process — critical cues, possible goals, expectancies and causal relationships, and potential responses — the factors affecting users can be identified as those relevant to

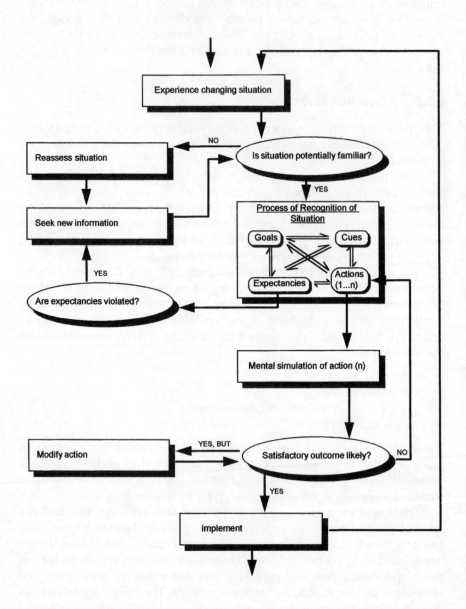

Figure 8.1 *Recognition-primed decision model (after [Klein 89])*

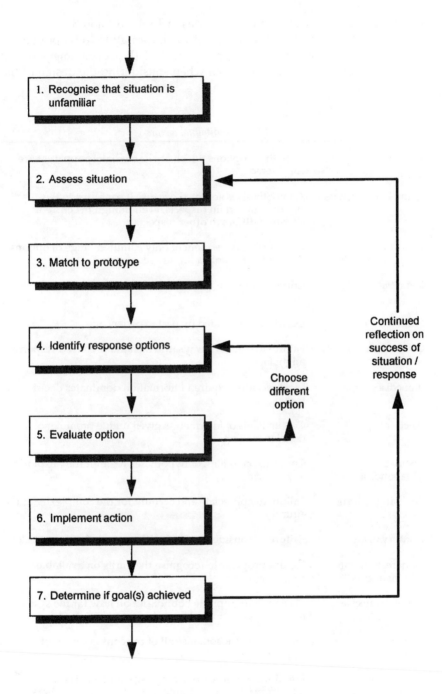

Figure 8.2 *Modified RPD Model applied to decision making in abnormal situations*

individuals and those relevant to teams, as in Tables 8.1 and 8.2.

Each factor, as well as the nature of the information to be presented, should be taken into account, in the context of the seven stages of the Modified RPD model (see Figure 8.2), when considering the design of the

Table 8.1 *Individual factors*

Availability	Only those options which are immediately apparent are considered
Representativeness	A hypothesis is accepted because it conforms to the situation in certain respects, even though the current situation differs in other respects
Fixation	Continual reassessment of only a limited range of options; accounting for only a limited set of data
Focusing	Concentration on only a limited set of features of the situation
Hindsight Bias	Decisions dominated by past experience
Over-confidence	Failure to consider ways in which the preferred response might fail
Immediacy	Most recently acquired information dominates decision making
Anchoring	Certain items of information given undue prominence in decision making
Thematic vagabonding	Continual changing of hypotheses without fully exploring any of them
Oversimplification	Failure to appreciate and account for the complexity of the situation
Conservatism	Failure to consider options with apparently higher risk
Underestimation of constraints	Failure properly to recognise the limits on available responses
Framing effects	Placing inappropriate boundaries on how far the system response is considered
Limited Imagination	Inability to conceptualise all of the consequences of a given response
Escape from responsibility	Identification of reasons for postponing decisions inappropriately, or for taking the decisions of others as binding
Knowledge decay	Failure to utilise all available knowledge

Table 8.2 *Team factors*

Failure to share information	Withholding of information from team colleagues
False consensus	Incorrect assumption that team is in agreement
Pluralistic ignorance	Incorrect assumption that someone in team has relevant information
Social pressures	Pressure to conform to social norms
Group think	Tendency for all team members to follow same inappropriate decision-making route
Group polarisation	Team divides into two or more groups with different, irreconcilable, views
Risky shift	Tendency of team to select more risky option than would any individual member of that team
Rejection of relevant information	Certain information not considered by team
Focus on shared information	Information held by few team members given less importance than information held by all
Individual conflict	Conflict between team members
Emergency situations - stress etc.	Effects on team cohesion of stress, etc.

interface for a safety-critical system:

(i)   Recognise abnormal or unfamiliar situation;
(ii)   Assess situation;
(iii)   Match to prototype;
(iv)   Identify response option;
(v)   Evaluate options;
(vi)   Implement action;
(vii)   Determine if goal(s) achieved.

Two further issues, 'stress' and 'panic', are relevant to describing the user's response to abnormal situations. Neither will be discussed in great detail as there often appears to be more myth than information available to describe their effects.

Stress is a subjective response to a situation, and hence is less amenable to quantification and control than might otherwise be the case. In essence, it may be considered as deriving from the inherent uncertainty of a situation and a lack of control as perceived by the individual concerned. Consequently, certain measures, such as reducing uncertainty and providing enhanced feedback of performance, may act to reduce stress. It is also the case that subjective reports of high levels of stress do not necessarily correlate with formal measures of performance.

Thus, while stress is not formally emphasised here, the recommendations contained in Section 8.4 include some which are oriented towards supporting the user in situations which might be considered to lead to high stress. The training recommendations, and in particular those associated with teamworking, should provide some defences against the performance effects of subjective stress.

Panic is even less well defined, although there tends to be a general acceptance of the term. However, when specific incidents are investigated, it is rare that any indication of panic can be identified. True panic would involve behaviour which is not goal directed, which might include aimless action or total inability to react. In almost all incidents, the behaviour exhibited by participants is clearly goal directed, although with hindsight or the benefit of additional information it may not have been the optimum behaviour. For example, where people are apparently scrambling over each other to leave a dangerous situation, this can be considered not as panic, but as a sensible strategy to hasten their own escape from danger. An operator in a nuclear power plant who ran through the plant to operate a valve manually was probably not panicking. Rather, he was attempting to respond as quickly as possible to a time-critical demand. Even though he was doubtlessly concerned, his actions were a reasonable response to the demands of the situation.

For this reason, panic also is not considered further in this chapter. However, it is clear that the more information which can be provided to a user, and the more of his knowledge which is based on training and experience, the less likely he is to suffer degraded performance due to what is commonly termed 'stress' or 'panic'.

8.3   SYSTEM-BASED SUPPORT

This section presents a series of recommendations for the provision of interface and procedural support to a user during abnormal or emergency situations. Section 8.4 considers the manner in which appropriate training must also be provided. The recommendations in both of these sections are based on the Modified RPD model shown in Figure 8.2. Clearly there are interactions between training, procedures, and systems, and hence any distinction drawn in this chapter is a somewhat arbitrary reflection of the need to impose a structure on the information. The importance of a coherent and integrated approach cannot be overemphasized.

8.3.1   Interface Support

The purpose of the interface is to support the user in:

- Obtaining the information required for the formulation of an appropriate course of action;
- Carrying out the identified course of action;
- Determining the effects of the actions taken in order to modify the responses as appropriate.

The following interface recommendations can be identified for each of the seven stages in the Modified RPD model of Figure 8.2.

(a)   Recognise an abnormal or unfamiliar situation

The interface must be capable of presenting all information relevant to the present situation, without overloading the user, and in a form which minimises the risk of selective access to information. The interface should also indicate expected values and should highlight interrelationships between parameters. The interface must be familiar to the user, and inherently consistent, irrespective of whether the situation is familiar or not (Chapter 6 discusses the factors which influence an effective interface design).

'Early fault detection' systems, which provide an indication of failures or sub-optimum system performance, can assist the user to recognise abnormality, but it must be ensured that such systems do not bias the user towards one particular scenario to the exclusion of consideration of alternative explanations of the observed data.

The interface is less effective than other support systems, such as procedures and training, at combating group tendencies. It may be appropriate for the interfaces for different team members to present information in different ways, in order to minimise the risk of selective sampling. However, such a strategy should be adopted with caution if users may be expected to fulfil different team roles at different times. Then, differences between interfaces may act to obstruct their users. In such circumstances users need to be familiar with all interfaces. Consistency of interface style and information presentation, both across interfaces and across situations, acts to reduce the likelihood of error.

(b)   Assess the situation

Relevant aspects of the system state must be clearly indicated to an appropriate level of detail (there is often a tendency to present information because it is available within the system rather than because it is of value to the user). This includes current plant status (operating state, availability, etc.), and the current values of parameters. The system should indicate trends so as to assist the user to determine the dynamic state. In addition, information should be available concerning not only critical parameters, but also potential threats to the maintenance of those parameters, such as pump availability, and the direction in which those parameters are moving if the rate of change is excessive.

(c)   Match to familiar prototype

Users will have a 'prototype' hypothesis against which they must match the perceived situation in order to 'recognise' it. To achieve a match, users consider salient aspects of the perceived situation, hypothesize future plant behaviour, and compare this with plant behaviour in previous situations. The interface must support this matching process by highlighting the salient features of the situation, including both current and predicted system behaviour. During normal operations, when the system appears to be deviating from the expected behaviour, but not to the extent of being abnormal, the interface should permit and assist the operator to assess the nature of the deviation. The interface should support the identification both of the similarities and the differences between the current situation and previously experienced situations from which the user's mental prototype will have been developed.

(d)   Identify response options

The ideal response might be to follow an appropriate procedure, where one is available. This requires that the user is able to identify the relevant procedure and assess its applicability. The interface must support this in two ways. First, the relevant procedure, or procedural information, must be readily available, with clear entry rules and concise indications of relevant plant parameters. In addition, the interface must present a clear and concise summary of the system state in order that the user can select the most appropriate procedure (whether a formal document or an established practice).

As the process of selecting response options is an iterative one, the interface must support the evaluation process by preventing premature closure. The interface itself should give an indication of potential options through the organisation of information so as to provide cues to areas of potential intervention.

(e)   Evaluate options

Evaluation consists of mental simulation in terms of applicability and potential for success. This is a key process in determining the effectiveness of the system response, and one which is affected by many of the factors identified in Section 8.2.3. The interface can support the simulation process through the provision of trend data and predictive information. In addition, it can support the group processes which might defend against such problems as mindset (fixation on inappropriate hypotheses to the exclusion of relevant data) by assuring effective communication and the development of diverse approaches to understanding the current situation.

(f)   Implement action

At this stage, assuming that the selected response option has been through the 'evaluation' process, a potential error is the underestimation of constraints, where the user fails properly to understand the restrictions on potential options. These might be time constraints or the unavailability of certain safety systems, and the interface can be valuable in highlighting them, if this is possible. Time constraints are often the most critical for the user. Appropriate trend displays can provide to the user an indication of the time available before the system reaches a certain state. System failures must be highlighted, and the means of confirming such failures provided.

(g)   Determine if goals have been achieved

The user must place the immediate goals in the context of the wider objectives of the task. The interface should ensure that these higher-level goals remain apparent, and that the relevance to, and relationships between, the immediate goals and the higher-level goals remain clear, together with immediate and unambiguous feedback, so as to allow a user to determine when they have been achieved. In the context of an abnormal situation, the higher-level goals will provide the assurance of safety and hence the attainment of a safe operating state. An intermediate goal might be the maintenance of coolant flow, and an immediate goal might be to start a pump or open a valve. The interface should make clear how the attainment of the immediate goal supports the higher level goals and hence, for example, whether the successful operation of a valve has moved the plant to a safer state.

8.3.2   Procedural Support

Current procedural guidelines tend to address the human factors issues associated with the presentation of procedural information, rather than the underlying nature and objectives of the procedural information itself. It is clearly important that procedures for use during abnormal or emergency situations conform to good practice, such as in terms of presentation, format and content.

Some abnormal situations may nevertheless be pre-planned, such as modified start-up sequences to account for plant item unavailability. In such circumstances the procedure must be available before the task commences and must highlight the potential difficulties and critical issues which the situation may present. It must indicate the anticipated system response, together with the relevant cues which will be available to the user to monitor the system.

For unplanned activities, such as responding to plant item trips, it is still possible for procedures to be available, although their level of detail may be reduced. Such procedures may provide decision-making guidance to the user. As with interface support, the procedures should be defined in the context of the seven stages in the modified RPD process (see Figure 8.2).

(a) Recognise an abnormal or unfamiliar situation

Typical event-based procedures frequently have clear entry rules, such as 'Pump A tripped'. By contrast, symptom-based procedures, which are considered to be more appropriate for emergency situations, require only that the user recognises the symptom (e.g., loss of flow or rise in temperature). The user does not need to know the cause. The procedures tend to be oriented towards the maintenance of critical safety functions rather than diagnosis and 'cure'. Such procedures tend not to have clear entry rules. It is therefore essential that the normal operating procedures and practices should continually encourage the user to assess the current situation and identify whether to change to different procedures (irrespective of the manner in which the procedural information is presented — whether paper-based or VDU-based). The interface must support the user in identifying that the situation is no longer as expected.

(b) Assess the situation

The procedure must support the user in identifying the nature of the situation, although this need not imply fault or failure diagnosis. Rather, it requires that the procedure should aid the user to determine sufficient key parameters of the situation to form a valid basis for matching it against previous experience and to recognise that it is abnormal or on its way to becoming so.

(c) Match to familiar prototype

The procedure must indicate the key criteria which should be considered by the user when attempting to match a situation against previous experience and thus to classify it. For example, in a chemical plant this might include the rate of change of temperature and the statuses of certain defined pumps and valves.

(d) Identify response options

Having classified the situation, however generally, the user is then disposed towards selecting response options on the basis of that classification. The procedure must recognise this tendency, and ensure that clear goals and objectives are presented to the user. The goals and objectives will form the basis on which the user selects response options. These goals may be safety-related, and as far as possible economic and efficient, and will be influenced not only by the system state but also by management, safety culture, and by the available options themselves. For example, in an aircraft in flight, the option of safely shutting down all engines is not available.

(e) Evaluate options

Prior to committing to a course of action, the user should evaluate each proposed response in order to identify the likelihood of achieving the required goals and objectives. By definition, where the situation is abnormal, the user may not be able to take full account of all relevant factors which might influence the successful outcome of the proposed actions. However, the procedure should attempt to support the evaluation process by presenting a structured approach to evaluation.

(f) Implement action

At this stage the range of actual responses available to the user is likely to be both limited and familiar (e.g., changing set-points or operating valves). However, the user may be unfamiliar with the anticipated magnitude or timescale of the system response to his actions. For example, the user might know that a particular switch, when held closed, causes an isolation valve to close. However, the user may never have needed to close that particular valve during normal operation, and, thus, may not know that it takes 20 seconds before the valve starts to close. Such a delay may lead the user prematurely to assume that the valve has failed and, hence, to abandon further attempts to close it.

The interface should provide accurate and appropriate information concerning the system response, such as indications first that a control signal has been sent and second that the plant item is responding. In the example given above, the interface should first provide an indication that the 'close' signal has been initiated, such as by way of an illuminated indicator, and then also provide positive feedback from the valve to indicate its changing position. Further, this latter indication should be derived not from the valve actuator, but from the valve spindle itself, in order to avoid misleading the user in the event of a failure in either the actuator or the valve.

The user's lack of familiarity with the system dynamics in the abnormal context must be recognised when considering the form and extent of feedback to be provided. Further, recognising the potential for the abnormal situation to present atypical indications to the user, where possible the feedback should include validity indications, although the difficulty in doing so for a truly abnormal situation is recognised.

It should be noted that these feedback issues apply equally to hardware systems and organisational characteristics, such as the extended communications delays which might arise when altered reporting routes have to be relied on during an abnormal event.

(g)   *Determine if goals have been achieved*

Given that the situation is abnormal, the user may not previously have completed the actions taken, and may have a reduced understanding of the system state. It may therefore be difficult for the user to determine precisely whether the goals have been achieved. The procedure must indicate clearly what criteria can be used for determining if goals have been achieved, what indications of successful goal attainment will exist within the system or at the interface, and how they can be detected.

In general terms, procedures should also:

- Provide checklists or reminders of key actions and system parameters;
- Avoid the need to predict the frequency of an event;
- Provide rate-of-change data;
- Identify clearly the objectives of each potential response option;
- Highlight predicted or anticipated values to assist discrepancy identification;
- Emphasise the need to consider system state history;
- Indicate action consequences;
- Highlight time and resource requirements;
- Identify achievable objectives.

In addition, procedures should recognise the various error tendencies discussed in Section 8.2. For example, to limit the effects on information processing which derive from *availability* (where only that information which is readily available is considered), a procedure should assist the user to identify features of the situation which distinguish it from other situations. A checklist approach may act to remind the user of the information and of indications which should be considered. It is not advisable to rely solely on the user's training during abnormal events.

Another error tendency which could be defended against by procedures is *thematic vagabonding*, where the user fails properly to address or test any one hypothesis, and instead constantly changes his model of the current situation. The procedures can control the tendency to switch from task to task by ensuring that there are clear objectives associated with each task, together with clear indications of the attainment of those objectives. The procedure should minimise the number of concurrent objectives, and should facilitate access to required data, perhaps by indicating where they are located.

8.4   TRAINING SUPPORT

Chapter 7 has considered training support in detail. The present section addresses the training issues which are relevant to responding to abnormal situations. Operation of many safety-critical systems occurs within the context of ill-defined problem domains. Difficulties for action may be compounded in abnormal or emergency situations because of changing patterns of learned behaviour, increased ambiguity and uncertainty, and the insertion of new difficulties and complexities. Training would normally include the use of drills and exercises to acquaint users with system requirements during abnormal or emergency situations. However, some training regimes may fall short of achieving these goals as a consequence of being based on simplified, unrealistic, or inadequate representations of reality, or because of an inability to predict all possible eventualities.

It is possible to develop training programmes which provide users with the bases for action in abnormal situations — not only those situations which are unfamiliar to the user but also those which may be unfamiliar to the industry. As has been noted, at the extreme the user is present to respond to situations which are unfamiliar to the industry, but it is inherently impossible to practise these [Bainbridge 92]. However, it is possible for users to gain experience in responding to situations which are unfamiliar to them. This can include training in problem solving. By allowing users to practise in unfamiliar situations, it is possible for them to acquire skills which can be applied to any unfamiliar situation, including those which derive from abnormal events.

Design of training programmes could be based, in part, on the RPD framework. Whereas training cannot guarantee success, well-designed programmes can provide significant gains in the development of skills, knowledge, and general abilities necessary for successful and reliable performance.

One other factor which is relevant to the planning of training is the selection of trainees. The value of selection techniques for ensuring that users are best able to respond to abnormal situations has not been considered here, for the evidence concerning the manner in which selection techniques can be used to support the aims of training is not clear. Furthermore, it is assumed that such selection techniques are by no means peculiar to safety-critical systems.

8.4.1   General Issues in Training Design

The effective training of both individuals and teams rests on common design requirements. In the design of training, the following should be considered:

- Transfer of training (skills acquired under one set of circumstances are applicable to another);
- Inert knowledge (knowledge acquired under one set of circumstances is not recognisable by the user as applicable under another);
- Fidelity of the training environment (the extent to which the training environment is equivalent to the real environment in which the skills and knowledge will be used);
- Feedback of performance during training (it should be timely, focused, appropriate, sequenced);
- Part-task training (acquisition of skills and knowledge appropriate to subsections of the task, which can be subsequently combined to

provide the competence to perform an entire, complex, task — see Section 7.6.1 of Chapter 7).

The integration of part- and whole-task training is critical. The effective transfer of training to real decision-making environments requires practice with whole tasks. Research results emphasise that both individual and team training methods are enhanced by the careful sequencing of part-task and whole-task training. One form of this process, termed 'scaffolding', describes the process of providing the trainee with readily understandable components of the whole task in a form which allows the gradual construction of a framework (scaffold) onto which can be hung the additional task information and skills which will be required for successful task performance. The availability of this framework allows the trainee to incorporate future information as it becomes available. During early stages of learning, the process avoids the danger of overloading the trainee with information which may not then be strictly necessary.

It should be noted that an effective safety culture will play an important role in supporting the user during an abnormal or emergency situation. Training has a part to play in the development of an effective safety culture, but such aspects of training are not relevant to the present discussion.

8.4.2   Individual Training

(a)   Domain-specific training

The following general guidance can be offered for the domain-specific training of individuals:

- Provide timely feedback of performance, with appropriate specificity and sequencing;
- Use simulators for repeated trials over a short period;
- Provide knowledge of process, system, and tasks in a hierarchical structure;
- Provide conceptual models to guide development of appropriate mental models;
- Provide repeated opportunities for pattern and problem representation and verification;
- Expose trainees to abnormal situations which require reconfiguration of skills and knowledge;
- Use techniques such as 'scaffolding' to reduce workload, complexity and stress during the early stages of training;
- Promote skill acquisition and skill and knowledge flexibility by using informed and self-controlled training techniques;
- Develop 'over-learning' (i.e., the acquisition of skills which become automatic) only for task components which will remain invariant between contexts; warn of effects of over-learning on flexibility and adaptability;

- Train for important sub-tasks that support problem solving (e.g., schematic reading);
- Introduce trainees to multiple tasks early in training; develop dual-tasking skills;
- Develop and provide strategies for minimising and monitoring workload;
- Provide opportunities for trainees to experience the actual 'stress' of unfamiliar situations, by adapting and altering the training regimes;
- Develop techniques and strategies for self-monitoring;
- Focus training on tasks which are particularly vulnerable to effects of stress, such as recognition of abnormality;
- Emphasise strategies for minimising framing effects (the manner in which the problem context affects the options which the user might consider), such as by highlighting the importance of considering similarities and differences;
- Practise responding to unfamiliar situations;
- Practise responding to emergency situations.

(b)   General training

In addition to domain-specific training, it is necessary to provide users with skills which can be employed in abnormal situations where the system is no longer behaving in an expected manner. The general skills which a user requires in such situations include the ability to generalise to new contexts, and to reflect on and regulate conscious cognitive processes. These abilities are often termed 'meta-cognition' [e.g., Redding 89] and are thought capable of enhancing user performance by assisting:

- The construction of appropriate mental models;
- Self-monitoring of cognitive limitations;
- The ability to modify behaviour when necessary;
- Application of general problem-solving strategies;
- Transfer and application of skills and knowledge acquired during training in one context to another (reduction of 'inert knowledge').

The following principles apply to meta-cognition training:

- Provide clear explanations of importance, rationale, and uses of heuristics and strategies;
- Show connections between old and new knowledge to improve access and utilisation;
- Maximise and highlight similarities between tasks and methods, rather than only providing bits of information;
- Focus on the development of problem space representation, including similarity and difference identification skills;
- Develop flexibility by encouraging experimentation and learning from feedback;

- Highlight the nature and consequences of cognitive heuristics and biases, such as fixation, anchoring and hindsight bias (see Table 8.1, and see also Section 2.4.7 in Chapter 2 for a discussion of biases in the classification of human error).

8.4.3 Team Training

Guidance for team training, in the context of abnormal situations, must recognise three principal areas:

- Individual skills for team performance;
- Teamworking (including coordination and communication);
- Team leadership.

Teams which appear effective during normal operations may break down during abnormal situations, yet it is during abnormal situations that the effective performance of the team may be most necessary (see Section 9.3 of Chapter 9 for a discussion of this). Consequently, it is essential that training provides the maximum amount of support for team-working.

While it is beyond the scope of this chapter to cover all aspects of team training, the following guidance highlights some of the issues most relevant to abnormal situations.

- Provide assertiveness training, emphasise the need to share information, encourage questioning of team members;
- Provide information about roles, needs and responsibilities of other team members, and the opportunity to sample other team roles;
- Train leaders to use effective leadership strategies and avoid ineffective ones;
- Develop clear and consistent communications skills;
- Train team leaders in conflict management; introduce uses of conflict (e.g., questioning assumptions) to enhance performance;
- Provide strategies for group reflection and monitoring;
- Expose team members to multiple-task environments which require coordination, sharing and sequencing;
- Expose teams to unfamiliar situations in meaningful contexts;
- Expose teams to unfamiliar situations which require reconfiguration of skills and knowledge;
- Provide opportunities to experience 'stressful' conditions in meaningful contexts;
- Sequence individual and team training;
- Train teams to develop mental models which are shared within the team;
- Begin team training after individuals have mastered sub-tasks;
- Provide feedback on performance as a team.

Most team-work training is generalisable across domains. Consequently, a substantial amount of team-work skills training required for responding to abnormal or unfamiliar situations can be provided in non-task-specific contexts, and in the context of normal situations.

8.4.4 Emergency Procedures Training

In order to ensure that users are prepared both to enter emergency procedures and to follow them to a successful conclusion, it is important that the training which they receive includes scenarios in which the emergency procedures are required. Such scenarios are an inherent part of the training programme suggested above.

However, in order to foster the skills necessary for dealing with abnormal or unfamiliar situations, it is essential that the training scenarios provide opportunities for exploring premature exits from the emergency procedures. The procedures must encourage the user to develop hypotheses concerning the current plant state, and provide them with the means for testing these hypotheses. This in turn requires that the interfaces provide appropriate information to support such hypothesis testing.

At the same time, the procedures must support the maintenance of critical safety functions during this diagnostic phase. This places demands on the system's ability to fulfil its intended function while supporting the user in obtaining necessary information. The procedures must jeopardise neither safety nor diagnostic accuracy.

It is therefore essential that the emergency procedures and training provide an integrated approach to event management.

8.5   CONCLUSIONS

When considering the design of safety-critical systems, and the incorporation of human factors principles into the design, it is essential that emergency operations are considered. The demands which the user may make on the system during abnormal or emergency situations are likely to differ from those made in normal circumstances. However, the system itself must still support the user, and must do so in a manner which recognises the particular demands of abnormal situations.

Human characteristics which influence the optimum design of systems and interfaces are consistent across both normal and emergency operations. However, many of those characteristics only become apparent during the latter situations. For this reason, the present chapter has outlined an approach to considering human decision making which emphasises the tendency of users to attempt to match the current situation to past experience, and then to use that match, however tenuous, to select response options.

A number of strategies, covering interface design, procedures and training, have been described to address the potential human failures which can arise but which also can be defended against.

Key recommendations are:

- Ensure that there is consistency of interfaces across normal and emergency operations;
- Ensure that goals and objectives are always clear to the user;
- Ensure that sufficient information is available to permit the user to identify the current system state to a sufficient level of validity;
- Ensure that the user can assess the consequences of response options;
- Ensure that the system provides sufficient feedback of action consequences for the user to determine if the response has been effective.

The fact that a safety-critical system must perform as required during abnormal situations means that the demands on the system, and the possible failures which can be incurred, must be addressed during the design process. Furthermore, as with the consideration of human factors for any system, the performance of the complete person-hardware system must be considered.

The interface, procedures and training all can influence the overall performance of the system, and the way in which certain deficiencies in the performance of the user can be identified and defended against. The design of the overall system can have a significant impact on the tendency for users to commit errors, or on their ability to detect and correct those errors. A particular message is that such errors are predictable and can be defended against even though the precise nature of the abnormal situation itself may not be easily predicted.

Finally, it must be emphasised that the overall performance of the system will be influenced by all of the factors described above, so effective system performance can best be guaranteed by ensuring that the interface, the procedures, and the training all form part of a coherent and integrated whole.

REFERENCES

[Bainbridge 92] Bainbridge L: *Unfamiliar Situations*. Unpublished manuscript. Department of Psychology, University of London, London, UK, 1992

[Cacciabue et al 89] Cacciabue P C, DeCortis F, Mancini G, Masson M and Nordvik J P: *A Cognitive Model in a Blackboard Architecture: Synergism of AI and Psychology*. In Proc. 2nd European Meeting on Cognitive Science Approaches to Process Control, Siena, Italy 24-27 October 1989

[Doerner 87] Doerner D: *On the Difficulties People Have Dealing with Complexity*. In Rasmussen J, Duncan K and Leplat J (eds): *New Technology and Human Error*. John Wiley and Sons, Chichester, UK, 1987

[Duncan 87] Duncan K D: *Reflections on Fault Diagnostic Expertise*. In: Rasmussen J, Duncan K and Leplat J (eds): *New Technology and Human Error*. John Wiley and Sons, Chichester, UK, 1987

[Klein 89] Klein G A: *Recognition-Primed Decisions*. Advances in Man-Machine Systems Research, 5, 47-92, 1989

[Rasmussen 86] Rasmussen J: *On Information Processing and Human-Machine Interaction: An Approach to Cognitive Engineering.* Elsevier, Amsterdam, Netherlands, 1986

[Reason 87] Reason J T: *A Preliminary Classification of Mistakes.* In Rasmussen J, Duncan K and Leplat J (eds): *New Technology and Human Error.* John Wiley and Sons, Chichester, UK, 1987

[Reason 88] Reason J T: *Framework models of human performance and error: a consumer guide.* In Goodstein L P, Anderson H B and Olsen S E (eds): *Tasks, Errors and Mental Models.* Taylor and Francis, London, UK, 1988

[Reason 90] Reason J T: *Human Error.* Cambridge University Press, 1990

[Redding 89] Redding R E: *Trainers teaching thinking skills: applications of recent research in metacognition to training.* In Proc. of Human Factors Society 33rd Annual Meeting — 1989. Human Factors Society, Santa Monica, Ca., USA, 1989

[Wagenaar et al 90] Wagenaar W A, Hudson P T and Reason J T: *Cognitive failures and accidents.* Applied Cognitive Psychology, 4, 273-294, 1990

3

Socio-technical considerations

9

Social factors in safety-critical systems

9.1 INTRODUCTION

In this chapter I will discuss the organisational aspects of human factors in safety-critical systems. Perhaps 90% of ergonomics literature concerns the individual operator and the individual machine. Yet in real life the relevant unit is often a group, organisation or network, and its associated socio-technical system. While there is a long tradition of ergonomics studies examining human error, the study of social factors in complex systems is a relatively new area. For example, when Perrow's *Normal Accidents* appeared in 1984 [Perrow 84], it was largely unconnected to the human factors literature as it then existed. Since that time, many more studies have begun to fill in the gaps, but basic literature on groups in human-machine systems is still in its formative period. In this chapter I will present an overview of the issues, using both others' studies and my own work, which largely concerns complex aviation and similar high-tech systems. While some of the areas have received much attention, we also need to note others as yet only dimly marked out.

Considering 'social factors' means that, in order to understand how the overall system works, the dynamics of the group, of whatever size, must be considered together with the dynamics of the technology. To provide safe operation, the group and its technology must work harmoniously. This seems straightforward enough. For the system designer or manager, however, 'social factors' often mean a step into the unknown. The psychology of the individual is far better understood than the psychology of the group, let alone the psychology of organisations and networks. Traditionally the system designer turned to the psychologist for help in putting together the human-machine interface. Now the person designing or managing a complex system must turn to the social psychologist or social scientist, who in turn may have little experience with the complexities of the technology.

While many ergonomists or human factors professionals are used to thinking in engineering terms, this is often a novel area for social scientists. In fact, the social scientists who have made major contributions to this area have had to immerse themselves in often unfamiliar jargon and concerns. But some have perservered, and out of their work have come some valuable contributions. I see these contributions as responses to basic questions in three key areas. These, and some of the questions, are described in the following paragraphs.

The first area comprises group processes in the initial design of systems. For each system studied, there is a unique context of design. In this context, how do man-machine systems come about? Who are the designers? What do designers do to ensure the safe functioning of the systems they design? What is the social system involved in conceiving and developing new technologies?

The second area is the effective operation of these systems. Unlike the ergonomist, the social scientist is likely to study operations in the field rather than the laboratory. Who are the operators? By what set of complex operations does the functioning of the system take place? How is authority distributed? How does decision making take place? What are the social dynamics of the team? Who trains the operators of these systems?

The third area is how these systems interface with other systems. Not only do the systems have complex dynamics themselves, but so do their interfaces. These interfaces have to be carefully managed. All systems are lodged within larger social and political bodies. How does the sponsorship of these larger systems shape their dynamics? And how do these systems shape the larger context in return?

Each of these topics will be explored below through a survey of relevant concepts and examples.

9.2   SOCIAL FACTORS IN THE DESIGN PROCESS

9.2.1   The Context of Design

How often in our lives we are driven to reflect on the adequacy of the design of the systems we use, from high heels and tin openers to aeroplanes and telephone systems! Every time I take a business trip, I find myself reflecting on the operation of the systems on whose adequate functioning I depend. Turn on the tap in the hotel room, and you get an immediate lesson in design adequacy. Enter the shower, turn on what you think is the hot water, and you find yourself playing Russian roulette. Yet somebody designed it all. Driving in my automobile, I have frequently cursed the designers of my glove compartment. I fumble for essential items while at a stop light; the light changes, I start off, and the contents of the glove compartment end up on the floor. Who designed this? Why haven't they learned to do it better? If designers can fail with such simple, often-designed and widely distributed items, it can be no surprise that they fail with larger, one-off or custom-built systems [Petroski 92].

Most studies of organisational interactions with technology take the design of the equipment for granted. However, it is important to include the designers as part of the whole system, as, intentionally or unintentionally, they shape the interactions which one is observing or participating in.

This leads to the question of what we know about the design process. The nature of design is an interdisciplinary area that could use much more study [Meister 71]. For instance, some years ago, the American Civil Aeronautics Board discovered that some aeroplanes were much more likely to crash than others, often as a result of poor human factors design [Bureau of Safety 67]. The process that led to the less than adequate designs, however, was not investigated. Often we find out about the design process only when things have gone so seriously wrong that a public inquiry unearths much of the relevant data. The Challenger space shuttle explosion [McConnell 87] is one example, but one can easily think of others. At the same time, a large literature exists in the industrial design area on the origin and the management of design, but little of this literature uses categorisation of a type that would facilitate comparison and the development of theory.

An examination of the strategies adopted in the design process give rise to two key questions. First, how do the designers determine what the users need? This is by no means a trivial question, as many systems are not designed with sufficient attention to what the users need (see also Chapter 4 for consideration of the user in design). One solution is to involve the users in the design of the system, as Boeing has done on its recently designed 777 airliner [O'Lone 92]. Second, once the design is released, how do the designers track the performance of their systems? This is important both to fix designs that are not working and to learn from previous experience. Unfortunately, very little attention of a systematic nature has been given to either one of these questions.

Let us take one of the more striking examples of poor product design, that of the M-16 rifle, and consider the context of its creation. Military systems — even those as small as a rifle — provide severe lessons in design. Even small glitches can show up poor design through casualty figures in training, in operation, and especially in combat. The M-16 rifle, widely used by the United States during the Vietnam era, reflects many of the issues here. Highly prone to jamming, it proved a serious hazard to Americans who used it in combat [McNaugher 84]. More remarkably, it represented a step backward from an initial design that was outstanding. It is an object lesson in the importance of the network of designers, improvers, and implementers who start with one idea, but shape and transform it into something else.

The M-16 began life as the AR-15 Stoner rifle, designed by Eugene Stoner of the Armalite Corporation. The AR-15 was designed in response to studies that showed the need for a rifle with a high rate of fire but lower accuracy than the traditional military rifle [Ezell 84]. It was not, however, requested specifically by the military services, and was therefore outside the official channels at birth — a 'technological orphan'. It was treated accordingly. It used non-standard ammunition and required non-standard tactics, and therefore violated many of the traditional habits of the community that needed it. It was at first refused, then imposed from above. Re-engineered into the M-16 to conform to inter-service requirements, it was produced in quantity and finally distributed widely as a combat weapon. Rumored to be a super weapon, impervious to dirt and grime, soldiers did not clean their M-16s often enough. However, even if they had wanted to, insufficient cleaning materials were provided to operational units. It was considered a tragic failure in combat, often jamming during operation, sometimes resulting in the death of the user, who was no longer able to return fire. The following quotation from a letter written by a US marine to his parents brings home the tragedy of faulty design.

> Dear ...: I got your letter today aboard ship. We've been in operation ever since the 21st of last month. I can just see the papers back home now —enemy casualties heavy, Marine casualties light. Let me give you some statistics and you decide if they were light. We left with close to 1,400 men in our battalion and came back with half. We left with 250 men in our company and came back with 107. We left with 72 men in our platoon and came back with 19. I knew I was pressing my luck. They finally got me. It wasn't bad, though, I just caught a little shrapnel. I wish I could say the same for all my buddies.
>
> ...Believe it or not, you know what killed most of us? Our own rifle. Before we left Okinawa, [we] were all issued this new rifle, the M-16. Practically every one of our dead was found with his rifle torn down next to him where he had been trying to fix it. There was a newspaper woman with us photographing all this and the Pentagon found out about it and won't let her publish the pictures. They say that they don't want to get the American people upset. Isn't that a laugh? (quoted in [Ezell 84], p. 208).

By the time a congressional investigation took place, it was evident that the rifle's design had been seriously compromised to meet inter-service requirements. Some even suspected that its failure was an intentional attempt at revenge by the rifle's enemies.

The (Ichord) Subcommittee's words were harsh. 'The much-troubled M-16 rifle is basically an excellent weapon whose problems were largely caused by Army mismanagement.' Addressing the management of the rifle programme, the staff of the Subcommittee decided that it had been operated in the most 'unbelievable' manner. 'The existing command structure was either inadequate or inoperative.' Moreover, their investigation indicated that 'the division of responsibility makes it impossible to pinpoint responsibility when mistakes are made'. Most damning was the Subcommittee's statement that there was 'substantial evidence of lack of activity on the part of responsible officials of the highest authority even when the problems of the M-16 and its ammunition came to their attention'. These bleak observations forced the Subcommittee to conclude 'that under the present system problems are too slowly recognized and reactions to problems are even slower' ([Ezell 84], p. 209). All these remarks point to basic difficulties with the nature of the design and the implementation network surrounding the weapon.

Ironically, the Special Forces, which had requested the original Stoner design and got it, found they had an excellent weapon. No better commentary on the failure to provide a sound product can be offered than the following case study, virtually a catalogue of what not to do.

9.2.2   The Ecology of Thought in the Design Process

Around every well-designed and well-fielded system is a protective envelope of human thought. This envelope comprises the technology's designers, manufacturers, operators, and evaluators. The technology's survival as well as its safety depends on the adequacy of this human shell. When the shell fails, the technology also fails, often with serious consequences for the users and for many others. Good systems come about because those involved in their creation and deployment think intelligently about the context of use. Most engineers are trained to think very well about the dynamics of electronic or mechanical systems. Few are trained to think about the social context into which the systems fit. Optimising the design for the technological system does not mean that the system will work effectively with human beings. Designing effectively for use means integrating people harmoniously with the hardware and software. This means that the human envelope needs effective design alongside the technology. I want to suggest, furthermore, that the sound design is the one that includes imagination, perception, and consideration for the user. This requires a willingness to move beyond the laboratory. If I am correct, then managers of the system design process need to think more about how to cultivate thought in their organisations. The literature on technological failure quickly reveals that when the design process doesn't provide a proper environment for thought, the systems often fail. Designers need, then, both more contact with users during the design process and more follow-up after the system has been released for use.

The M-16 saga illustrates the dangers created when such a comprehensive system of inquiry is lacking. These dangers appeared at several stages of the system's life cycle. Lack of realistic testing of the rifle's re-engineered design is the first point. The inadequate training and cleaning supplies given to soldiers is another point. Delayed detection of its failures in performance during combat operations is the third point. The rifle failed because it lacked the protective envelope of thought present in every well-engineered system. Once it left the hands of Stoner, no one seemed to pay enough attention to it

Table 9.1 *Types of organisational climate (from [Westrum 90])*

Pathological	Bureaucratic	Generative
The organisation does not know what to do	The organisation may not find out	The organisation actively seeks information
Messingers are shot	Messengers are listened to if they arrive	Messengers are trained
Responsibility is avoided	Responsibility is compartmentalised	Responsibility is shared
Bridging is discouraged	Bridging is allowed but neglected	Bridging is rewarded
Failure is punished or covered up	The organisation is just not merciful	The organisation initiates inquiry and redirection
New ideas are crushed	New ideas present problems	New ideas are welcomed

at the key points, to provide either a solid initial design or a quick and reliable fix when things started going wrong. There was no monitoring process to ensure that things didn't go wrong, or to fix them when they did. Large systems need even more urgently an attentive and nurturing environment to ensure design adequacy.

The Sidewinder missile provides an interesting case. Sidewinder was carefully designed to exacting user requirements by an outstanding government laboratory [Westrum 89] and constantly upgraded to provide a better weapon. While Sidewinder's designers paid careful attention to what pilots would need in an air-to-air missile, its design was still not good enough for the real world. Moreover, it was introduced without adequate training in the relevant tactics for the pilots who used it. Even though Sidewinder was better than its competitor missiles, Sparrow and Falcon, it initially did very poorly in combat because its capabilities and limitations were not properly appreciated by the users — a failure of implementation [Wilcox 90], which is an issue explored further in the following section.

Table 9.1 (after [Westrum 93]) illustrates some of the kinds of environments in which design can occur. It suggests that one critical feature is the nature of the supportive environment that surrounds the technological system. All too often, the leadership necessary to ensure a generative environment is lacking. 'Failure to inquire' and to think critically can lead to groupthink [Janis 72] and to a false sense of security.

9.2.3   Inadequate Study of Users' Needs

One source of safety problems is a lack of contact with the realities of the tasks faced by users. The design engineer often seems to be living in an ivory tower, and the context of design is often far removed from the context of use [Perrow 83] (see also Chapter 4). The users are remote in space and time, and not 'psychologically present' to the designers. This has often let architects design attractive buildings that turn out to serve their users very badly [e.g., Sommer 74]. Similarly, car designers may sacrifice safety for good looks. This happened with the 1963 Corvette Stingray, when a strip down the rear window, put in by General Motors' head of styling Bill Mitchell, interfered with the vision of the driver. A great many 1963 Corvette owners had the windows removed at their own expense. It was left out of subsequent models [Ludwigsen 78, p. 160].

All too frequently, when engineers do leave the laboratory and take their first field trip, they are shocked when faced with unfiltered reactions from users. Designers are surprised that instructions are seldom read, that users take short cuts, that jury-rigs or quick fixes compromise or cancel safety features. When Ingersoll-Rand designed a new grinder in 1989, an interdisciplinary team toured the country to get reactions and ideas from users and dealers. They discovered that users had covered their grinders with tape to protect themselves. They were shocked when they discovered that users' hands got black and blue from using the previous model. They were surprised when dealers reacted with comments such as: 'We doubted our input would mean anything. We thought once again Ingersoll-Rand was going to design something in a dark room and ram it down our throats' [Kleinfield 90, p. 6]. The net result of the field studies, however, was a grinder that did not leave workers' hands hurting at the end of the day. One of the prototypes given out to potential customers for test was hidden, and the worker refused to give it up [Peters 92, p. 80].

When large safety-relevant systems are involved, the consequences of failure to study users' needs are more serious. Clearly 'designing in the dark' led to the infamous case of the US Navy's 1200-psi steam propulsion plant [Naval Personnel 82]. Hazards created by this ship power plant included lethal high pressure steam leaks, unlabelled and often difficult-to-read gauges, and a complex start-up operation that was virtually guaranteed to cause accidents. A lack of skilled personnel and a high turnover also contributed to making the system risky. As a result, at least 37 boiler explosions occurred [ibid, p. 5]. While all boiler explosions involved deviations from standard procedure, the complexity of the procedures encouraged such violations (see Chapter 11 for a discussion of violations and their causes and Chapter 7 for guidance on the design of prcoedures to minimise them). Defective design, then, is symptomatic of an organisational environment which does not encourage the requisite inquiry needed for thorough product development. Safety is often one of the neglected features.

9.2.4   Failure to Follow up Product or System Release

Sir Karl Popper suggested that good scientific investigation should be conducted by making bold hypotheses, then subjecting the hypotheses to tough tests [Popper 61]. In the same way, engineers need to innovate boldly, but also to test thoroughly. When such testing is not done, unsatisfactory or unready products reach users. And testing should extend to the users. Product use and experience need to be monitored far beyond the laboratory. When such monitoring is carried out, unexpected negative effects are frequently detected. These effects need immediate action. Failure to follow up can mean that potentially disastrous conditions may be ignored until it is too late. For those who grew up during the 1960s, the Thalidomide tragedy is a perfect example of an organisation's post-release carelessness [Knightley 79].

Design typically takes place in a commercial environment, and this is very important when safety-relevant systems are involved. Safety-critical systems must sell if their creators are to stay in business, yet safety may be an incidental rather than a focal part of the design process. A bad safety record can affect profits, but so can many other product features. Only with the liability revolution has product safety become central to business survival. Furthermore, large complicated systems can always be defended on the grounds that inadequate training or 'human error' caused an accident, rather than fundamental design problems.

Contemporary examples are readily apparent, as illustrated by two recent examples from the medical field. The first is the Bjork-Shiley heart valve, a widely distributed prosethesis whose failure cost over 400 people their lives [Subcommittee 90]. The strut on this valve was highly prone to breakage, partly as the result of bad design, and partly as the result of poorly trained welders on the assembly line. However, even taking these two problems into consideration, the Shiley company failed to act to investigate and elucidate reports of valve failure once they began pouring in. Failure, both to investigate rapidly and to communicate with the public, is symptomatic of an environment that does not encourage thought and communication.

The second medical example comprises the Therac-25 deaths [Leveson 92]. The Therac-25 radiation machines had a software 'bug' that could turn them into medical death rays, apparently responsible for killing several people. Although the internal processes involved in design and development of the Therac-25 are not clear, there are several indications that the ecology of thought surrounding the product was not good. Apparently the software was written by a single person, but the documentation for design is inadequate, making it difficult to retrace the design process. When fatalities occurred, the company's responses to indications of serious problems were slow, ineffective, and less than helpful to users, who were typically hospital nuclear medicine departments. Responses that were 'too little, too late' indicate that corporate management found itself unable to focus effectively on the kind of design revision that might have cured the problem once and for all.

9.3   THE CONDUCT OF OPERATIONS

9.3.1   Small Mistakes Lead to Big Accidents

Large systems are vulnerable in a host of ways to small errors [Reason 90]. Consider, for instance, what something as simple as a rocket firing circuit may do if it is badly designed or manufactured. A Zuni air-to-ground rocket is about 9 feet long and weighs roughly 68 kilograms. On 29 July 1967, one was fired prematurely while on a plane on the deck of *USS Forrestal*, an aircraft supercarrier off Vietnam. The rocket hit the gas tank of another plane, and started a huge fire that engulfed the deck. By the time the fire was over, some 21 aircraft were destroyed and another 43 damaged. The death toll was 134, with another 62 crew members injured. The carrier was put out of action and had to return to the United States [Wilcox 90, p. 71]. The events leading to the accident were multiple. The immediate cause of the accident was a short circuit, probably created during the manufacturing of the missile, but another key factor was failure to give the crew adequate training in missile handling. Well-trained sailors know that you don't connect the umbilical cords of live rockets until the plane is on the catapult, pointing out to sea. But the sailor who connected the rocket to the plane had not had such training.

The *Forrestal* fire is a good example of the need for orchestration. Operating a complicated system requires the complex orchestration of technology and people. The epitome would be something like a manned spacecraft launch, certainly one of the most complex operations ever attempted by humans [Murray 89]. But even in something as small as a two-pilot aircraft, teamwork is still important, as it also is on a hospital surgery team. Large or small, the group must be coordinated, led, and administered, and how this is achieved is far from an exact science. Further, as already noted, our grasp of social dynamics tends to decline in proportion to an increase in the size of the group. Conceptualisation and precision become more difficult as we go from the dyad to the larger group, and to the organisation and the network.

Highly automated systems in particular tend to be challenging to groups, as often very small inputs can cause large disasters. And because few inputs are needed to change the state of the system, automation tends to amplify the importance of the human factor (see Section 2.3.2 of Chapter 2 for a discussion of some of the impacts of automation). Consider the destruction of Korean Airlines Flight 007, generally considered to be caused by a faulty autopilot instruction, which brought the plane into hostile airspace. Or consider the plight of the flight deck crew faced with an autopilot whose unpredictable output nearly sent its 747 airliner into the ground [Carley 93]. These situations place a high premium on effective communications, training, and supervision.

To consider all aspects of the operations of technological systems by groups would make this review impossibly long. So I will concentrate on one particular aspect, decision making, and provide illustrations of some key issues.

9.3.2   Making Full Use of Intellectual Resources

A system is only as good as the intelligence directing it. Basically, this means using the 'full brainpower' of those in the system. But how is that to be done? Evidently it is not easy, as studies documenting technological accidents readily illustrate. All too often, the information is in the system, but it is not used. Sometimes the system must rely on information passed by informal means of communication, as discovered during observations made in a London Underground control room (see Chapter 4, Section 4.5).

The 'known problem' happens when one or more of the crew knows exactly what is going on, but communication barriers or mindsets prevent corrective action. A United Airlines accident during a landing in 1978 at Portland, Oregon, is a fascinating example of this category of accidents. During the landing, the aircraft's first officer and the flight engineer were increasingly uncomfortable with the low fuel state, and advised the captain, with increasing urgency, of the need to land. The plane had a gear-up problem and was loitering over Portland to consume fuel. When the captain headed outbound (away from the airport) to begin his final approach, the two other crew members screamed at him that there wasn't enough fuel left to make the planned approach, and to turn back. The captain did not respond and continued with his plan, which resulted in a flame-out and crash a mile or two from the airport while on final approach. The pilot, who had a reputation as a dunderhead, was quickly retired following the accident [Doyle 93]. Such accidents usually occur because the full intellectual resources of a system are not used. This problem has led, in the United States and elsewhere, to a form of corrective action which began as 'Cockpit Resource Management' but has since been generalised to 'Crew Resource Management' [Wiener 93].

In about 1979 studies by several key individuals in the United States suggested problems with interaction on airliner flight decks [Aerospace Medical Association 91]. These problems included failure to pass on key information, refusal to listen to other crew members, and loss of situational awareness. In simulations and in operational experience, such failures of leadership and communications had led to serious accidents. Often pilots refused to listen to the suggestions and observations of co-pilots or other members of the flight crew during dangerous operations. This was sometimes the result of overconfidence or an arrogant attitude on the part of the captain. More generally, however, the problem was a failure of the crew to interact so as to make the best decisions based on the information available. As a result of these studies, two forms of training for better use of cockpit intellectual resources came about: cockpit resource management training (CRM) and line-oriented flight training (LOFT); the latter involves using simulators to provide practical training in the former. CRM is becoming the generic label for this style of group effectiveness, and it is widely used by airlines in their training programmes for cabin and ground crews as well as flight deck personnel [Bovier 93]. Already CRM has helped to save dozens of passengers who might otherwise have died in the Sioux City crash [Fitch 91]. While CRM has helped to improve information flow in aeroplane cockpits, the same principles are obviously useful for other team-decision areas, such as air traffic control, nuclear power plant control, and even anaesthesiology in hospital operating rooms [Howard 92].

The need for CRM training is apparent. A lack of CRM led to fatalities in the only crash of a B-52 bomber in the Gulf War [Karam 91]. Interestingly, the crew had experienced the same failure during a simulation. In each case a co-pilot's suggestion that the crew eject was ignored. When the simulator 'crashed', crew members were disturbed; when the real B-52 crashed, crew members died. Hospitals probably need CRM (in this context it is called 'crisis resource management') even more than cockpits [e.g., Stein 67, Howard 92]. And no doubt the US and Royal Navies will eventually adopt 'Bridge Resource Management'. They certainly need it, as witness the circumstances leading to the destroyer *Spruance* running aground [Associated Press 89]. The lieutenant in charge ignored warnings from his quartermaster and a junior lieutenant, and ran the ship onto a reef. Note also the running aground of the huge ocean liner *Queen Elizabeth II* from similar causes [NTSB 93].

Use of intellectual resources is thus a problem for all large and complex man-machine systems. Airliners provide only one location where such issues arise, but they are a good place to start. Beginning with the dyad or triad on an airliner flight deck, and then gradually moving out to consider larger groups, we can see the kinds of factors that affect a group's use of its full intellectual resources. These issues have been highlighted in an outstanding series of studies carried out by Todd Laporte, Gene Rochlin, and Karlene Roberts of the University of California's Institute for Governmental Studies. This group has studied aircraft carrier landings, nuclear power control, and air traffic control. These studies have focused on how systems can be both highly complex and highly reliable. They have helped to clarify how really complex the management of large systems is, and in the process the Berkeley team has brought to light some remarkable ways in which high-performance systems operate [Rochlin 87, Roberts 93]. Not only do high-reliability systems use intellectual resources better, but they do so through a variety of informal practices and cultural norms that other systems might do well to emulate.

9.3.3   Coordinate Leadership

One of the interesting results of the Berkeley group's study of aircraft carrier landings is the recognition of the importance of what I have called 'co-ordinate leadership' [Westrum 91, p. 227]. 'Coordinate leadership' means that the leadership role shifts to the person who currently has the answer to the problem in hand. During a crisis, this style of system operation puts greater emphasis on knowledge than on organisational position. Mary Parker Follett, one of the founders of management theory, spoke of the 'law of the situation' [Metcalf 40] — the need to do what the situation requires.

This may be typical of a multi-disciplinary team, where different knowledge may be required at different steps in a process. It may also be possible in situations in which there is a common understanding of the situation, so that a person with a critical observation or understanding may be empowered to take action without checking with the formal leader. For instance, during a night landing on an aircraft carrier, the naval hierarchy on the ship recedes until the plane is landed and halted on the deck. While supervision actually intensifies during the landing, its character changes from directing to monitoring. The lowest seaman in the system can overrule an admiral if he knows the key facts about what must happen at that moment. Any person with the answer has a right, indeed a duty, to do the right thing regardless of rank. The moment the plane is down, however, the hierarchy reasserts itself [Rochlin 87]. Thus, one of the important features in the management of complex but unstable systems is the ability of the team to shift its structure to provide for safe operation.

Coordinate leadership is important not only because it ensures that the best answers are used but also because it facilitates the management of the complexities of a situation by making sure that no one is overtasked. The dangers of leaders' excessive mental workloads were brought out in a study of flight leads and mechanical failures [McKinney 93]. The flight 'lead' is the pilot in charge of a group of two or more planes. In principle, when anything happens to the lead's plane, the lead should pass the torch over to a wingman. In real life, however, leads actually made worse decisions when their planes developed mechanical failure than wingmen did when faced with the same situations. Although flight leads probably had superior skills to their wingmen, they appeared reluctant to engage in the same information-searching that the wingmen did, an effect that worsened with the pilot's level of experience! The study suggests the dangers of leadership which cannot shift into a different mode. When a plane develops a problem, the ability to draw on the expertise and ideas of others is very important, since the best ideas may not come from the person in charge.

In this regard, one of the most remarkable and effective uses of full intellectual resources took place when one of the Apollo (moon mission) spacecraft developed a serious problem due to an internal explosion on 13 April, 1970. Intellectual resources inside and outside the space programme were tapped to provide an answer for the plight of the crew on Apollo 13. The search for answers was so wide that it went well beyond the boundaries of the United States. Fortunately, these resources proved adequate, and the technological fix allowed the spacecraft to return safely to earth [Murray 89].

The management of group and individual consciousness thus emerges as an important feature of the effective group. In the aircraft carrier studies, the Berkeley group found that considerable attention was paid to the level of situational awareness of higher officers. 'Having the bubble' signified that the commanding officer had an appropriately complete picture of the disposition of the various ships in the task-group fleet [Roberts 90]. If the commander started to lose the bubble, it needed to be transferred to another officer, so that someone had an adequate grasp of operations. Transferring the bubble meant the transfer of authority to others. Consciousness is typically slighted in ordinary system studies, but Roberts found that 'having the bubble' was a phenomenon very important to fleet safety. The operational 'maestro,' whether Air Boss or otherwise, had to have a sense for what was happening over a broad area.

The situational awareness about which Roberts wrote is related to broader issues. One of these is 'groupthink,' identified in 1971 by Irving Janis as pressure for consensus in situations that are dangerous [Janis 72]. In many respects groupthink also represents a failure to maintain situational awareness, although it is typically applied to strategic rather than tactical situations. In groupthink, an organisation, due to conformity and complacency, fails to pay attention to how its plans might go wrong. As a result of the pressure for consensus, the group may ignore alternative courses of action and other intellectual resources, including the stifled critiques of group members. Janis showed how groupthink has often been involved in situations that led to strategic fiascos. Groupthink forces below the surface the information that needs to flow freely. As in crew resource management studies, the failure to encourage communication leads the group not to examine all options, and to take courses of action that are sub-optimal. For a discussion of how biases such as 'groupthink' affect human decision making, see Section 2.4.7 in Chapter 2 and Table 8.1 in Chapter 8.

9.3.4   Tinkering

Complex systems often come complete with extensive training, rules, manuals, and procedures. In real life, however, the rules and procedures are frequently bent or broken, and not always for bad reasons (see Chapter 11 for a discussion of such 'violations'). Pick up any technical manual and try to use the technology to which it applies without any previous knowledge, and you will find that frequently key steps have been assumed and left out. Alternatively, the manual itself may be wrong, as one of the manuals was for the fire-control circuit wiring in a Boeing airliner [Fitzgerald 89]. Often the technology just does not do what it is supposed to do. This is just as much the case with a nuclear power plant as a home videotape player. Additional (tacit) knowledge is frequently needed to get the system to operate. Sometimes, as with the 1200-psi power plant (see Section 9.2.3 above), the technology is just not user-friendly and needs to be made responsive. In other cases the system's objectives simply cannot be achieved by adhering to the rules. So users tinker: they add on additional controls, they reconfigure the system, they adjust the system to their needs.

Sometimes their tinkering is ingenious. In Vietnam, soldiers taped two M-16 magazines bottom to top, so that when one was empty, the other magazine could be readily inserted. Sociologists have discovered that workers in factories frequently know how to make production go faster or better; these secrets are seldom disclosed to management, but they can often allow rapid finishing of a production run when there is a deadline. Similarly, a study of mobile aircraft electronics checkout systems on aircraft carriers showed considerable 'customisation' of work procedures, contrary to rules and regulations [Kmetz 84].

Tinkering can also be dangerous. Workers' knowledge of the machinery may be extensive, but this does not mean that they really understand the dynamics of the systems with which they work. Taking short cuts may expose them to unexpected dangers. Chemical reaction vessels may explode when used incorrectly. Alarms which offend may provide vital signals, but workers at nuclear installations have been known to turn them off if they sound once too often. This is particularly true of the undertrained worker. A common cause of technological accidents is the use of persons who are temporary workers, or who have had insufficient training [Fitzgerald 89, Schneider 91]. They often do not understand the potentially serious consequences of failing to follow the rules. In a classic essay on mistakes in building, Riemer showed how one mistake in a building can lead to a series of successive re-adjustments in the way the building is actually put together [Riemer 79]. This kind of tinkering can lead to dangerously underbuilt systems, as in the Hiatt Regency Hotel Disaster in Kansas City [Petroski 85].

9.3.5   The Role of Maestros

I have previously referred to the orchestration of resources in socio-technical systems. Now I wish to consider the conductors of the orchestra. The complexity of safety-critical systems means that somehow the various elements involved must be brought into harmony: the people, the hardware, the software, and the various interfaces, including political and financial support. To achieve this harmony requires people who Arthur Squires has designated as 'maestros' [Squires 86]. Simply put, a maestro is a person who makes the system work in harmony [Westrum 93]. Maestro abilities include some combination of the following traits:

- Technical virtuosity;
- A high energy level;
- A broad attention span;
- High standards;
- An ability to focus on key questions.

There are different kinds of maestros — research and development maestros, operational maestros, problem-solving maestros, and so forth. These types may overlap, but the common element is the ability to act as the manager of complexity. Every successful project requires a maestro. Some projects may require many maestros. A study of the NASA Apollo project uncovers too many maestros to make a definitive list, but names such as Hugh Dryden, Wernher von Braun, Max Faget, Joe Shea, Harrison 'Stormy' Storms, Walter Gilruth and George Low come immediately to mind; and there has been an outstanding series of NASA Flight Directors, many of whom became legends.

One has to look hard to find this level of leadership in NASA today. A maestro's presence does not guarantee safety, as the Apollo fire showed, but a maestro's absence in a safety-critical system can spell disaster. When maestros are missing, bad things happen.

Bridge-building provides some object lessons in the presence or absence of maestros, and safe bridge design requires maestro leadership [Westrum 93]. When maestros are present, there is proper attention to key questions, even down to small details. Petroski, in a remarkable book about engineering design [Petroski 94], has shown that in designing for the future maestros often pay significant attention to what has gone wrong in the past. Safety issues are carefully explored, and valuable safeguards are installed. When problems arise, they quickly become the focal points of inquiry and their solutions tend to be thorough. The absence of a maestro doomed the first Quebec River Bridge and allowed an exceedingly dangerous situation to develop into a fully fledged disaster [Tarkov 86].

Probably the single most important function of the maestro is 'having the bubble' in the larger sense. In large projects, some of the component teams may have an undue concentration on their own tasks. Without someone who has a larger vision, this concentration can lead to sub-optimisation or even to serious imbalances that imperil safety. Often, for instance, the interstices between teams are neglected, and no one is doing the 'organic management' necessary to ensure coordination [Burns 61]. It is in the interstices that many accidents originate. The maestro must empower internal champions to solve major problems, especially when these problems cross organisational boundaries. It is also the maestro, who must force project teams to attack the build-up of 'latent pathogens', that James Reason has so ably described [Reason 90]. The maestro must have a vision of the entire process and focus effort on items with the most critical priorities.

Often the maestro will be the exemplar for ensuring fast and adequate communications flow. The famous anecdote about von Braun sending a bottle of champagne to an engineer who reported his own error is a case in point. When a Redstone missile went out of control, the rocket team at Huntsville, Alabama, was at a loss as to what to do until one engineer reported his suspicion that he had caused a short circuit. Checking proved the engineer right, and von Braun sent him a bottle of champagne [McCurdy 93, p. 70]. An organisation willing to confront its own errors is likely to manage a technology far more safely than one afraid to admit such problems.

At first glance, the ideas of a maestro and coordinate leadership might seem utterly opposed. With a strong leader, so the argument goes, less initiative is needed from the troops. Actually, however, one of the results of maestro leadership is precisely an empowerment of others to act. The high standards of maestros tend to create the pressure for action when something goes wrong. In another study on which the author is at work, for instance, a constant refrain was 'I couldn't let the chief down'. Maestros encourage prompt action, not passing the buck. To give the reader some idea of what I mean, I will quote a colleague who was lucky enough to work for one of the great maestros of technology, Charles Franklin 'Boss' Kettering. In a 1946 speech to some of his Electromotive Division engineers, Kettering said something like the following:

> You men are now Mr. General Motors. You are not just representing the Electromotive Division but all of the GM organisation. I want you to stay at good hotels, eat at good restaurants, be punctual, take taxis if public transportation is not suitable. If someone you contact is having trouble with a GM automobile or a truck diesel, get in touch with the District or Regional Manager and get his problem solved. We're giving you special uniforms and full expenses. I want you to show people that you are representing a first class organisation. [Blaise 91]

While it is true that believing that one belongs to a 'first-class' organisation can produce groupthink, often instead it can provide a not-so-subtle pressure to perform to a higher standard.

9.4   MANAGING THE INTERFACES

9.4.1   Across the Boundaries

Every system has boundaries, and this fact is important because things (ideas, supplies, people) move across the boundaries. Sometimes the boundaries are spatial, as when an airliner moves from one air traffic control territory to another, or a hospital patient is taken from one unit to another. Sometimes they are temporal, as when one team in a factory hands over to another at a shift-change, or when a new crew comes on board a ship. Across the boundaries of systems move pieces of information, people and, of course, other systems. How these transitions are managed is a subject appropriate for more detailed study.

For instance, in a fascinating study of the world of the ambulance, Metz described the ambiguity of the passage of the patient from the ambulance to the hospital [Metz 81]. As the patient leaves the dedicated but poorly trained hands of the ambulance paramedics for the highly-trained personnel of the hospital emergency room, the information gathered by the former is seldom well transmitted to the latter. There are two reasons for this. One is the attitude of the hospital personnel, who see themselves as superior to the ambulance attendants. The other is the real ignorance of the attendants, often working for minimum wage, whose knowledge is only occasionally deepened by feedback from the hospital personnel. One result is that the patient is not accompanied by a complete file of information, important for diagnosis and treatment. Another is that the ambulance personnel are not 'educated' for their next encounter with the same problem.

In an early study by Patterson, accidents on an airbase led a sociologist into an analysis of factors contributing to them. In very short order, Patterson discovered that one of the problems was that the pilots and ground controllers did not understand or trust each other [Patterson 55]. To provide cross-training, he put pilots in ground centres and gave controllers simulated flying experience in link trainers. By managing the interfaces (and group reorientation), and working across the organisation's boundaries, Patterson was able to get the accident rate down and move the system towards a safe mode of functioning. This kind of 'organic management' [Burns 61] provided the necessary linkages.

In the study by Rochlin, LaPorte and Roberts [Rochlin 87], one of the important features of the Navy culture in which the carriers floated was the subculture of the bosuns themselves. While the crews of the carrier changed constantly, the bosuns did not, and they provided a mobile source of organisational learning about how arresting gear and other technical systems worked. Interestingly, the bosuns exchanged this information without any previous planning by the Navy. Rochlin and others have worried that many of these informal mechanisms might be compromised by a naval administration that was unaware of the human underpinnings of technological reliability. A similar phenomenon was observed in a London Underground control room, already referred to in Section 4.5 of Chapter 4.

Systems originate from, and serve the purposes of, larger groups or organisations. I have suggested the concept of a human envelope of thought around every socio-technical system. But the safety provided by the envelope can be compromised in a variety of ways. Higher management may peel away resources for more 'urgent' projects, leaving the system undermanned and exposed to accidents. This happened at Bhopal [Shrivastava 87]. Political pressures may intrude upon technological decisions, encouraging actions with high risk, as in the Challenger case [McConnell 87]. Scheduling pressures or resource constraints may discourage the necessary pre-operation checks, as with the Comet airliners, three of which crashed before the organisation realised that a problem existed in the basic design [Petroski 85]. Omitting the checks and balances also was a factor in the failure of the Space Telescope [Lerner 91].

Organisational instability and conflict can also be a significant source of accidents. A study by Little et al [Little 90] showed that the financial instability of an airline might be related to its accident rate. Another study, by Boeing Aircraft, questioned corporate management pilots about their perceptions of the 'safety culture' of the flight department [Lautman 87]. Most felt that the department's culture was strongly shaped by the top management of the larger corporation; expansion of the sample size validated the early conclusions.

When some parts of an organisation become unwilling to share information with other parts, danger signals may not be transmitted. This is particularly the case when the person who spots the problem belongs to 'the wrong department'. Wise organisations, through a process I have called 'pop-out,' encourage individuals anywhere in the organisation who spot problems to make them known [Westrum 93]. But often 'not invented here' is a far more powerful factor.

Other organisations outside the organisation in question may also play a role. Many find their work caught up in a complex network of sponsors, contractors, and regulators. Responsibility may not be clearly identified. How the network 'hangs together' may well shape whether an accident emerges in a given context. When the network is acting as an integrated whole, as during the Apollo 13 crisis, there is substantial additional protection. When the network is the subject of commercial or political pressures, as in the failure of the Hubble space telescope, protection is seriously compromised. In the Challenger case, independent organisations supposed to be acting as monitors did not fulfil this role [Vaughn 90]. In the Bjork-Shiley heart valve scandal, the Food and Drug Administration was not forceful enough to prevent hundreds of deaths as Shiley continued to ship and market risky heart valves. These observations suggest that inter-organisational bridging activities need more study than they have received.

9.4.2   Case Study: Air-to-Air Guided Missiles

At this point I would like to develop an extended example of the kinds of issues that crop up when one begins to investigate interesting real-world problems involving human factors. It concerns the development and use of air-to-air guided missiles, and extends from the 1940s into the 1970s. Some parts of this story are well known, but others are obscure. This example is presented in detail, because it illustrates many of the key points that I have previously mentioned.

The story begins shortly after the end of the Second World War, when it became obvious to military and scientific personnel in many countries that air-to-air combat would soon involve guided weapons. In the Second World War, planes shot down other planes with machine guns, but the advent of jet engines meant that machine guns, or even the more effective aerial cannon, were likely to become obsolete. Their range and rate of fire would be limitations in the coming era of jet combat. Some kind of self-guided weapon was necessary to achieve the required distance, and to follow and intercept the target aeroplane. The coming of jet bombers with atomic bombs made the self-guided weapon even more imperative.

In the United States, both radar- and infrared-guided missiles were developed by the early 1950s. Three missiles in particular, the Falcon and Sparrow (both radar guided) and the Sidewinder (infrared guided), became the weapons of choice. By about 1955 all three had been given a variety of tests and released for use. They were soon distributed to the Navy, Air Force and Marines, and were made available to friendly governments. The first combat use of Sidewinder by the Nationalist Chinese Air Force in 1958 seemed to indicate that the missiles would be successful in air combat.

Let me digress for a moment to show that testing of at least one of these missiles was by no means cursory, but involved a series of fairly difficult hurdles. Sidewinder went through many tests by its own research and development team at China Lake in the Mojave Desert (California) before release. Then the missile was given an extensive test by the Navy's Bureau of Ordnance. This test was followed by more tests by the Navy's Operational Test and Evaluation Forces. When the missile was released to the fleet, still further tests were carried out. At each stage more improvements and refinements went into the missile. One test, for instance, showed that putting Sidewinder together on shipboard was difficult, so the parts were redesigned to make this easier. Ships were specially designed so that Sidewinders could be assembled in the mess hall.

All these tests and evaluations, as well as other general analyses, seemed to show that the missiles would work well in combat and create a virtual revolution in air warfare, making the machine gun and dogfighting obsolete. New planes were designed without aerial cannon, and air combat training for pilots was discontinued. Most pilots were forbidden to engage in dogfighting manoeuvres during training since they caused so many accidents. The new air war was to be an electronic, push-button war, in which distant planes would be identified and shot down without the 'eyeball-to-eyeball' manoeuvres so familiar from the world wars. The American armed services entered Vietnam trained for exactly this kind of remote-control air war.

But in combat the missiles typically failed; they missed their targets, failed to explode or simply plunged into the ground. Only one missile in ten shot down the plane at which it was fired [Ault 89]. Contrary to doctrine, dogfights still took place, and planes without aerial guns were at a disadvantage. The first phase of the Vietnam air war was a victory, but it was far too costly. Something was clearly wrong.

One person who wanted to do something about it was an aircraft carrier commander named Captain Frank Ault. He had written a series of letters to Washington about the problems, but largely without result. Later, when Ault had been reassigned to the Pentagon, he was called in by higher officers and given an assignment to fix the problem. Ault was given authorisation to get the personnel he needed and go wherever he needed to go. What followed was one of the most comprehensive examinations of system failure in the history of operations research, an example that belongs in every textbook.

In each of the disciplines involved in guided missiles, there were maestros, people recognised as the highest practitioners of their arts. Ault got five of them to serve as the heads of his investigation teams. In each of the five areas — design, manufacture, maintenance, logistics and training — the system was examined from beginning to end. Missiles were followed from design, through manufacture, to shore installations, on board ships and into the skies. In every area there were problems. Missiles were designed without sufficient concern for users; they were badly manufactured; they were abused in the handling processes on shore and on ships; their firing procedures were not user-friendly; and most of all, pilots did not have sufficient training to use them well. For some missile shots, it was the first time either the pilot or the plane had fired a missile.

Ault believed in getting industry involved, and forced those who manufactured the missiles to consider how they would be used. In one case,

Ault and one of the industrialists he was 'educating' were on board an aircraft carrier.

Escorting a Raytheon Vice President through a Yankee Station carrier work area, Ault and the VP watched an F-4 squadron maintenance crewman lay out twelve feet of schematics across the hangar deck trying to figure out what was wrong with the Phantom radar on which he was working. He hardly got it all laid out when a forklift full of bombs ran right across the middle of it. Ault said to the VP, 'The trouble with you guys is you put on your white smocks back there in Massachusetts and don't understand what happens when your system gets into the hands of the great American sailor. This isn't your five micron lab atmosphere. Probably the circuit that kid wants is right under that big greasy tire track right there'. [Wilcox 90, p. 106]

Ault's report, after some nine months, 100 man-years of research and about $3 million at today's value, showed that there were 241 problems that needed solving. In most cases, the solutions had already been identified, but were not being used. So, much of Ault's effort was to establish a network of people who either knew the answers or could discover if someone else knew them. This network, the supportive context, was critical in later implementation.

Ault's report resulted in many changes, but let us follow only one of them, the creation of a school to teach air combat manoeuvreing. Even after the Ault report, it was very difficult to get such an effort going. For the first two years, the 'Top Gun' school at Miramar operated on funds and material whose origins were irregular, to say the least [Wilcox 90]. When Top Gun later became official and successful, a great deal of publicity was given to its triumphs. The bible of this school, of course, was the Ault report, which served as a blueprint for the kind of changes that were needed.

In the second phase of the Vietnam air war, the 'kill ratio' went up enormously. Missiles, now redesigned, functioned much better. They were better handled by the ships and shore installations, and pilots now understood their limitations and the best way to use them. Fewer accidents, fewer 'duds' and fewer dead American pilots were the result. The Ault investigation was enormously successful. But why was it necessary?

At the heart of the missile failures was the absence of the supportive context that would track and correct the problems as they arose. The missiles had originally been designed to shoot down bombers; they were used against fighter aircraft. No one was paying attention to this change in mission. Dogfighting was supposed to be obsolete; it was not. Aerial cannon were still needed (and were retrofitted on aircraft) for short-range engagements. Manufacturing was supposed to produce high reliability rates, but reality was otherwise. Guided missiles did not work as they were supposed to, but then in real life things never do. Problems with a system are inevitable and the key to coping with them is to provide an inquiring system that will identify and fix them as soon as possible. Therefore, it is incumbent on higher management to provide an environment in which there is a free flow of communication, both inside the organisation and between the organisation and those who use its products. Taking action on the identified problems will increase this flow of information.

9.5   CONCLUSION

Safety demands a comprehensive envelope of human thought surrounding the process of design, operation, and implementation. There is simply no substitute for asking all the questions that need to be asked, and insisting on getting good answers (and corrections!) all the way through the process. Often this demands that individuals or groups go beyond formal organisational boundaries to make sure that things are working as they should. Managers at all levels who know how to conduct such imaginative inquiries are of great value to safe operation. We have seen in this chapter what happens when such people are lacking or left without the power to enforce needed changes.

Similarly, the safe system is one whose design is carefully thought through and tested, typically in conjunction with the users. Its design takes into account the activities and habits of social groups that are going to use the system. It brings these groups into the design process, either through field studies or by placing them on design committees. Even the best design requires thorough training for the users and an open line of communication between the users and the designers.

Once the system is operational in the field, it needs to be monitored by the design group. Some kinds of hardware and software problems will become apparent only through use in the field. The problems that arise need to be carefully studied and rapidly corrected. But above all, somebody needs to be paying attention. There are many systems that would benefit from the kind of inquiry that Captain Frank Ault made of the air-to-air missiles. This kind of thoughtful care and support of products and systems is the best guarantee of safety.

It is evident that many of the areas covered in this chapter might profit from more profound study by researchers interested in safety questions. It can only be hoped that this formulation will encourage the filling in of the many gaps this survey has exposed.

REFERENCES

[Aerospace Medical Association 91] Aerospace Medical Association:*Cockpit Resource Management*. In: *Aviation, Space, and Environmental Medicine*, March 1991, 268-271

[Associated Press 89] Associated Press: *Navy Says Officer Ignored Advice and Caused Destroyer Grounding*. New York Times, 30 November 1989

[Ault 89]   Ault F: *The Ault Report Revisited*. In The Hook, Spring, 1989

[Blaise 91] Blaise R A: Personal communication, 1991

[Bovier 93] Bovier C: *Teamwork: The Heart of an Airline*. In Training, June 1993, 53-56

[Bureau of Safety 67] Bureau of Safety, Civil Aeronautics Board: *Aircraft Design-Induced Pilot Error*. National Transportation Safety Board, Washington, DC, June 1967

[Burns 61] Burns T and Stalker G M: *The Management of Innovation*. Tavistock Publications, London, 1961

[Carley 93] Carley W M: *Mystery in the Sky: Jet's Near-Crash Shows 747s May Be At Risk of Autopilot Failure*. Wall Street Journal, 26 April, 1993, A1-A6

[Doyle 93] Doyle T: Personal communication, 1993

[Ezell 84] Ezell E C: *The Great Rifle Controversy*. Stackpole Books, Harrisburg, Penn, 1984

[Fitch 91] Fitch D E: *Can CRM Affect the Outcome?* In: 44 International Aviation Systems Symposium, Singapore, 1991

[Fitzgerald 89] Fitzgerald K: *Probing Boeing's Crossed Connections*. IEEE Spectrum, May 1989, 30-35

[Howard 92] Howard S K, Gaba D, Fish K J, Yang G, and Sarnquist F H: *Anesthesia Crisis Resource Management Training: Teaching Anesthesiologists to Handle Critical Incidents*. Aviation, Space and Environmental Medicine, Vol. 63, No. 9, September 1992, 763-770

[Janis 72] Janis I L: *Victims of Groupthink*. Houghton Mifflin, Boston, 1972

[Karam 91] Karam R: *Gulf War's Only B-52 Crash Eerily Echoed Simulation*. in Detroit Free Press, 7 July, 1991, page 8E

[Kleinfield 90] Kleinfield N R: *How Strykeforce Beat the Clock*. New York Times Business Section, 25 March, 1990, 1-6

[Kmetz 84] Kmetz J L: *An Information-Processing Study of a Complex Workflow in Aircraft Electronics Repair*. Administrative Science Quarterly, Vol. 29, No. 2, June 1984, 255-280

[Knightley 79] Knightley P, Evans H, Potter E and Wallace M: *Suffer the Children: The Story of Thalidomide*. Viking Press, New York, 1979

[Lautman 87] Lautman L G and Gallimore, P L: *Control of the Crew-Caused Accident*. FSF Flight Safety Digest, June 1987, 1-7

[Lerner 91] Lerner E: *What Happened to Hubble?* Aerospace America, February 1991, 18-24

[Leveson 92] Leveson N G and Turner C S: *An Investigation of the Therac-25 Accidents*. Computer, July 1993

[Little 90] Little F L, Gaffney I C, Rosen K H and Bender M M: *Corporate Instability Is Related to Airline Pilots' Stress Syndromes*. Aviation, Space, and Environmental Medicine, Vol. 61, No. 11, November 1990, 977-982

[Ludwigsen 78] Ludwigsen K: *Corvette: America's Star-Spangled Sports Car -*

The Complete History. Princeton Publishing, Princeton, New Jersey, 1978

[McConnell 87] McConnell M: *Challenger: A Major Malfunction*. Doubleday, Garden City, New York, 1987

[McCurdy 93] McCurdy H: *Inside NASA: High Technology and Organizational Change in the U.S. Space Program*. Johns Hopkins University Press, Baltimore, Maryland, 1993

[McKinney 93] McKinney E: *Flight Leads and Crisis Decision-Making*. Aviation, Space, and Environmental Medicine, Vol. 64, No. 5, May 1993, 359-362

[McNaugher 84] McNaugher T L: *The M-16 Controversies: Military Organizations and Weapons Acquisition*. Praeger, New York, 1984

[Meister 71] Meister D: *Human Factors Psychology*. John Wiley & Sons, New York, 1971

[Metcalf 40] Metcalf H C and Urwick L: *Dynamic Administration: The Collected Papers of Mary Parker Follett*. Harper and Row, New York, 1940

[Metz 81] Metz D L: *Running Hot: Structure and Stress in Ambulance Work*. Abt Books, Cambridge, Massachusetts, 1981

[Murray 89] Murray C and Cox C B: *Apollo: The Race to the Moon*. Simon and Schuster, New York, 1989

[Naval Personnel 82] Naval Personnel Research and Development Center: *Problems in Operating the 1200 psi System Propulsion Plant: An Investigation*. NPRDC 82-25, San Diego, May 1982

[NTSB 93] National Transportation Safety Board: *Bridge Communication Failures Led to QE2 Grounding*. In NTSB News Digest, Vol. 12, No. 4, 1993

[O'Lone 92] O'Lone R G: *777 Design Shows Benefits of Early Input from Airlines*. Aviation Week and Space Technology, Vol. 137, 12 October, 1992

[Patterson 55] Patterson T T: *Morale in Work and War*. Max Parrish, London, 1955

[Perrow 83] Perrow C: *The Organizational Context of Human Factors Engineering*. Administrative Science Quarterly, Vol. 28, No. 4, 1983, 521-541

[Perrow 84] Perrow C: *Normal Accidents: Living with High-Risk Technologies*. Basic Books, New York, 1984

[Peters 92] Peters T: *Liberation Management: Necessary Disorganization for the Nanosecond Nineties*. Alfred A. Knopf, New York, 1992

[Petroski 85] Petroski H: *To Engineer is Human*. St Martins, New York, 1985

[Petroski 92] Petroski H: *The Evolution of Useful Things*. Alfred A Knopf, New York, 1992

[Petroski 94] Petroski H: *Design Paradigms: Case Histories of Error and Judgment in Engineering*. Cambridge University Press, New York, 1994

[Popper 61] Popper K: *The Logic of Scientific Discovery*. Science Editions, New York, 1961

[Reason 90] Reason J: *Human Error*. Cambridge University Press, Cambridge,

UK, 1990

[Riemer 79] Riemer J: *Worker Mistakes.* In: Hard Hats: The Work World of Construction Workers. Sage Publications, Beverly Hills, California, 1979

[Roberts 90] Roberts K H and Rousseau D M: *Research in Nearly Error-Free, High Reliability Organizations: Having the Bubble.* IEEE Transactions, 36, 1990, 132-139

[Roberts 93] Roberts K H and Weick K: *Group Mind: Heedful Interaction on Aircraft Carrier Flight Decks.* Administrative Science Quarterly, Vol. 38, No. 4, September 1993

[Rochlin 87] Rochlin G I, LaPorte T R and Roberts K H: *The Self-Designing High-Reliability Organization: Aircraft Flight Operations at Sea.* Naval War College Review, Autumn 1987, 76-91

[Schneider 91] Schneider K: *Study Finds Link Between Chemical Plant Accidents and Contract Workers.* New York Times, 30 July, 1991, A10

[Shrivastava 87] Shrivastava P: *Bhopal: Anatomy of a Crisis.* Ballinger, Cambridge, Massachusetts, 1987

[Sommer 74] Sommer R: *Tight Spaces: Hard Architecture and How to Humanize It.* Prentice-Hall, Englewood Cliffs, New Jersey, 1974

[Squires 86] Squires A M: *The Tender Ship: Government Management of Technological Change.* Birkhauser, Boston, 1986

[Stein 67] Stein L I: *The Doctor-Nurse Game.* Archives of General Psychiatry, Vol. 16, 1967, 699-703

[Subcommittee 90] Subcommittee on Oversight and Investigations, US House Committee on Energy and Commerce: *The Bjork-Shiley Heart Valve: 'Earn As You Learn'.* US Government Printing Office, Washington DC, 1990

[Tarkov 86] Tarkov J: *A Disaster in the Making.* American Heritage of Invention and Technology, Spring 1986

[Vaughn 90] Vaughn D: *Autonomy, Interdependence, and Social Control: NASA and the Space Shuttle Challenger.* Administrative Science Quarterly, Vol. 35, No. 2, June 1990, 225-257

[Westrum 89] Westrum R and Wilcox H A: *Sidewinder.* American Heritage of Invention and Technology, Fall 1989, 56-63

[Westrum 91] Westrum R: *Technologies and Society: The Shaping of People and Things.* Wadsworth, Belmont, California, 1991

[Westrum 93] Westrum R: *Cultures with Requisite Imagination.* In Wise J, Stager P and Hopkin J: *Verification and Validation in Complex Man-Machine Systems.* Springer-Verlag, New York, 1993

[Wiener 93] Wiener E, Kanki B G and Helmreich R L: *Cockpit Resource Management.* Academic Press, New York, 1993

[Wilcox 90] Wilcox R K: *Scream of Eagles: The Creation of Top Gun and the U.S. Air Victory in Vietnam.* John Wiley & Sons, New York, 1990

10

Learning from incidents at work

10.1 INTRODUCTION

Accidents and near misses do occur once in a while, and mostly when they are least expected. The adverse consequences of accidents, such as injured persons, fatalities, and various categories of losses, are always undesired. This is especially the case in systems in which people work to reach preset goals. If an accident occurs, these system goals are jeopardised, and rehabilitation of the system to the pre-accident level of functioning incurs costs additional to the initial damage. Accidents result from unplanned deviations in system operations. These deviations initiate an undesired process which, if not stopped, will lead to an accident [Hirschfeld 63, Kjellén 83a, Hale 87, Hendrick 87]. A near miss reveals such a process without actual damage being done. By early detection and identification of the deviation process, corrective action can restore operations to the desired modes with minimal costs.

Analysis of near misses is particularly relevant to a better understanding of human interaction with risk [Schaaf 91]. People are often more willing to talk about their actions and errors in relation to a near miss or minor accident than a major one; the feelings of guilt are less. In addition, near misses can give clues to the human recovery behaviour that stopped them becoming serious accidents.

In safety-critical systems, analysis of serious incidents is normally focused on finding causal factors of system failures in order to improve the system and, thus, to prevent such deviations in future. The effectiveness of the analysis depends very much on the objectives of the investigation and the methods used [Benner 85]. Accident analysis in systems which are not (officially) classified as 'safety-critical' is too seldom aimed at learning lessons to improve the overall system performance. In such systems, accident data serve hardly more than simple statistical purposes. There is a strong belief that for a single company too many accidents are needed to draw from the analysis of the collected data sensible, statistically founded conclusions regarding system weaknesses. Also, resources for in-depth analysis of minor incidents are limited.

In safety-critical systems, the need to learn from single accidents is probably obvious. Less evident is which of many kinds of lessons can usefully be learned from a single accident. The potential is considerable. In-depth investigation, often by independent committees, can spell out lessons with respect to failure of equipment or operational procedures and to human interventions as well as to organisational issues; it can lead to redesign of equipment, rewriting of procedures, retraining of operators, restructuring of the organisation, or replacement of management.

Powerful investigation and analysis methods have been developed in the last decades for analysing serious accidents in technologically complex, safety-critical systems. The emphasis on post-accident corrective measures has shifted to assurance of safe functioning prior to the commissioning of the safety-critical system under review [Hendrick 87]. For example, it could not be accepted that on the first flight to the moon astronauts would be killed during the mission while the whole world was watching. Early pre-commissioning risk analysis techniques like Sneak Circuit Analysis [Boeing 70, Rankin 73] or hazard and operability studies (HAZOP) [Kletz 92] focus on the system hardware. With the commissioning of unique, first-time systems, it became apparent that the system's mission as well as the assurance of safe functioning required a full systems approach. Operational readiness of a complex and safety-critical operation calls for a perfect match of equipment, human operators, operational procedures and managerial functions [SSDC 75, SSDC 78a]. The scope of risk assessment and risk control, and also of accident analysis, has been broadened to encompass the safety management system of the operation. Among the main results of this change to a focus on the assurance of safety in highly critical systems was the development of the Management Oversight and Risk Tree (MORT) philosophy and its accompanying concepts and toolbox, including the MORT chart [Johnson 80]. The analytical tools help to bridge the gap between the harmful sequences of events leading to the accident and the organisational functions which are needed to ensure and maintain stable conditions for running safety-critical operations safely.

Many of the accident analysis techniques which have been developed for highly technological and safety-critical systems can also be used for the analysis of near misses and minor accidents. Usually, and due to limited resources, such analysis techniques are not applied if the incident-related risk is classified as minor. Also, powerful tools can be too time-consuming to apply to resolve minor problems. On the other hand, one single, serious accident may be foreshadowed by more than ten times as many minor accidents and by an even greater number of near misses having similar causes. Iceberg models have been published to illustrate this point, from the perspective of both accident ratios and loss control — for example, see Bird et al [Bird 76]. While these ratio studies can sometimes be criticised for being too simplistic (ratios vary widely between types of incident, and many minor incidents have no potential for being major), they do indicate that minor accidents and near misses which have the potential for being serious should be analysed on the same principles as are used in the analyses of serious accidents. Such analyses of minor incidents would considerably increase the opportunities for learning lessons from accidents and, as a consequence, the control of the operations under review could better be ensured.

This chapter describes the tools for learning lessons from such potentially or actually serious incidents. A number of accident analysis models and methods will be reviewed as well as their merits with respect to the feedback mechanism within the organisation.

After a brief introduction of accident models in Section 10.2 and accident analysis methods in Section 10.3, a PC-based system, ISA (Intelligent Safety Assistant), will be described. In this, the generic MORT method has been applied as a structured-knowledge framework for the collection of data and the automated diagnosis of accidents at work. The diagnostic interpretation of task performance errors will also be discussed, with the aim of showing the application of techniques which result in systems which work with increased efficiency because of incident analysis. Thus, accident reporting and analysis is seen as a special case of the feedback loop required in quality management systems [ISO 94] to produce constant improvements in system quality.

In this chapter, the 'system to be kept under control' is the enterprise which is running one or more safety-critical processes, e.g., plants at at least one site. It includes the organisational and management structures, and the human and other resources which are relevant to the primary processes in which hazards are to be kept under control. Specific knowledge about technological subsystems is considered to be part of the relevant resources. Techniques for performing an accident investigation, e.g., interview methods, evidence collection, or reconstruction techniques, will not be described. For this, readers are referred to texts such as Ferry [Ferry 88, SSDC 85a]. The ISA application discussed below was tested in the mining, tobacco, chemical and aerospace industries, and further work is being carried out to adapt it to hospital departments.

10.2   ACCIDENT MODELS

In the past, many theories or models about accidents and their causes have been developed and used [Hendrick 87, Johnson 80]. A number of them remain valuable to meet the objective of learning lessons to improve system performance. These fall into three categories, namely models which:

i)   Start from system processes with intrinsic hazards to be kept under control;
ii) Describe conditions of human error which may contribute to system failure;
iii) Describe organisational conditions which are prerequisites for system failure in a safety-critical operation.

See [Benner 83, 85] for a more detailed overview and assessment of approaches to accident analysis. In the next sections only those which are most relevant to learning lessons from accidents will be briefly described.

10.2.1 Accident Models Focused on 'System Under Control'

Accident models whose focus is the system to be kept under control presume that intended activities and operations entail hazards which, if not controlled, may initiate a harmful process. In such a process, energy within the system is wasted, or transferred to exposed objects and, if it is in an amount above the limit that the recipient element can absorb, converted into their harmful deformation or degradation.

The idea of energy transfer underlies several accident models [Haddon 66, McFarland 67]. Its simplest form is the 'energy trace and barrier' model shown in Figure 10.1.

Any accident can be described as one or more sequences of energy transfers, influenced by more or less successful barriers. This model also contains a strategic concept about energy management [Haddon 73, Johnson 80], namely that the elimination of a hazard source (the reduction or elimination of an energy source) is more effective than providing a vulnerable object with its own protective barrier, as long as this elimination or reduction does not threaten the system goals. If sufficient energy is present, the

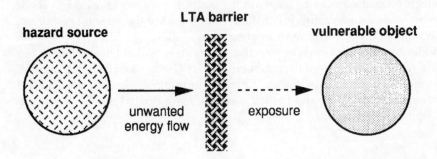

LTA barrier

hazard source   **vulnerable object**

unwanted
energy flow   exposure

Figure 10.1 *Energy trace and barrier accident model*

prerequisites for an accident are present. If a vulnerable object is exposed to an energy flow without sufficient barriers, then the accident will become a fact. There is a near-miss incident if an unwanted energy flow occurs without hitting a vulnerable target. The distinction between an accident and a near-miss incident is, therefore, marginal. Moreover, a barrier analysis will reveal which energy flows — as precursors of the accident — should have been interrupted, where and when [SSDC 85b].

MacDonald's and Kjellén's accident sequence models [MacDonald 72, Kjellén 83b] start from the idea that it is possible to define a stable, normal operating state of a system in which processes run according to plan and hazards are kept under control. Harm or damage only occurs if the (designed) control is lost and not recovered before the energy comes into contact with the vulnerable system element. Thus, an accident is a process of deviation from the desired, normal mode of operation. These models describe time-sequenced process phases, although not in terms of detailed event chains. They provide a basis for a classification system for the types of prevention measures which are possible, by identifying the stages in the accident process where intervention is possible.

Figure 10.2 combines the deviation model with the energy trace and barrier model and makes the link to prevention and system design explicit.

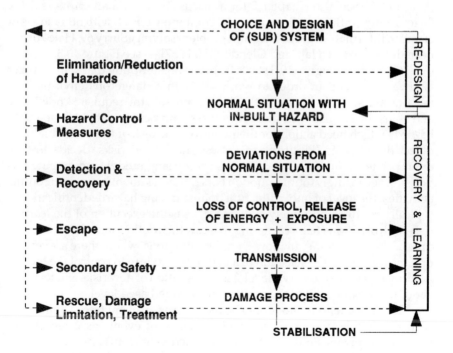

Figure 10.2 Accident deviation sequence model

The types of interventions can be grouped under the following headings:

(i)   Elimination (e.g., substitution of toxic chemicals);
(ii)   Built-in control functions (e.g., automatic process parameter regulator);
(iii)   Detection and recovery measures (e.g., preventive maintenance, alarm systems);
(iv)   Escape and exposure limitation (e.g., emergency exits, automation of dangerous processes);
(v)   Secondary safety barriers (e.g., protective clothing);
(vi)   Damage limitation (e.g., fire fighting, rescue and treatment measures).

The model also points out that redesign of the system is necessary if the desired stable mode cannot be regained and maintained by detection and recovery measures.

10.2.2 Accident Models Focused on Human Behaviour in the Control of Danger

More systematic assessment of human behaviour in the control of danger was made possible by the work of Rasmussen [Rasmussen 80], who developed a taxonomy of decision-making errors by control-room operators from which Reason developed his Generic Error Modelling System (GEMS) [Reason 87]. See also Chapter 2 for a discussion of human errors and the factors which influence them. A model combining GEMS with other ideas of errors in information processing and problem solving [Surry 69, Hale 70] has been developed by Hale and Glendon [Hale 87] — see Figure 10.3.

The model of Figure 10.3 starts from the presumption that some danger is always present. In order to work safely in a dangerous environment, persons at risk must recognise the danger, possess the requisite knowledge and skills, be familiar with the procedures, and be motivated to cope with hazards which need action for their control. The task of the individual is to detect the hazards (input mode), assess the related risks, decide to take appropriate action to reduce the risks (processing mode), and execute this decision accordingly (output mode). For common known hazards in routine activities, the level of individual functioning during hazard identification is typically routine and largely skill-based, as is the execution of the learned risk-control actions (e.g., cornering in a car, the routine shut-down or start-up of a chemical plant). In more complex situations, where there are several options for action, the selection of a correct procedure to control the hazard is normally at the rule-based level. The recognition of new risks and learning about new responsibilities are at the knowledge-based level.

The Hale-Glendon model can be redrawn as an event tree [Thomas 95], so it is possible to plot an accident sequence of events as categories of behavioural errors. Solutions to the identified behavioural problems differ according to the type of erroneous behaviour and the phase of the accident process, but can, in most cases, be embedded in the system under review. The model enables its user to distinguish slips and lapses on the one hand (where,

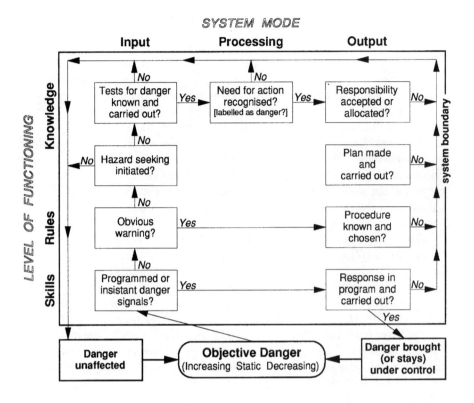

Figure 10.3 *Behaviour in the face of danger*

for example, two routines are confused with each other, or an individual is distracted and loses his place in the sequence), from misdiagnoses (in which, for example, the wrong routine is chosen on the basis of confusing information — say, due to wrong expectations or misleading warnings), and from mistakes at the knowledge-based level (where the mental model or understanding of the task in hand is a basic cause). It also distinguishes between input problems (e.g., failure to notice or understand warning signals, lack of knowledge that something is hazardous), processing problems (e.g., unclear allocation of responsibility, lack of knowledge of preventive actions) and output problems (e.g., clumsiness in carrying out an action, or failure to respond rapidly enough to an emergency).

10.2.3 Accident Models Focused on Safety Management Systems

When analysing the nature and causes of accidents, the intuitive approach has been to concentrate on the physical harm process (see Section 10.2.1) and the actors directly involved (see Section 10.2.2). A third viewpoint, which has developed with the increased emphasis on management responsibility and on self-regulation [Robens 72, DMSA 93], is to focus on the organisational mechanisms and systems which were meant to control and maintain the intended system operations, but which allowed the occurrence of an accident or near miss. Models which reflect this viewpoint start from the idea that the owner of the safety-critical operation can assure full control of the process-related risks. Full control means that all risks have been identified and either have been brought under operational control or have been accepted (as residual risks). In order to realise this level of risk control, safety functions and responsibilities which are linked to departments or specific people within the organisation must be defined. If such a safety function fails with respect to the safety-critical process, the accident process has already begun (in terms of the deviation models in Section 10.2.1) and damage will occur sooner or later.

The Management Oversight and Risk Tree (MORT) [Johnson 80, Benner 85, Hendrick 87] is one of the very few examples in this category of accident models. It includes a model of the safety management system needed to ensure safe operations and encompasses the energy trace and barrier model described in Section 10.2.1. Figure 10.4 shows the top tiers of the MORT chart (see Section 10.3 for a further discussion of it). This tree model has a top-event 'T' and three main branches, the S-, the M- and the R-branch. The model says that losses due to an accident (SA1) either result from failure of the safety management system or fall within the limits of a predicted residual risk (R)

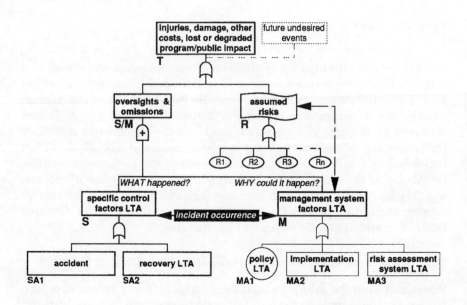

Figure 10.4 The MORT accident model

which has been accepted by appropriate management. The cornerstones of the idealised safety management system are worked out in the programme elements: 'policy' (MA1), 'policy implementation' (MA2), and 'risk assessment system' (MA3). The specific operational control functions which failed to prevent the accident are worked out in the S-branch.

One important problem in this accident model is to establish whether the risks, as they have manifested themselves in an accident, are really those which were known, assessed and accepted as residual risks when the safety-critical operations were started. The initial assumption, though, is that accidental losses are due to failures in the safety management system unless otherwise proved by the evaluation of the accident-specific control factors. The model enables the identification of the relevant parts of the safety management system, as well as specific control functions which need to be improved in order to realise the desired level of risk control.

10.3   ACCIDENT ANALYSIS TECHNIQUES

Criteria for accident analysis models and investigation methodologies were defined in the early 1980s: a model should be realistic, supportive, definitive and satisfying, it should impose discipline, and it should be comprehensive, direct, functional and visible [Benner 83]. An accident investigation methodology should encourage participation, be independent and non-blaming, support personal initiatives to come up with positive suggestions, discover safety and health problems, increase competence, systematically define countermeasure options, support the enforcement of effective standards, encourage governmental bodies to take responsibility, and help test the accuracy of investigation outputs, and be compatible with pre-investigation methodologies. It is outside the scope of this chapter to elaborate on these criteria; see Benner and Hendrick for further reading [Benner 83, Benner 85, Hendrick 87].

On the basis of these criteria, Benner reviewed all the accident analysis approaches used by 18 governmental agencies in the USA [Benner 85]. He found 14 ways of accident modelling and 17 methodologies for accident analysis. Very few methods and investigation methodologies can meet all the stated requirements — from which Benner constructed a table of criteria and a scoring system with a possible 20 points. The assessment results indicate that the modelling of an accident as a process with multiple series of events offers the best approach for the analysis of safety problems (score: 19 points out of 20). The second-best scoring approach (18 points) uses the energy trace model to consider the accident as a process in which harmful energy is released and converted into losses. Of all the methodologies evaluated, events analysis, using the Sequentially Timed Events Plotting technique (STEP) [Hendrick 87] and the MORT analysis system scored best with 18 points out of a possible 20. Fault tree analysis, used as an approach for accident modelling, scored 14 points, and as an investigation method 16 points, the third place overall in Benner's table. Single event and causal factors chaining scored only 1 point with respect to accident modelling and just 4 points when it came to the analysis of the accident. In the next sections, the accident investigation techniques STEP and MORT will be discussed.

10.3.1 STEP: Sequentially Timed Events Plotting

A principal objective of the STEP accident investigation technique is to reconstruct the harm process by disciplined charting of elementary events which were part of, or which contributed to, the accident. This creates a trustworthy 'mental movie' of the accident process: what happened, when and where. Key concepts in STEP are the idea of the deviation of a process from a desired mode of homeostasis into an unstable mode, the concept of actors, which can be either human beings or objects, and elementary event building blocks. During any process, an actor acts on other things or other people who or which, in turn, react or counteract. An event can be thought of in terms of a simple formula: *1 actor + 1 action = 1 event.*

When a process is planned, the events are predicted and defined in advance. The accident process *begins* when an event or change is undetected, undesired, or does not fit within the limits of tolerance (or recoverability of equilibrium) which can be accommodated by all the actors involved without disrupting the process. The accident process *ends* when it has achieved its outcome (harm, loss), whereby the last harmful outcome is linked directly through successive events to the first undesired or unaccommodated change that disrupted the planned process.

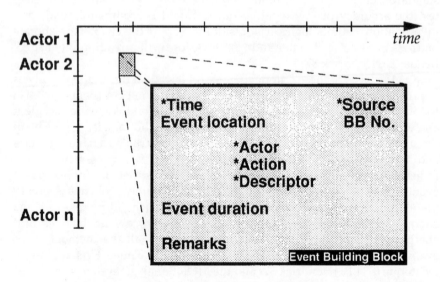

Figure 10.5 STEP worksheet (first stage)

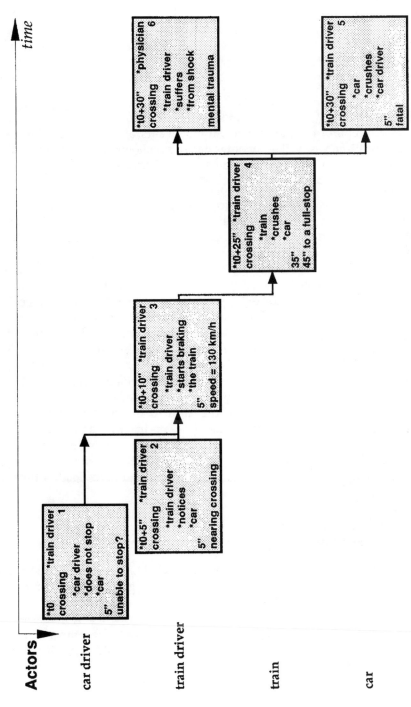

Figure 10.6 STEP worksheet of a fictitious train-car accident

The core activity in a STEP investigation of an accident is setting up the STEP worksheet by documenting all relevant events that occurred during the accident process, using event building blocks. These building blocks are ordered in rows (one row per actor) and in time. Figure 10.5 shows a generic design of a STEP work sheet in the first stage, that is, before the events have been linked together [Hendrick 87].

The second step is to interconnect all events using the following conventions:

(i)   Each arrow should point forwards in time;
(ii)   Except for the starting and ending events, all events should have an incoming as well as an outgoing arrow showing precede and follow relationships between events.

In Figure 10.6, a STEP worksheet of a (fictitious) train-car accident is shown in which these connection rules for linking events have been applied. See Figure 10.5 for the layout of the building blocks. Within each building block t0 (+ x") gives the start time in seconds relative to the beginning of the accident process. In the lower part, y" represents the time duration of the event. The text at the top represents the reference source. The central text gives the event, and the comments at the bottom give additional data.

Thus, the identification of the starting and ending events of the accident process follows out of the plotting procedure. Safety problems are identified in terms of causal links between events which occurred in practice and so propagated the accident, but which could have been prevented, or links to preventive events which were missing, but which also could have diverted the accident process. From there, relevant recommendations can be developed.

STEP can demonstrate clearly the complexity of an accident in terms of the number of actors involved and the nature of interactions. This permits a new definition of large- or small-scale accidents in terms of their complexity. The larger the scale of the accident, the more event building blocks are to be processed, the larger and more complex the STEP worksheet will become and, usually therefore, the more investigators and investigation time are needed.

10.3.2   The MORT Chart:
Main Structure and Use in Accident Analysis

The Management Oversight and Risk Tree (MORT) is a powerful investigation tool. The MORT (single page) chart covers 98 generic safety problems, decomposed into over 1500 basic 'causes' within the safety system of the process owner [Elsea 83]. Combining principles from the fields of management and safety and using fault tree methodology, the MORT tree aims at helping the investigator discover what happened and why. Both specific control factors and management system factors are analysed for their contributions to the accident. People, procedures, and plant hardware are considered, first separately, then together, as key system safety elements [SSDC 92]. A great

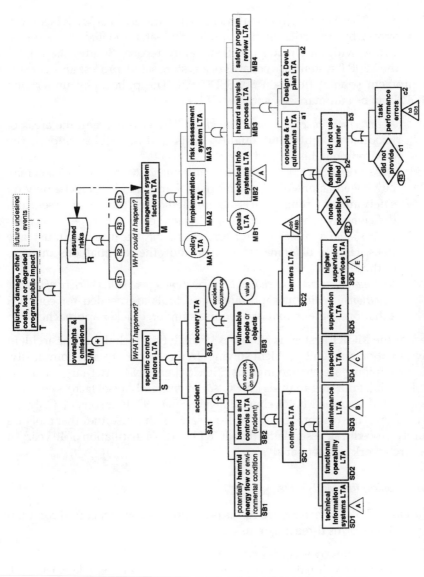

Figure 10.7 The basic management oversight and risk tree (simplified)

advantage of the MORT chart is that it is already drawn out and ready for use. It can be applied whenever a harmful energy flow or hazardous environmental condition can be identified. In the case of a major accident, the investigation can turn into a safety programme review by linking the whole M-branch (on management system factors) with the incident under review (see Figure 10.7).

In the 1970s, William Johnson and his team conceived of MORT as a way to decrease, by an order of magnitude, the safety-related losses of an organisation with an already excellent safety record. Besides the MORT chart, the MORT system also consists of a system safety process and a means of problem solving [Johnson 80, SSDC 92]. The philosophy underlying MORT can be summarised as follows [Elsea 83]:

(i)   *Management takes risks of many kinds.* These risks are in the areas of product quantity and quality, costs and schedules, and of the environment, health and safety.

(ii)   *Risks in one area affect operations in other areas.* Management's job may be viewed as one of balancing risks. For example, to focus only on safety and environmental issues might increase the risk of losses from deficiencies, schedule delays, and related costs.

(iii)   *Risks should be made explicit where practicable.* Because they must take risks, it is helpful for management to know the potential consequences of those risks.

(iv)   *Risk management tools should be flexible enough to suit a variety of diverse situations.* While in-depth analytical tools are needed for complex situations, other situations require simpler, quicker approaches.

Part of the MORT system is a toolbox of analytical techniques for accident investigation aimed at problem solving. In increasing order of complexity, these techniques are: change analysis [SSDC 81], energy-barrier-target analysis [SSDC 85b], fault tree analysis [SSDC 82], events and causal factors charting [SSDC 78b], and the MORT chart analysis [SSDC 92, Johnson 80]. Only the MORT chart has comprehensive built-in (safety) expertise; the other tools are purely procedural. Essentially, the MORT chart is a distillation of 40 years of research work in safety analysis.

(a)   Using the MORT Tree

The basic MORT tree in Figure 10.7 is best described as 'structured common sense'. It performs three major tasks:

*   It identifies oversights and omissions;
*   It enables residual risks to be identified and, where possible, calculated;
*   It enables the analyst, the employer, or whoever is using the chart, to put the effort and the finance where it is most needed.

The MORT chart looks like a fault tree, but rules for going through the tree are modified: in addition to the fault tree logic (AND and OR gates) the user must work not only from top to bottom, as in a fault tree, but also from left to right [Conger 84]. In many branches, a tree event to the left is a prerequisite to the one under review; e.g., in the barrier sub-tree (SC2), it would not make sense to investigate whether a prescribed, specific barrier meets its required specifications if such a barrier could not be applied under the actual pre-incident circumstances.

In incident analysis, the MORT chart says that adverse outcomes or losses are the results of oversights or omissions concerning relevant risks which management should have known of, or the anticipated consequence of explicit acceptance of an identified, assessed risk. This 'accepted' risk should have been evaluated by a person who had authority delegated by management to assume that (residual) risk, on the grounds that it was considered to be:

- Tolerably low (minor) in frequency or consequence;
- High in consequence but impossible to eliminate (e.g., a hurricane);
- Simply too expensive to correct when weighed against the risk consequences.

The overall losses form the top event (T) of the tree. Analysis of the specific-factors branch (S-branch) will show whether any specific relevant risks have been accepted by a management authority or not. If this decision has been made adequately within the context of the specific problem area, the risk is transferred to an oval with code R# in the R-branch just below the top of the tree. If the acceptance decision has not been correctly made, this remains an oversight or an omission and, therefore, indicates a management problem.

Moving down the tree from the top leads to the S/M-branch (oversights and omissions) and thence to the first AND gate. This gate indicates that, in any accident, specific control factors as well as management system factors will always be present in a state which can be assessed qualitatively as 'less than adequate' (LTA). The important principle in MORT is that specific failures alone cannot be responsible for accidents; the possibility of these specific failures should have been predicted by management and, for those risks which have not been accepted, systems should have been put in place to prevent or detect and correct them.

Moving down further and to the left leads to the accident (SA1) and recovery (SA2) branches. Rescue implies fire fighting, prevention of secondary accidents, public relations, and rehabilitation of victims and operations.

After assessment of the specific control factors, the management system factors may have to be evaluated by going through the M-branch. This is only cost effective in the case of very serious incidents with a high loss or loss potential or of a series of incidents with a pattern of similar specific failures.

The structure of the MORT chart below the accident event, SA1, is based on the energy trace and barrier accident model of Figure 10.1. SB1 represents the harmful energy flow or environmental condition, SB3 stands for the vulnerable targets, and SB2 for the barriers between the two as well as for energy flow control functions. The AND gate here says that there will be no accident if at least one of the specific control factors — SB1, SB2 or SB3 — functions according to adequate plans and specifications. Obviously, the safety margin is increased if all three, or at least two, are present and functioning, although, strictly, only one needs to be brought fully under control in order to prevent future accidents with similar underlying problems.

Post-accident recommendations will focus on resolution of the problems discovered at the level below SB1, SB2 and SB3. The main problem areas will be found by analysing the tree event SB2: barriers and controls 'less than adequate'. This applies even when a vulnerable target was not hit by the energy flow, either by chance or because there was, also by chance, no vulnerable target around. The incident then is a near-miss occurrence.

(b)   Specific control factor SB2

The tree event SB2 sits above two branches in Figure 10.7. The first one, starting with SC1, is used to evaluate the factors which condition the actual operations. These are:

- Technical information systems (SD1) with respect to the hazard: information on the work processes, communication systems, monitoring systems, data collection and analysis, triggers for initiating hazard analysis, audit outcomes;
- Facility functional operability testing (SD2), especially of procedures, man-machine interfaces, and operational skills — relevant at initial start-up of a plant or other safety-critical operation, or at the first start-up after a major plant modification;
- Maintenance (SD3);
- Inspection (SD4);
- First-line supervision (SD5);
- Support of the first-line supervisor by higher supervision services (SD6).

All these factors are aimed at the realisation of an undisturbed process as intended in the system design stage, whereby inherent hazards are kept under control by design features as well as by real-time control and adjustments. Problem areas under SC1 may, therefore, point back to the design of the plant. More often, identified problems call for measures regarding internal communication about known risks and risk-control options, as well as about supervisory improvements in daily operations.

The barrier sub-tree SC2 is relevant whenever a harmful energy flow might reach a vulnerable, valuable target, especially a human being. Barriers are things which separate a harmful energy flow or a harmful environmental condition from a vulnerable target, in time or space, either completely or by reducing the energy to a harmless level before the target is exposed to it. This sub-tree should be used for each relevant barrier successively. Thus, the relevant barriers must be identified first. A list of relevant barriers can efficiently be prepared by checking each of the four barrier categories (which have been omitted for clarity in Figure 10.7), which are:

- On the hazard source (a1);
- Between the hazard and a vulnerable target (a2);
- On the target (a3);
- As separation in time or space (a4).

By way of example, we will follow the barrier sub-tree further in order to discuss how a MORT-tree analysis works. The first item is tree event b1, 'none possible'. For the barrier under review, the question is whether this type of barrier was possible in the (pre-incident) situation. The answer to MORT questions is always qualitative and requires a judgement by the investigator. The options are: *adequate, less than adequate (LTA), more information needed*, or *not relevant*. If the barrier under review in the case in question was not possible under the pre-incident circumstances, the answer would be: LTA. This means that a relevant barrier could not be applied where it was needed. This leads to the question of whether the organisation went through the proper process for accepting the risk of not having the relevant barrier in place. This is denoted in Figure 10.7 by the small oval R2 on the left below the tree event SC2-b1. If this risk was accepted in a proper way at the appropriate level of responsible management (R2 = adequate), then we have an assumed risk R2 which can be noted in the R-branch just below the top of the MORT chart. In that case the other sub-tree events b2 and b3 can be ruled out as being not relevant with respect to the barrier under review.

The circular basic tree event b2, 'barrier failed', relates to the issue of whether the barrier worked as intended, i.e., conformed to its specification. If not, then b2 scores LTA, thus indicating a basic 'cause' of the harm process in the specific accident. Next is b3, 'did not use barrier'. The meaning of this can be explored by going one tier deeper: b3 results from either c1, 'did not provide', or c2, 'task performance error'. If the barrier was not provided, then it could not be used (c1 = LTA). Again, the issue of acceptance of an identified risk (R3) by appropriate management is raised for assessment at this point. If the barrier was provided (c1 = adequate), then it might have been used wrongly or not at all (c2 = LTA). This can be assessed in detail by entering the task performance sub-tree section on the chart (SD5-b3), denoted by the triangular transfer symbol beneath tree event SD5. This branch is not detailed in Figure 10.7, but the topic will be discussed in the next section. Note that the analysis of a specific barrier stops when each of the subtree events, b1, b2 and b3, has been assessed. The approach outlined above applies to the whole MORT tree.

(c)   Task performance errors in the MORT chart analysis

The role of human operators is consistently reviewed in relation to the specific task or activity on which they were working, either during the incident or at some earlier time relevant to the incident, for example a maintenance task which was not performed according to plan. If a task performance error is identified, e.g., in relation to maintenance, inspection, supervision, the use of a specific barrier, or during initial recovery action, then an extensive sub-tree section (SD5-b3: not shown in Figure 10.7) is used. This branch of the MORT chart is employed to evaluate underlying management factors, such as the functional situation of the operator, training, the role of the supervisor, motivation (including issues like social climate), time pressure, job interest building, group norms, and physical hindrances. Management system factors underlying task performance errors are analysed in greater detail (for more detailed discussion of the role of such factors see Chapters 7, 8 and 9).

There are seven main problem areas with respect to task performance errors, of which only one, namely 'personnel performance discrepancy' (a sub-problem of 'employee motivation'), permits the user to indicate that deviating behaviour by an individual operator beyond management control was a determinant of the accident. This underlines an important principle within the MORT system, that the analyst should be discouraged from pointing the finger too hastily at the operator as a scapegoat when it comes to task performance errors in a safety-critical system: the bottom line is not to *blame* somebody but to *gain* improved control over safety-critical systems.

(d)   Learning lessons from daily accidents using the MORT chart

The structure of the MORT chart enables its user to derive recommendations to be submitted to management for improved control of the safety-critical process from those tree events which have been evaluated as 'LTΛ' or which lack conclusive data for a sound assessment, i.e., have been classified as 'more information needed'. The problem areas which are directly related to an incident are found by performing an S-branch analysis. Once found, recommendations to management can be generated by moving upwards in the tree (see Figure 10.7). Then:

- SB1 concerns the control of hazard sources (post-design);
- SC1 also concerns the control of hazard sources, but now with respect to design and development, system implementation, maintenance and management of operational use;
- SC2 considers exposure reduction so as to prevent harm or damage to vulnerable people or objects;
- SB3 relates to control of, and evasive action by, a vulnerable person or object.

Recommendations may easily go beyond the span of control of the first-line supervisor. For instance, task performance errors might call for improvements in personnel selection, training, or management concern (vigorous personal action) for safety displayed by top executives.

10.4   THE ISA SYSTEM

The knowledge of management system factors underlying incidents as structured in the MORT chart has been used to build a PC-based support system, ISA (Intelligent Safety Assistant), for operational safety staff. This has been done by:

- Using selected parts of the S-branch, tailored on the basis of adverse outcome potential, i.e., incidents with only minor potential require less detailed assessment than those with more serious potential;
- Providing diagnostic rules reflecting the idealised safety management system from the MORT chart, with the additional inclusion of supplementary diagnostic rules derived from other expert domains, e.g., psychology with respect to skill-, rule- and knowledge-based human errors;
- Generating generic diagnostic reports, which must be interpreted by the safety practitioner in terms of local reality, for feedback to relevant managers.

The ISA system has been tested by local operational safety staff in a large-scale pilot project for the analysis of minor accidents at work. The staff were advisers to management but they had not received any previous training in the MORT system. The project extended across the mining, tobacco, chemical and aerospace industries. Small-scale trials in patient-critical hospital departments are in progress and indicate promising results.

10.4.1 Analysing Minor Accidents at Work

The opportunities to learn about system factors that need improvement for increased safety are more frequent if small-scale accidents and near misses at work are analysed as well as major incidents. In particular, humans in the system play a control role in the processes leading both to deviation and recovery. They make errors and miss opportunities for recovery or improvement because of the way in which the system within which they work is designed and organised. The enterprise has an interest in system factors that contribute to a decrease in the incidence of such human error. Such errors can often be more easily studied in minor accidents or near misses because those concerned are less inclined to cover up their errors or to have forgotten because of the trauma of a major accident.

The assumption is often made, and we make it here, that minor accidents and near misses, if analysed, would give useful information about the weak points in a safety system, comparable to the results of the analysis of major accidents. The theory of Heinrich [Heinrich 80] (see also [Bird 76]) is based on the assumption that the underlying causes are identical. We would not ascribe to such a theory unreservedly, as it ignores the fact that people and organisations in general take more care to avoid major accidents than minor ones. For us, the question is an open one which should be resolved by empirical research. Meanwhile, we place only one clear limitation on the accidents and incidents which it is worth analysing with ISA: care must be taken to ensure that incidents and errors which could never have led to significant harm or losses do not receive too much attention, which would be a waste of resources.

Where the analysis of the 'everyday' incidents is undertaken by line management or operational safety staff, the feedback of lessons to effect improvements in operational safety measures is often good. Additionally, in order to bridge the gap back to the design or redesign of the system, it is vital to provide specific communication channels by involving either operational safety staff in design reviews or design staff in incident reviews. The MORT chart provides structured expert knowledge to do this, but using the full chart for investigation of a small-scale incident is like taking a sledgehammer to crack a nut. For occupational accidents, the use of parts of the specific factors branch is normally sufficient [Conger 84].

Principles of knowledge-based registration and analysis of accident data as outlined above have been implemented in the expert system ISA. The initial trigger for the development was the realisation, during one project involving accidents to stevedores and others in two chemical companies [Hale 91], that identification of accident scenarios on the basis of pattern recognition from an extensive classification of superficial factual data was doomed to fail because of the lack of sufficient numbers of accidents within a reasonable time. As an alternative, the ISA system, based on the MORT chart, has been developed as a tool for the investigation of all incidents with respect to organisational factors.

10.4.2 Description of the ISA System

ISA is designed as a supporting tool for monitoring incidents in daily safety-critical operations in order to provide feedback and improved control of the process (see Figure 10.8). After notification of an incident, the investigator feeds data into ISA, guided by the system. The system determines the type and severity of the incident on the basis of early answers and, from there, navigates only to topics which are relevant to the analysis of the incident. Terminating criteria are based on the type and severity of the incident, as well as on previous answers by the investigator.

Unnecessary questions are avoided. For instance, in the case of a very minor incident (i.e., no injuries or lost time and only negligible economic losses), questions will be asked only as far as the top events at the SB-level (SB1-SB3) in the MORT tree (see Figure 10.7). On the other hand, if the incident is a major one, the system will alert the investigator to initiate an in-depth investigation, going beyond the bounds of the ISA system, which is limited to learning from minor incidents. In addition, further data will be asked for if that is required for other purposes, e.g., for government reporting. Entry data consist of facts describing the incident and judgements by the investigator regarding MORT issues. All data are kept in a database for

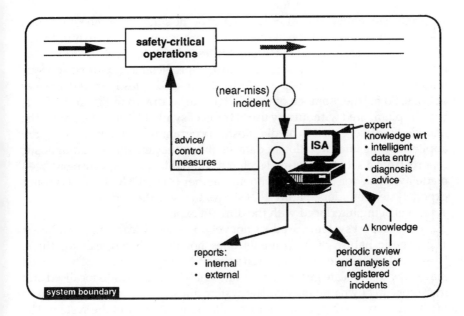

Figure 10.8 ISA system - real-time incident monitoring and feedback

future retrieval.

During data entry, support to the investigator is given by means of context-sensitive help pages. Especially, before a judgement on a specific MORT issue is entered, the help function provides arguments to assess the topic either as adequate, LTA, more information needed, or not relevant. After entry of all relevant data, the in-built analysis module generates a report about the organisational, causal factors underlying the incident, including advice for improvement of the safety management system with respect to the operations under review. The investigator has to translate these analysis results into options for concrete measures, tailor-made for the company as process owner.

ISA also contains a user-access control mechanism which can be used to control navigation through the system (e.g., determining which parts of the program should be entered), database management functions, and a pseudo-language programming environment for the creation and editing of modules to control data entry and to carry out analysis. Besides the report generator, additional functions include a data import-export module and a simple statistical module for frequency tables and two-dimensional cross-tabulations with a 'chi-squared test' utility. Thus, the investigator of daily accidents needs to use only one system for fulfilling all basic incident-reporting needs. The ISA system can handle different incident categories and it produces reports in different formats in order to help the investigator to concentrate on the issues that matter.

In the current version of ISA, one part of the contents of the knowledge base is determined by the navigational programmes which mirror the MORT-tree structure. It includes 'stop' rules to limit the level of depth of the investigation, additional questions regarding predictable traps for erroneous human behaviour, and help pages to the related questions and answer options. So far the S-branch of the MORT tree, as drawn in Figure 10.7, has been implemented without any use of transfer symbols. The other part of the knowledge-base consists of diagnostic programmes in which rules are applied for the interpretation of data, enabling the generation of diagnostic reports from a set of predefined statements. In this way, expertise has been made available for end users who are neither trained MORT analysts nor expert system builders. Figure 10.9 shows the way the end-user of the ISA system is being interfaced with the domain experts.

As an expert system, ISA does not yet possess self-learning capabilities. The generic nature of its contents does not immediately call for them. However, once implemented within a specific company, site or plant, a need may grow to translate generic issues and analysis reports into localised and more specifically named items. The system allows such modifications to be made, but care needs to be taken to maintain the integrity of the system. The generic core of the system enables the validation of navigational and diagnostic rules and of the generated analysis reports at meta-company level, i.e., on the

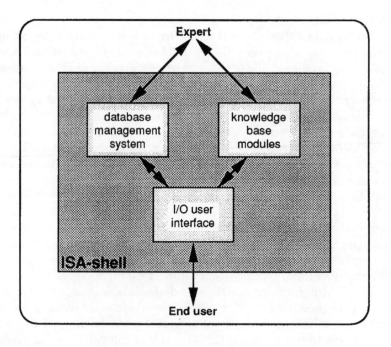

Figure 10.9 *Expert end-user interfaces in the ISA system*

basis of combined incident data from different companies in dissimilar industrial sectors. At this level, effective learning can be organised by system users, supported by experts, in order to improve and extend the knowledge base and other functions of the ISA system.

10.4.3 Example of an Accident Analysis using ISA

In the following paragraphs, ISA is illustrated by way of an example. The example is of a fatal accident which occurred during maintenance of a heating installation in an apartment building. The text below provides a transcription of the investigator's typed input to the system, the system's output following analysis, and a running commentary.

The incident description, as initially available for the ISA analysis, is given in free text format in order to facilitate understanding of the subsequent incident process:

During dismantling work in a heating installation, two employees were busy cutting off a pump with a blowlamp. They got an electric shock because the body of the pump was energised. The electrician switched off the cut-off at the switch board, but the conductors were still visibly connected. At the switch board, the electrical wiring was connected wrongly: the earth line was connected to the live phase. The switch had been installed a few weeks earlier and had never been tested.

The losses due to the incident are entered first. These were:

Results: 1 fatality. Work stopped for 3 days.

The user was then required to indicate the mode of operation of the production process which is normally to be expected at the place of the incident. Here it is a complete shut-down for maintenance. The assessment is used for the generation of diagnostic statements and recommendations. Then the classification of the most relevant hazard must be entered.

Additional incident data:

Production phase: unusual conditions were applicable.

Hazard type: electricity (< 1 kV).

After the investigator has assessed and entered into ISA all the relevant issues, and before the ISA system generates its automated analysis results, he is asked by the system to state his professional conclusions and recommended preventive actions with respect to the incident.

Conclusions by the investigator:

1. *Leads not disconnected.*
2. *Wrongly connected electrical bridge switch.*
3. *No test if current was actually switched off.*

Preventive measures suggested by the investigator:

1. *Check electrical installation.*
2. *Retrain employees.*
3. *Provide better supervision of infrequently performed tasks.*

In this way the professional opinion of the ISA-user can be judged against the analysis outcomes generated.

The following ISA elements were activated in the accident assessment:

- The severity of the incident implies that a medium set of questions should be asked.
- The mode of operation of the heating system was set to a complete shut-down during the maintenance activities.
- Non-functional presence of a hazard, i.e., electricity at the heating installation, has been established.
- Maintenance was the only activity and contributed as such to the incident.
- Task performance errors have been scored during maintenance execution (during the previous activity on the electrical bridge switch).
- Task performance errors regarding the use of the cut-off switch (barrier) have been identified.
- Regarding training or reinstruction, it has been noted that both employees were competent, but did not do this task regularly.
- In such a case, the supervisor's role is to increase workers' alertness, e.g., by oral briefing, before they start the job; work site supervision was LTA in this respect.
- With respect to this maintenance job, the technical information system of the maintance work contractor is to be assessed as LTA.
- A barrier (i.e., cut-off switch) that is not functioning according to specification is a trap for workers.

Based on these diagnostic issues, the following analysis results were generated by the ISA-system:

REPORT OF AN ACCIDENT category: incident with serious loss potential

- With respect to the hazard, a problem has been identified with the technical information system. A review of the following issues is recommended: technical information (knowledge and/or communication), monitoring systems, data collection + analysis, hazard analysis process (HAP) triggers, safety system appraisal.
- The incident appears to be linked with maintenance activities. A review of the maintenance plan and execution is called for.
- Non-functional presence of a relevant hazard source has been demonstrated. The presence of the identified, non-functional conditions calls for closer attention with respect to 'housekeeping', permit-to-work or access procedures, and supervision of adequate application.
- First-line supervision plays a key role in work-site safety. The following supervisory problem area(s) has been found:
  - the first-line supervisor did not adequately fulfil his supervisory duties. This normally indicates a lack of help and training for the supervisor, lack of time, inadequate plan for responsibility transfer, LTA detection and correction of hazards, and/or performance errors. This supervision problem should be looked into;
  - supervision on appropriate task execution during rarely occurring

tasks or task circumstances is called for.

- Periodic re-instruction or re-training is recommended for tasks which are seldom performed by the employee in question.
- Frequently occurring task performance errors indicate a general management problem regarding:
  - the analysis of job- or task-related risks;
  - the selection, training and motivation of personnel;
  - work-site supervision of personnel.

  Further analysis of this problem area is called for.
- Task performance errors have been scored with respect to:
  - maintenance;
  - the use of barriers.

The categories of loss potential are based on the assessment of a combination of the human loss potential (from no injury, through permanent injury or sickness, to many fatalities) and the economic loss potential (from none to disastrous). This assessed loss potential is not the same as the actual losses suffered: it represents the possible range, showing the maximum that realistically could occur.

ADDITIONAL DIAGNOSTIC CONCLUSIONS

- The technical information system does not meet the work process needs. Assess priorities for improvements (see MA3-MB2).
- Planned barriers did not function according to specifications. This is a serious quality assurance failure which must be looked into immediately.

The additional diagnostic conclusions are generated as a summary of the main report to highlight safety problems which require the immediate attention of higher management.

When this sort of computer-generated diagnosis was presented to those concerned in a pilot project (see Section 10.4.4) their conclusion was that it was very relevant to the incident and revealed a greater number of fundamental problem areas for improvement than had the recommendations of the original investigators. The text above also illustrates that the report which is generated needs to be translated and made more specific before it can be presented to people, such as managers, staff or operators, who are not familiar with the concepts of the MORT tree.

10.4.4 Trial of ISA in the ARA Project

In 1993, a large-scale pilot project, ARA (Accident Registration and Analysis), built around ISA, was launched in Poland as part of the technology transfer programme between the Netherlands and Poland, with the collaboration of government, the labour inspectorate, a trade union, companies and safety institutes. The main objective of the ARA project was to develop a consistent and unified accident registration and analysis system for preventive purposes at company level. The participating companies were from several industrial sectors: mining, tobacco, chemical and aerospace. During the project, their safety staff and those of the labour inspectorate learned to use ISA.

The field trials showed that in each of the companies taking part the ISA system led to improved quality of the post-accident recommendations made by occupational health and safety staff, thus increasing their professional image. It also showed that the recommendations were produced in less time than previously. Typically, the time needed for entry of one small-scale accident using ISA is 1–2 hours, and in this time not only were diagnostic reports on safety problems available for transformation into concrete options for management action, but also statutory and other compulsory reports were prepared. In the existing Polish situation, without ISA, the time needed to generate only the required statutory accident reports averaged 5–7 hours. Thus, during the pilot project, improved efficiency in the use of resources on the investigation of minor accidents has been realised.

At the national level, data from the different companies are compatible, because the same knowledge base has been applied, so it will be possible to create a database of cases combining verified accident data from all companies. Such a case base is necessary for the validation and improvement of the knowledge base in ISA, which could then be redistributed as an updated version of the system, enabling the sharing of improved expert knowledge. In order to facilitate and steer such incremental development of the ISA system, an ISA User Club Poland has been founded. Supported by scientific institutes, this will control future releases of the ISA system in Poland.

ISA's future development needs, to make it more flexible and user-friendly, include the following. First, more experienced ISA users want to be allowed to shortcut the entry of known data of frequently occurring incidents of a specific type. Second, minor incidents at work will normally be perceived from angles which do not directly match those of the MORT-based knowledge framework of the ISA system. For instance, it might be easier for a supervisor, an operator or an inspector to start an incident investigation with the identification of an operator's error which has led to an operational failure than to identify the relevant harmful energy flow or a hazardous condition to which a vulnerable person or object has been exposed and which needs to be investigated. If such different perspectives of ISA use exist, solutions for the perceptional problems have to be found, e.g., by implementing user-dependent navigation schemes which start from the user's framework of incident perception before the user is guided into the core of ISA.

10.5   DISCUSSION

In Section 10.3, STEP and MORT were introduced as the two most powerful examples of accident investigation techniques, the latter offering a content-laden and the former a content-free framework. With these techniques, incidents can be analysed effectively in order to provide the feedback loop needed for the monitoring and continuous improvement of a safety management system. However, general safety staff and managers need the support of built-in expertise on the organisational control factors which underlie accidents in safety-critical operations, in order to learn from them. Minor incidents at work can reveal safety problems if they are analysed in the right way, but analysis requires the discipline imposed by a consistent framework that aims at improved risk management. The MORT tree provides this without any prerequisites with respect to the way in which the process owner has organised his safety-critical operations.

Other content-laden techniques exist (e.g., Root Cause Analysis [SSDC 89] which has been derived from the MORT chart) but they offer a less consistent framework when it comes to automated generation of analysis outcomes and recommendations to management at relevant levels. The development of ISA and its field trial in the ARA project have shown that the MORT structure, coupled with insights from cognitive theory, can be translated into a series of diagnostic rules, which can help the non-expert to identify the relevant features underlying an accident and to home in on the possible weaknesses of the safety management system. Even one accident can be used to generate these signals, which alert the organisation to the need to look more closely at particular elements of the system and examine their overall functioning. This is an important breakthrough in changing from an attitude of allocation of blame and monocausal thinking to one of searching for generic lessons at the system management level.

The ISA system is designed to accommodate the user's needs in the reporting, to internal as well as external parties, of different kinds of anomalies in daily operations. Once the reporting criteria and formats have been defined, the system can be tailored to local needs without jeopardising its generic core. Thus, the end-user only needs one anomaly registration system to handle all kinds of incidents, without having to bother about the classification of the anomaly. The necessary reports and paper work are generated by the system, so the user's time can be devoted to answering the questions posed by the system according to the diagnosed type and severity of the anomaly. In practice, this may lead to a shortcut of existing administrative notification procedures — which are often time consuming and seldom provide useful feedback to the daily operations.

In addition, the system produces a diagnostic report about problem areas in the safety management system of the enterprise. This is designed to appeal to the professional judgement of supervisors and higher management and to be used as feedback for improving operational control. By its nature, it closes the door on scapegoat hunters because the recommendations address organisational controls of shop-floor operations. Thus, after implementation of the recommendations, operators are enabled to function better, knowing that adequate organisational controls are in place to prevent human errors from leading to unacceptable process deviation which may result in an accident.

In practice, the likelihood of acceptance of the system's diagnostic output is boosted by the disciplined and objective way in which the system asks questions, regardless of the persons who might be involved in the incident. Further, even after one problem has been identified during data entry, the system does not ignore other types of potential safety problems.

System users at different levels within the organisation can be allocated to using the system. For example, they may be at shop-floor level where operational anomalies become visible and where incidents occur or at higher line-management levels where decisions are to be made and where supporting staff may be available. The ISA system demands a controlled, multi-staged entry of incident data. Thus, for instance, the shop-floor worker could feed the system with the initial descriptive facts such as notification data, a trained staff member could enter the answers to system questions, and a dedicated risk manager could translate the diagnostic reports of the system into decision options for managers at appropriate organisational levels.

The system is constantly being improved as the result of observations by its designers and feedback from its users.

10.6   ACKNOWLEDGEMENT

The authors wish to thank Dorian Conger [Conger 84] and Ken Elsea [Elsea 85] for the introduction to MORT ten years ago, as well as Allan St John Holt for making us aware of this powerful approach in 1980 [Holt 80].

REFERENCES

[Benner 83] Benner L Jr: *Task report no. 1: Accident Models and Investigation Methodologies employed by U.S. Government Agencies in Accident Investigation Programs*. OSHA, Washington DC, 1983

[Benner 85] Benner L Jr:*Rating Accident Models and Investigation methodologies*. In: Journal of Safety Research, 16(3) 105-126, Fall 1985

[Bird 76] Bird F E Jr and Loftus R G: *Loss Control Management*. ILCI, Institute Press, Loganville GA, 1976

[Boeing 70] *Sneak Circuit Analysis Handbook*. Boeing Aerospace Company, Space Systems Division, Doc. No. D2-118341-1, Houston TX, 1970

[Conger 84] Conger D S: *MORT Course Notes*. Leuven (B)/Woodstock GA, 1984

[DMSA 93] *Memorie van Toelichting: Aangepaste Arbowet 1993*. (Explanatory Memorandum: Revised Working Conditions Law 1993.) Dutch Ministry of Social Affairs. Staatsuitgeverij, Den Haag, 1993

[Elsea 83] Elsea K J and Conger D S:*Management Oversight and Risk Tree*. The Risk Report, Vol. VI, No. 2. International Risk Management Institute, 1983

[Elsea 85]   Elsea K J: *Accident Investigation Workshop*. Course notes. Delft (NL)/ Woodstock GA, 1985

[Ferry 88]   Ferry T S: *Modern Accident Investigation and Analysis*. (2nd edition) John Wiley & Sons, New York, 1988

[Haddon 66]   Haddon W Jr: *The Prevention of Accidents*. Preventive Medicine, Little Brown, Boston, 1973

[Haddon 73]   Haddon W Jr: *Energy Damage and the Ten Counter-Measure Strategies*. Human Factors Journal, August 1973

[Hale 70]   Hale A R and Hale M: *Accidents in Perspective*. Occupational Psychology, 44, 115-122, 1970

[Hale 87]   Hale A R and Glendon A I: *Individual Behaviour in the Control of Danger*. Elsevier Science Publishers BV, Amsterdam, 1987

[Hale 91]   Hale A R, Karczewski J, Koornneef F and Otto, E: *IDA: an interactive Program for the Collection and Processing of Accident Data*. In Schaaf T W et al (eds): Near Miss Reporting as a Safety Tool. Butterworth-Heinemann, 1992

[Heinrich 80]   Heinrich H W, Petersen D and Roos N: *Industrial Accident Prevention: A Safety Management Approach* (5th edition). McGraw-Hill, New York, 1980

[Hendrick 87]   Hendrick K and Benner L Jr: *Investigating Accidents with STEP*. Marcel Dekker, Inc., New York and Basel, 1987

[Hirschfeld 63]   Hirschfeld A H and Behan R C: *The Accident Process*. Journal of the American Medical Association, 186, 193-199, 19 October 1963

[Holt 80]   Holt A J St John: *The Integration of System Safety Principles into UK Education Programmes*. Proceedings First European System Safety Conference, 3SF, Bordeaux, 1980

[ISO 94]   *Quality Management and Quality System Elements. Guidelines*. ISO 9004, CEN, Brussels, 1987

[Johnson 80]   Johnson W G: *MORT Safety Assurance Systems*. Marcel Dekker, Inc., New York and Basel, 1980

[Kjellén 83a]   Kjellén U: *Analysis and Development of Corporate Practices for Accident Control*. OARU, Royal Institute of Technology, Stockholm, 1983

[Kjellén 83b]   Kjellén U: *The Deviation Concept in Occupational Accident Control Theory and Method*. OARU, Royal Institute of Technology, Stockholm, 1983

[Kletz 92]   Kletz T: *HAZOP and HAZAN: Identifying and Assessing Process Industry Hazards*. Hemisphere Publishing Corporation, Bristol, USA, 1992

[MacDonald 72]   MacDonald G L: *The Involvement of Tractor Design in Accidents*. Research Report 3/72, Dept. of Mechanical Engineering, University of Queensland, St Lucia, USA, 1972

[McFarland 67] McFarland R A: *Application of Human Factors Engineering to Safety Engineering Problems.* National Safety Congress Transactions, Chicago, 1967

[Rankin 73] Rankin J P: *Sneak Circuit Analysis.* Nuclear Safety, Vol. 14, No. 5, 1973

[Rasmussen 80] Rasmussen J: *What can be learned from human Error Reports?* In Duncan et al (eds): *Changes in Working Life,* John Wiley & Sons, UK, 1980

[Reason 87] Reason J T: *A Framework for Classifying Errors.* In Rasmussen et al (eds): *New Technology and Human Errors.* John Wiley & Sons, UK, 1987

[Robens 72] Lord Robens: *Safety and Health at Work: Report of the Committee 1970-1972.* HMSO, London, 1972

[Schaaf 91] Schaaf T W van der, Lucas D A and Hale A R (eds): *Near Miss Reporting as a Safety Tool.* Butterworth-Heinemann, 1991

[SSDC 75] Nertney R J, Clark J L and Eicher R W: *Occupancy-use Readiness Manual.* ANC, ERDA-76/45-1 SSDC-1, Idaho Falls, 1975

[SSDC 78a] Nertney R J:*Management Factors in Accident and Incident Prevention.* DOE76/45-13 SSDC-13, Idaho Falls, 1978

[SSDC 78b] Buys J R and Clark J L:*Events and Causal Factors Charting.* EG&G, DOE76/45-14 SSDC-14 (Rev. 1), Idaho Falls, 1978

[SSDC 81] Bullock M G: *Change Control and Analysis.* DOE-76/45-21 SSDC-21, Idaho Falls, 1981

[SSDC 82] Crosetti P A: *Reliability and Fault Tree Analysis Guide.* DOE-76/45-22 SSDC22, Idaho Falls, 1982

[SSDC 85a] *Accident/Incident Investigation Manual: Second Edition.* SSDC-27, Department of Energy (DOE), Washington DC, 1985

[SSDC 85b] Trost W A and Nertney R J: *Energy Trace and Barrier Analysis.* SSDC-29, Department of Energy (DOE), Washington DC, 1985

[SSDC 89] Cornelison J D: *MORT Based Root Cause Analysis.* SSDC WP-27, Idaho Falls, 1989

[SSDC 92] Knox N W and Eicher R W: *MORT Users Manual. For Use with the Management Oversight and Risk Tree Analytical Logic Diagram.* DOE-76/45-4 SSDC-4 (Rev. 3), Idaho Falls, 1992

[Surry 69] Surry J:*Industrial Accident Research: a Human Engineering Appraisal.* University of Toronto, Department of Industrial Engineering, Toronto, 1969

[Thomas 95]Thomas M T: *An Analysis of the Causal Factors Leading to Fatal Accidents in Agriculture in Great Britain.* Journal of Health and Safety, No. 12, 1995

11

Procedural violations - causes, costs and cures

11.1   INTRODUCTION

The costs to international industry of incidents arising from unsafe working are significant. For example, the UK Health and Safety Executive (HSE) has estimated that the annual cost to UK industry from such incidents is between £11 billion and £16 billion.

Studies by the HSE's Accident Prevention Advisory Unit and others have also suggested that human error is a major contributory cause of 90% of accidents, of which 70% could have been prevented by management actions. Further, it is estimated that most of these costs are uninsured — total accident costs are typically between 8 and 36 times the insured costs. The effects of accident costs on industry can therefore be crippling. A study has shown that the accident costs of one industry represented 14% of its potential output. In another industry accident costs were shown to represent 37% of its profits. Outgoings on this scale are staggering and it is management's responsibility alone to reduce them.

Whereas most industries have gained a good reputation for controlling physical risks, relatively few have applied the same degree of control to the risks associated with human error. Despite the attention paid to research into factors influencing human error, there is growing criticism over the balance between research which increases our knowledge of causes and that which enables the practical application of that knowledge [Meister 92]. Although much is published on the theories underlying human error, very little published work addresses the practical means of increasing human reliability.

Human error study has usually focused on the inadvertent or accidental errors which are made by operators, for example failing to notice an unacceptable trend on an indicator in a control room (see Chapter 2). There is, however, a class of human error that has largely escaped attention, and yet has a most significant potential impact on accidents and lost production. This is a human failure type called 'violations'.

As violations can have a major effect on unplanned losses, the challenge is for industries to achieve the same success in controlling the human risk (from both inadvertent and intentional errors) that has been achieved in the control of the physical risks.

11.1.1   What are Violations?

Violations can be defined as any deliberate deviations from the rules, procedures, instructions or regulations introduced for the safe or efficient operation and maintenance of equipment. This applies to all levels, from an operator through to high-level management.

Violations can occur for many reasons and are seldom wilful acts of sabotage or vandalism. The case of someone driving a car through a red traffic light is clearly a traffic violation, but it may have occurred as a result of the driver being distracted by children shouting in the back seat or trying to re-tune the radio. In this instance the human failure is unintentional and not, therefore, a violation in the human-factors sense. Violations are restricted to deliberate breaches of rules and procedures, and would include a driver deliberately crossing red lights to get home early to watch golf on the television. In this instance the driver may be assumed to have made a conscious decision based on the perceived probability of the risk of accident and of detection and its consequences, as opposed to personal gain.

What can appear, at least initially, as a simple violation can hide complex causal factors. Within the area of deliberate rule violations come a number of subtle variations. Some rules, such as the 70 mile per hour speed limit on motorways, are regularly broken by a large proportion of the population, in this case drivers. Other rules are only broken under certain circumstances — because the rule is, or is perceived as being, impractical under those conditions. For example, it may be a safety requirement that a person should only enter a vessel when a second person is positioned outside in case of problems. Because of a practical difficulty in getting a second person allocated to the job on certain shifts, a maintenance worker may decide to take the risk and work alone. He may also believe (correctly or otherwise) that his boss would expect him to work in this fashion under these circumstances.

Management therefore requires a degree of flexibility in the approaches that need to be adopted to reduce such behaviour [e.g., Harper 91].

11.1.2 The Significance of Violations

There is a considerable body of evidence to suggest that violations are a relatively frequent occurrence, both in work situations and in general life. Many accidents and injuries arise partially or wholly from violations, such as removing guards on machinery. In addition, some recent disasters have shown that violations can become the normal way of working, e.g., the work practices and lack of supervision involved in the Clapham Junction rail tragedy [Hidden 89]. There is also a growing concern about the potential impact of rule violations in the 'high-hazard' industries — see for example [Munipov 92] and [ACSNI 93].

The exact proportion that violations contribute to the total costs of behavioural accidents and other losses varies between industries. Certain industries have, however, estimated that violations of rules and procedures were a significant contributor to approximately 70% of their total accidents.

The safety impact of violations (e.g., compensation payments and lost time through injuries) is relatively easy to see from the available evidence, but there are other less obvious impacts. These include lost production and the resulting poor quality of work, perhaps with reworking costs. There can also be deliberate deviations from maintenance rules and procedures which incur additional costs, for example as a result of ineffective maintenance and subsequent poor equipment reliability. In addition, violations can lead to plant being damaged.

The cost implications of violations can therefore be considerable and are frequently not fully appreciated by either the workforce or the management. Such a lack of appreciation may in itself be a factor which increases the potential for violations and causes management to overlook the relative ease with which they could be overcome.

11.1.3 Direct Motivators and Behaviour Modifiers

There is a variety of factors which can influence a person's decision knowingly to break rules [Mason 92] and these can be considered at two levels. The first is of factors which directly motivate the operators, maintenance craftsmen or line management to break agreed rules or procedures. These factors will be called the 'direct motivators'. The second level is of the supplementary factors which could increase or reduce the probability of any individual deciding to commit a violation. These will be called the 'behaviour modifiers'. For example, being late home may be a direct motive for wanting to break a speed limit when driving, but the absence of any effective detection by the police in that area may be a behavioural modifier which increases the probability that the violation would occur.

A list of the more important direct motivators and behaviour modifiers are given in Table 11.1 and the explanation beow. It should be noted that the direct motivators and behaviour modifiers are not mutually exclusive but may overlap. For example, financial gain for others could also be combined

Table 11.1 *Direct motivators and behaviour modifiers*

The Direct Motivators

Making life easier

Financial gain

Saving time

Impractical safety procedures

Unrealistic operating instructions or maintenance schedules

Demonstrating skill and enhancing self-esteem

There could also be:

Real and perceived pressure from the 'boss' to cut corners

Real and perceived pressure from the workforce:
(a)   to break rules
(b)   to work safely

Behaviour Modifiers

Poor perception of the safety risks

Enhanced perception of the benefits

Low perceptions of resulting injury or damage to plant

Inadequate management and supervisory attitudes

Low chance of detection due to inadequate supervision

Poor Management or supervisory style

Poor accountability

Complacency caused by accident free environments

Ineffective disciplinary procedures

Inadequate positive rewards for adopting approved work pract

with peer group pressure. A low chance of detection could also result from poor accountability. Ideally, it should be possible to identify those direct motivators and behaviour modifiers which have contributed to major incidents. This is, however, difficult, as most inquiries fail to elicit the deeper causal factors which lie behind the decision to break rules and procedures. In many cases the underlying causes can only be speculated upon.

For example, a ferry's bow door which if left open could cause an accident, as at Zeebrugge [Steel 87], could knowingly be left open for a number of reasons. The direct motives could range from the increased air flow clearing the exhaust fumes quickly (making life easier for the crew) to allowing quicker ferry turn-round times (saving time for the company and getting the job done more quickly). There could also be the direct motivator of real or perceived 'pressure from the boss' for short cuts to speed up the turn-around time. The behaviour modifiers could include inadequate perception by the crew of the risks of leaving the door open and a low chance of detection — especially if the captain had no facilities on the bridge to monitor the situation.

The Chernobyl incident provides a good example of inadequate identification and reporting of likely underlying causes behind the explosion at the nuclear power plant. A review by the HSE of the official Russian [USSR 86], and numerous other, reports concluded that it was noteworthy that in spite of the pre-eminence of human error in the accident, not one report gave a substantial analysis of the possible underlying causes of the errors.

It perhaps comes as no surprise to discover that violations are highly susceptible to management influences — since many of the underlying causes of violations are created, often inadvertently, by management itself. Management is therefore in an ideal position to influence the behavioural modifiers, as well as some of the direct motivators.

11.2   DIRECT MOTIVATORS

11.2.1 Making Life Easier

Making life easier for themselves is a prime motivator for many people to cut corners. An unwillingness to carry the appropriate tools and safety equipment necessary to perform all tasks in the agreed manner is a common problem. Equipment which is badly maintained, or a lack of suitable equipment, could result in alternative and unsafe procedures being adopted. In fact, the Health and Safety Executive has shown that failure to set up and use safe methods of work is a significant cause of maintenance-related accidents [HSE 85]. Better design which makes equipment easy to maintain plays a major role in encouraging safe working practices.

For example, a routine component inspection may require the maintenance worker to crawl under a machine and work in a very confined space, or it may require spillage clearing or the removal of several heavy covers. With such 'added' difficulties on top of the 'required' task, the likelihood of this component inspection being neglected may be expected to be much higher than if the component being inspected was readily at hand and access to it was easy.

11.2.2 Financial Gain

Bonus schemes are often linked purely to the throughput of product. The costs of poor workmanship on quality or costs resulting from incidents or accidents are usually not included. Anything which speeds up a job will result in financial gain for the individual. Such bonus schemes can therefore be seen to encourage rule violation, especially in those situations where supervisors are not present. In addition, such financial gain by individuals often results in financial loss for the company due to poor quality, customer dissatisfaction, and perhaps accidents, though they may only be recognised in the context of increased quality awareness.

11.2.3 Saving Time

Finishing a task early is an important motivator, especially if the task is unpleasant or if its early completion gives a person more time to spend on a rewarding aspect of the job.

A tyre fitter working on a lorry tractor unit, at a roll-on-roll-off ferry terminal, was crushed as the tractor axle fell off the supporting jacks. He had been using a sledge hammer on the wheel disc and had neglected to lock the vehicle wheels, use an axle stand, or lower the trailer legs. The fitter did not want to delay the driver's departure [HSE 85].

11.2.4 Impractical Safety Procedures

The deliberate breaking of safety rules often occurs as a result of the rules being genuinely impractical or perceived as being impractical. In such instances the motive to violate may be to help the company to get the job done, as opposed to any personal gain.

Safety procedures are often continually amended to prohibit actions that have been implicated in recent accidents or incidents. The resulting effect is that safety rules can become increasingly restrictive [Reason et al 95].

Furthermore, it is all too easy for management hastily to introduce rules and procedures, perhaps following an incident, without the full practical implications being considered. In such circumstances, rules and procedures can be thought of as being introduced simply to 'protect the backs of management'. For example, a study of Dutch railways showed that 80% of the workforce considered that the rules were mainly concerned with pinning blame, and 95% thought that work could not be finished on time if the rules were all followed (as reported in [HSE 95]). None of the respondents could remember ever having referred to the rules in a practical situation.

Such overspecification can result in permitted actions which are far less than those necessary under anything other than ideal conditions [Reason et al 95]. Rule violations will often therefore be necessary, or be perceived as necessary by the workforce.

Although work practices seem easy to write, understanding and following them can cause difficulties. Some work practices have been shown to be impossible to adopt while others have not been fully understood. Maintenance workers may be asked to report back to management whenever they consider any working methods to be difficult to comply with, but in practice many workers would attempt to get the job completed using improvised methods.

For example, a code of practice stated that no person should enter a bunker unless all material adhering to its sides had been removed above the point where the work had to be performed [Rushworth et al 86]. This was a requirement to prevent vibrations causing adhering material to fall onto the workers below. Despite the obvious importance of this requirement, men ignored it. It was only when the question *why?* was asked that it became apparent that there were no practical means of satisfying the stated requirement. Men therefore chose to 'risk it' in order to get the job done. Changes in bunker top design have since been introduced which allow workers to free adhering material much more easily than was originally possible.

Likewise, work plans must not be seen as inviolate. Such an attitude encourages the workforce, if faced by a problem, to improvise rather than to inform management. In such situations it should generally be recommended that management and the workforce collectively develop better, more realistic plans or codes of practice to cope with events as they arise. The following guidelines are appropriate:

- Safety practices should be reviewed regularly to check that they are practical under all foreseeable conditions and achievable with the equipment which is normally available;
- Safety practices must be well communicated and understood throughout the workforce — with one of the best ways of achieving this being to involve the workforce in writing the rules and procedures;
- Working practices should be rigorously monitored by supervisors;
- The workforce should be encouraged to approach management whenever the safety procedures are difficult to implement.

11.2.5 Unrealistic Operating Instructions or Maintenance Schedules

Operating instructions or maintenance schedules which place excessive demands on the users are likely to be rejected and alternative improvised methods adopted. For example, the 'pre-shift' checks specified by equipment manufacturers can be excessive. In one instance the pre-shift checks for a piece of equipment were estimated to take a fitter longer than the shift itself! Such unrealistic procedures can cause a gradual unwillingness of the workforce to adopt certain working practices. Over time the erosion of the rules gradually becomes accepted practice by most workers. In such instances, the unofficial work practices come to be regarded as the norm and may even be taught to new recruits. The design of procedures and operating instructions are discussed in detail in Chapter 7.

Breakdowns are often found to be associated with, or to result directly from, a lack of routine maintenance or an abuse of equipment, rather than from poor engineering. During breakdowns, a strong motive can exist to accelerate repair in order to return to production. Any corners which are cut can lead to increased safety risks as well as risks of subsequent plant damage. In such circumstances, improving equipment reliability may not be the most cost-effective route to overall system reliability. Improvements are likely to have more impact if management direct their attention towards the human reliability aspects of routine and breakdown maintenance operations by ensuring that procedures are not only practical but also understood and accepted by the workforce.

11.2.6   Demonstrating Skill and Enhancing Self-esteem

There is some evidence that people's behaviour is influenced by social image, and that the adoption of *ad hoc* procedures as a demonstration of their skills is a direct motivator. This motive is therefore likely to be strong with groups of individuals who consider themselves as possessing higher than normal levels of work skills, especially where they are employed in jobs with low inherent demands on their abilities (see Section 11.5).

11.2.7   Real or Perceived Pressure from the 'Boss' to Cut Corners

Repeated management expressions of concern over production or machine availability can easily create an impression that they are more concerned with production than with safety. Operators and maintenance crews alike may therefore think that management would want them to cut corners and take risks if it speeds up the job. Of course, this may in fact be the case. In order to 'look good' to management (e.g., to get considered for recognition, promotion and overtime) some workers are willing to violate rules.

This problem is probably more widespread than commonly realised. Several large groups of electricians were recently asked to complete a risk and hazard awareness questionnaire [Mason et al 94]. As part of one question they were asked to agree or disagree with the statement, 'Management never put me under pressure'. Up to 93% of the recipients in certain work groups disagreed. The same groups of electricians were also asked to agree or disagree with the statement, 'Management expect us to stick to safety rules'. Up to 76% of the members of the groups agreed.

In order to stimulate a safety culture, management should:

- Consistently show commitment to and an active interest in safety issues at all levels in the organisation;
- Continually be on the look-out for any situation which is not up to the expected safety standard and then always undertake to remedy the situation or, in exceptional circumstances, to discipline individuals concerned;
- Praise individuals when jobs have been completed in the safe manner

(this has been repeatedly shown to be a strong motivator to comply with safe practice).

11.2.8 Real and Perceived Pressure from the Workforce

(a) To break rules

Repeated studies [for example, Levine et al 76] have shown that individuals are strongly influenced by the behaviour of their workmates, even if they realise the behaviour is unsafe. There is a strong need to be seen by the group to fit in with its methods and values. This peer pressure to fit in (which may be either explicit or implicit) is likely to be especially strong for new young entrants into teams or during temporary replacements as a result of absenteeism.

In the attitude survey mentioned above, the same groups of electricians were asked to comment on the statement, 'Workmen never put me under pressure'. Between 73% and 91% of the members of the groups disagreed with the statement. Pressure from members of the work groups and from the rest of the workforce who may have interests in keeping production going, for example to gain bonuses, should not be overlooked by management.

Newly qualified operators, craftsmen and apprentices should ideally be placed only with teams with good safety records

(b) To work safely

The above direct motivators can cause people knowingly to adopt working methods which break agreed working practices. The same direct motivators could encourage or motivate a person to comply with the agreed safe working methods.

By emphasising that violations of safety rules by other people are likely to have a direct influence on their own safety, peer pressure can be exerted by the group to make individuals work safely. Groups may be set safety targets and goals by which performance is measured — and there is real pressure on indivduals not to 'let the side down' in their achievement. As an example, giving non-electricians some basic awareness training in the safety risks associated with electrical maintenance may provide positive peer pressure. An electrician may be less 'comfortable' breaking safety rules by working 'live' on equipment in a potentially gaseous atmosphere if he knows that nearby workmen would be able to identify that he is putting his life and theirs at risk. Equally, the nearby workmen would be less likely to put pressure on the electrician to cut corners to reduce down-time if they were more aware of the risks.

The disapproval of team members increases personal discomfort for those not complying with the recommended safe practices. One means of achieving this is to increase the visibility of a violation of safety rules.

Workmen should be able to determine whether other workmen are in breach of safety procedures and be made aware of the possible risks to themselves. For example, colour coding for approved lifting equipment can be used to increase the visibility of men using 'out-of-date' or inappropriate equipment

11.3   BEHAVIOUR MODIFIERS

Direct motivators are factors which may influence behaviour. The probability that a worker will knowingly deviate from the agreed procedures is affected by the presence or absence of the following behavioural modifiers.

11.3.1   Poor Perception of the Safety Risks

An individual often consciously balances the risk against the benefits before deciding to commit a violation. It is therefore instructive to determine how an individual determines the degree of risk, especially when faced with new experiences.

If operators perceive the risks to be less than they really are, they may be willing to take a risk which would normally be considered unacceptable. Risk perception tends to be influenced both by an individual's own experience and by formal and informal training. Real-life experiences impact on an individual's perception of risk and therefore act to modify behaviour. However, living through the experiences of major incidents will (it is to be hoped) be rare and therefore cannot be relied on to show up examples of risk. Where such experiences have been gained, the effect on risk perception and associated behaviour in the long term appears to diminish over a 3-5 year period. The ability to synthesise and evaluate novel situations and their associated risks must therefore depend on formal training as well as on an individual's own experience. There is a need to control both these factors if an organisation is to manage the risk-taking behaviour of its workforce.

Additional information provided by training or a deliberate programme of extended work experiences may change the perception of certain practices, thought of as safe, to being unsafe and hence posing an unacceptable risk. Risk, and an understanding of it, has two dimensions, as discussed earlier in Chapter 1. The first is an accurate understanding of the likelihood of the accident. The second is an accurate understanding of the consequences — e.g., the nature of potential injury, or the effects of long-term exposure to high-level noise or environmental pollutants. It is important that information about consequences is effectively communicated to all personnel.

People have to be aware of a hazard before they can make any judgement as to the associated risks to themselves and others. How people perceive this risk is central to the concern over deliberate breaches of safety rules and procedures.

An analysis of the Three Mile Island and Chernobyl disasters [Ahearne 87] showed that in both cases 'mechanical systems were defeated by operators who did not understand what they were doing and took actions that deliberately overrode safety systems'. This point was further emphasised by the Director of the HSE in their 1987/88 Annual Report [Rimington 88]: 'a great many accidents happen through ignorance not only of proper precautions, but even of the existence of hazards'.

Hazard awareness and risk perception can be assessed using a variety of questionnaires. An early use of this method in the mining industry addressed the attitudes of people who worked in bunkers or silos [Rushworth et al 86]. Questionnaires were given to men in seven job categories: trainers, experienced bunker workers, inexperienced bunker workers, skilled men, supervisory staff, shaftsmen, and the area shaft team. Participants were asked to rate 22 different activities on a dual scale of 'riskiness'. One scale was benchmarked with day-to-day risk activities, such as smoking 20 cigarettes a day, with the end points of 'certain death' and 'no risk of death'. The second scale covered the numerical 'rough odds of dying each year' (from 1 in 10 to 1 in 10000000). Table 11.2 shows the mean ratings of riskiness for all activities — converted, for simplicity, to a ratio referenced to the risk of being run over on the road. For example, a score of 10 equates to ten times more likely than being run over on the road.

The table clearly shows the large differences in risk ratings between the groups, with the more experienced groups rating most activities as relatively safe. This is probably explained by familiarity with the task and by the lack of any accidents being used as proof that the various working practices in the questionnaire were in fact safe. There were equally large differences between the risk ratings for the 22 individual tasks.

Table 11.2 Risk ratings for 'all activities' performed in bunkers/silos

Job Category	Risk Rating
Inexperienced bunker workers — newcomers to such work	240
Trainers — Instructors in methods of working in bunkers	112
Skilled men — welders, blacksmiths etc. who occasionally apply their trade in bunkers	3.0
Shaftsmen — experienced workers who normally work in shafts but less in bunkers	2.1
Supervisory staff — charged with planning and monitoring work in bunkers	2.0
Experienced bunker workers — experienced at conducting a variety of jobs in bunkers	1.9
Area shaft team — very experienced, regularly work in shafts and bunkers	1.3

Hazard awareness and risk assessment measures can be used to assess training courses. When administered before and after training, such measures quickly establish where training is ineffective for the majority of the group and hence where a syllabus needs changing. Should the results identify instances where only a small number of individuals have poor post-test hazard awareness or risk perceptions, it may be assumed that the training was probably adequate but not fully understood by the low performers and that special attention may only be needed for them.

11.3.2   Enhanced Perception of the Benefits

The personal benefits from working to agreed safety procedures are often non-existent in an organisation. By comparison, individuals may see many benefits of breaking rules and procedures, in addition to the financial gains which can often arise.

Instances are often reported of people being praised for their initiatives if they improvise to 'get the job done' under difficult conditions. The violating of rules and procedures which this involves may appear to have been condoned as long as they did not result in unwanted events such as injury or plant damage.

An attitude often develops where strictly adhering to the rules is seen as conflicting with the objectives of the organisation — e.g., where working to rule is an industrial or commercial threat.

11.3.3   Low Perception of the Likelihood of Injury or Damage to Plant

An operator or maintenance worker may be more willing to deviate from a prescribed working practice if he or she considers it unlikely to result in injury or equipment damage. For example, the perceived level of importance of a maintenance requirement to the reliability or functioning of an item of equipment would influence the probability that the maintenance would be ignored or carried out badly.

Maintenance which is excessively difficult to conduct, or a situation where the crew has limited time available to conduct a number of operations, are both conditions where tasks may be omitted or not completed to the desired standard, especially if they are thought to be of little importance.

A recent survey of machine reliability identified dramatically different reliability figures for essentially the same machine at different locations. Subsequent analysis showed that a major contributor was the relative performance of the maintenance crews. A study was therefore conducted into the attitudes of the maintenance crews [Mason and Rushworth 92]. Of special interest were the methods that fitters used to overcome overheating on diesel powered vehicles, as this had previously been identified as a major cause of breakdown. Many of these methods were in strict violation of rules and procedures. Table 11.3 summarises the responses from 100 fitters on

Figure 11.3 Fitters' estimates of likelihood of engine damage

Action	No chance of engine damage %	Chance of engine damage %	Strong chance of engine damage %
Changing air filter too soon	97	3	0
Leaving air filters too long	1	65	34
Changing oil filters too soon	95	5	0
Leaving oil filters too long	1	42	57
Removing filter in tank to speed filling	5	55	40
Running with low levels in conditioner box	43	46	11
Cooling a hot engine by hosing it	1	12	87
Cooling a hot engine by hosing the radiator	28	50	22
Cooling a hot engine by removing the thermostat	26	50	24
Leaving out chemicals in conditioner box	55	31	14

whether they thought each action would cause damage to the engine units.

The results clearly showed that there was a relatively poor agreement between fitters for most effective actions which could be taken to reduce overheating. There was only good agreement on the strong chance of damage caused by hosing a hot engine, and no chance of damage by changing air filters too soon.

By studying the risk perceptions of the individuals, using purpose-designed questionnaires, problem areas can be readily identified and training tailored exactly to redress misperceptions, thereby helping to reduce the sources of induced maintenance costs.

11.3.4 Inadequate Management and Supervisory Attitudes

Many of the above issues are influenced by management. In a production environment, where the emphasis in on output, there can be a tendency to regard safety procedures and maintenance schedules as undesirable — a cause of lost production. Operators may cut corners. Pressure to get equipment operating as soon as possible (following both routine and corrective maintenance) can result in imperfect repairs and subsequent failures which would not otherwise have occurred.

The influence of this perceived pressure can be strongly affected by the perception which the workforce holds of management's (including immediate supervision's) attitude towards safety. An apt example was recently shown in a study of electricians working in the coal industry [Mason et al 94]. The attitudes of 'craftsman-grade' electricians to safety, their acceptance of the need for safety rules and procedures, and their compliance with safety rules were strongly linked to the perceptions they had of management's commitment towards safety.

11.3.5   Low Chance of Detection due to Inadequate Supervision

There are a number of factors which can reduce the chance of a violation being detected. Organisational factors may result in supervision only being present during predictable periods — when violations may be committed with minimum risk of detection. In such situations it is particularly important that rule violations can easily be detected subsequently.

Supervisors may not have been given sufficiently detailed training to identify the occurrence of certain specific violations. For example, a supervisor with no electrical training may be unaware of even the most basic safety infringements being made by electricians. For this reason, the training of supervisors in the principles which are basic to the areas of their supervision is desirable.

Supervisors may not have been given, or may choose not to operate on, a sufficiently wide remit to perform the necessary range of safety checks. They often concentrate only on the presence of unsafe conditions and not on unsafe behaviour. This may be for a whole range of reasons, for example it is easier to detect such conditions and also to rectify them, than to deal with the underlying cultural issues. Alternatively they may choose to focus on the operators and not check for violations by maintenance crews or by management.

The absence, inability or unwillingness of supervisors effectively to identify breaches in approved working methods removes a significant safeguard for ensuring compliance with working practices. Thus, for many reasons a low chance of detection can greatly increase the likelihood of violations occurring.

11.3.6   Poor Management or Supervisory Style

Day-to-day production pressures may lead to many managers and supervisors failing to develop an environment which encourages staff to adopt safe working practices. Unsafe practices may frequently go unnoticed or uncorrected. Whenever a manager encounters a situation in which rules or codes of practice are not being strictly adhered to but fails to take corrective action, he is seen by many to be condoning the breach. Probably without realising it, the manager has demonstrated to the workforce that he is not totally committed to safety.

If managers fail to stop 'bad' behaviour, it will gradually drift into becoming 'normal' or 'accepted' behaviour. In addition, if immediate management does not prevent this from happening, it is unlikely that anyone else within the management structure will discover it and stop it. It may be useful to note that this may be in spite of safety polices and handbooks and the stated — and often genuinely intended — commitment of upper management. Studies have also shown that many workers model their behaviour on that of key individuals. Their immediate boss is usually implied, but good workers or senior managers may also be the role models. It is vital that any key individuals who may not always strictly adhere to prescribed safety behaviour are identified and their behaviour changed. Exposing them to the effects they have on the unsafe behaviour of others may be sufficient for them to alter their behaviour.

A first requirement may therefore be to determine the hazard awareness and risk perceptions of the supervisors and managers. Those who consider that accidents are an inevitable part of the operation will exacerbate the problems.

A total commitment to safety by top management is also an important precursor to improved safety attitudes throughout the organisation. This is borne out by a number of recent public disaster inquiries.

The Piper Alpha oil rig fire inquiry [Cullen 90] stated that, 'I am convinced from the evidence ... that the quality of safety management ... is fundamental to off-shore safety. No amount of detailed regulations for safety improvements could make up for deficiencies in the way that safety is managed.'

The important role that management plays was also highlighted by the inquiry into the capsizing of the *Herald of Free Enterprise* [Steel 87]. The inquiry concluded that 'all concerned in management, from the members of the Board of Directors down to the junior superintendents, were guilty of fault in that all must be regarded as sharing responsibility for the failure of management. From the top to the bottom the body corporate was infected with the disease of sloppiness.'

In considering how management can set an effective example in safety issues, the following are relevant:

- Management must consistently lead by example in safety matters;
- Management must be seen to be actively interested in safety — by continually communicating their own commitment to it;
- Management should ensure that personnel do not perceive them to give production priority over safety;
- Management should be directly involved with trainees in training and work settings;
- Behaviour of 'key individuals' must be targeted as a priority for ensuring total compliance with safety procedures;
- Regular compliance checks should be made by supervisors and management to determine the degree of working to the prescribed rules and procedures — the results of such checks, when fed back to the workforce has been shown to influence behaviour.

11.3.7 Poor Accountability

Any lack of clarity in roles, responsibility, accountability and authority creates an organisational environment which can encourage violations. Individuals with the authority to decide work methods, but without any direct or immediate responsibility for safety consequences, are more likely to be motivated to commit violations than those who have been given responsibility for safety.

Machinery used on shift work may be taken as an example. Routine maintenance could be undertaken by different craftsmen. In such situations a machine failure caused through poor routine maintenance could not be linked to any individual craftsman, thus removing a major incentive to adopt good maintenance standards. Making individual workers directly accountable for items of equipment can increase the motivation to maintain plant in good order. A sense of pride at the 'ownership' of the maintenance of important pieces of equipment is a positive influence here. To summarise:

*   Individuals who can influence safety-related decisions or actions should have clear accountability for their own actions;
*   Accountability for any specific action or decision should never be shared between individuals or functions;
*   Maintenance duties should be allocated (wherever possible) so that each maintenance worker can be held accountable for the condition of all of 'his' items of plant.

11.3.8 Complacency Caused by Accident-free Environments

The absence of an accident to someone who constantly and knowingly breaks safety rules is 'proof' to him (and others) that his methods are safe. For example, a fitter may have got into the habit of omitting certain safety preparations before starting the job if it makes his life easier. It would not be until a catastrophic accident, or a string of serious accidents to himself, or others, that he would voluntarily choose to reconsider his perceptions of the risks or hazards associated with his job and to modify his behaviour.

Several commentators on the Chernobyl accident observed that operators and management were complacent as a result of their good operating record [Howieson and Snell 87]. A trade union leader also commented that 'we had become complacent, Chernobyl punished us for it'. It must, however, be added that there had apparently been a number of near misses at the plant, notably in 1982 and 1985, when elementary safety rules had been ignored without supervisors being alerted. There are also further reports of violations of safety regulations [Hornick 87]. This suggests that relatively short accident-free periods can generate complacency.

The consequences of complacency can further be judged from comments made by the Vice President for Health and Safety of Union Carbide. Commenting on whether a similar problem to that at their Bophal plant in

1984 could occur in their US Plants '. . . Could the same thing happen here? . . . We said "no" in December of 1984 based on our . . . understanding of the process, and our confidence in our safety systems and procedures. Now, after the investigations, we are even more certain of our answer based on a comprehensive analysis of what happened in the tank in India. We can confidently say: "it can't happen here".' In August 1985 an escape of process chemicals, including methyl isocyanate, caused 150 casualties amongst the employees of Union Carbide's West Virginia plant and the local population.

Similar conclusions were drawn from the inquiry into the Clapham tragedy [Hidden 89]: 'A belief in a positive safety culture . . . which is sincerely and repeatedly expressed but which is not carried through into action is as much protection from danger as no concern at all.'

To aid the avoidance of complacency, managers and supervisors must continually make workers aware of accidents happening at other locations.

11.3.9 Ineffective Disciplinary Procedures

People are not born with an attitude. Attitudes are shaped and developed by both the individuals and the circumstances surrounding them. If changes are made to the behaviour or the circumstances which shaped the original attitudes, then the new circumstances will shape new attitudes.

An example of this philosophy is the enforcement of correct behaviour through high-profile policing of procedures and effective disciplinary action. It is argued [Krause et al 84, 90, Krause and Hidley 92] that this subsequently shapes attitudes to the point where the desired behaviour will remain even when the high-profile policing has been removed.

An analogy can be drawn from the campaign in the UK for the compulsory wearing of seat belts. Initially many drivers were against this on the grounds of removal of personal freedom or that they felt safer if they weren't strapped in after an accident. Voluntary campaigns had limited effect. Compulsory requirements to wear seat belts, accompanied with a very high police profile, subsequently greatly increased compliance with the rules. It is suggested that attitudes to seat belt wearing have now changed so much that even if the wearing of belts was no longer compulsory the vast majority of people would continue to use them.

Effective disciplinary procedures are essential if this approach is to work. Very often, first-line supervision does not discipline a worker for violating a safety rule for fear of the effect it would have on their working relationship. In this way, small offences may become common practices which allow the development of greater offences. Thus:

- Supervisors should be taught the inter-personal skills needed for two-way face-to-face disciplinary action — for example, they should not criticise the employee but the unsafe practice;
- The organisation should have an unambiguous policy regarding the expected safety-related behaviour of employees;

- The policy concerning the use of punishment is most effective if it has been formulated with the involvement of employee representatives;
- The reason for punishment should always be clearly given; punishment should be administered soon after the offence and should not be unduly harsh;
- Supervisors should never allow an unsafe practice to be used under certain conditions but not others.

11.3.10 Inadequate Positive Rewards for Adopting Approved Work Practices

The benefits of breaking safety rules in most organisations can be said to be faster completion of the task, higher earnings, praise from others, and less effort. The natural 'punishers' for breaking safety rules are, however, usually weak and infrequent — injury, damage and disapproval from other workers. Likewise the natural reward for any individual working safely are usually non-existent. It is rare for anyone to receive praise for working strictly to the safety rules. In fact, working to rules is often viewed as an industrial threat. However, when small rewards have been offered they have been shown to be effective.

One case study showed that the company costs from accidents following use of a positive reward incentive scheme were reduced from $294,000 to $29,000. Such savings were said to justify a bonus system in which around 50% of this saving was returned to the workforce. Evidence has shown that returns as low as 1:13 or even 1:30 will work. However, many managers take exception to having to pay people to work safely — i.e., fully adhering to the safety rules which management introduce or impose. Praise from the boss has been said to be the strongest reward of all. The following are useful guidelines:

- Forms of reward for achieving high safety standards should be devised — they can include better promotion prospects, clear acknowledgement of the individual's personal achievement among the workforce, praise from management, and forms of financial reward;
- The safety culture must accord high status to safe workers.

11.4　A WORK DESIGN APPROACH

A second framework for understanding violations began from studies on the influence of new technologies on the jobs of mineworkers [Mason and Simpson 92]. It is based on some of the traditional motivational theories which have been well documented for the design of jobs. These can be used to investigate potential organisational links with factors likely to cause procedural violations. In the framework, violations are considered to be caused primarily by the following motive types:

Motive A: that which gives financial reward;

Motive B: that which leads to group acceptance, affiliation, or social benefits and which can result in pressure being placed on individuals to conform (requiring sociological solutions);

Motive C: that which gives ego- or self-fulfilment, power, or achievement advantages (requiring psychological solutions).

Maslow developed a 'hierarchy of needs' model which suggested that there were a number of motivators, but that only one operated at a given time and that they occurred in a set order [Maslow 54]. According to this model, the most basic motivator, at the lowest level of the hierarchy, consists of the 'physiological needs' of food, etc. Only when this is fulfilled does the second-level motivator of 'safety' become active. Again, only once an individual considers himself safe does the third level of motivation apply. This is the 'social needs', or the need to be accepted by a work group. The fourth level of motivation consists of the 'ego needs', and the final level the 'self-fulfilment needs' — i.e., the need to be better tomorrow than you are today. There are many instances where people are operating only at the third level, with the fourth and fifth levels being inapplicable.

Using this well-known theory, it may be expected that people likely to commit Motive C violations would be operating higher in the Maslow hierarchy (levels 4 or 5) than those committing Motive A and Motive B violations. For example, those operating at the 'social level' may be exposed to group norms which encourage Motive B violations in order to fit in and to stop them being dissociated from their workmates. Senior staff who are more likely to be working at the higher levels of the hierarchy may be more likely to commit Motive C violations because of their needs to 'achieve' results or gain power.

This concept could be taken further by utilising the differences between individuals who are classed as 'motivation seekers' and 'maintenance seekers'. Motivation seekers can be looked on as those individuals who are primarily enthused by the challenges of a job. By contrast the maintenance seekers are those who are primarily interested in status aspects. Only the former type of individual would be motivated by the intrinsic characteristics of a job. It could be expected therefore that only motivation seekers (including management) are likely to commit Motive C violations, in order to take risks to show themselves as good performers. Maintenance seekers may commit only Motive A and B violations.

A complication arises if senior staff are seen to violate rules and are still judged to be successful. This could then create a social need for others to copy this type of behaviour. Motive C violations by senior motivation seekers could therefore introduce Motive B violations in maintenance seekers.

It is possible that individuals seeking the 'ego needs' and 'self-fulfilment' benefits from violations are more likely to work within an organisation where they cannot demonstrate superiority and receive acknowledgement for it by always keeping to the rules. Therefore, there may be a conflict between the demands of the employer and those of the employee. Using the 'infant-adult' dimensions offered by Argyris, it is likely that such motives to violate would be more likely to affect those motivation seekers working within organisations which do not fully stretch the individual in terms of challenging work and rewards [Argyris 60]. These people still need to demonstrate their superiority within an organisational framework which limits their involvement, and so may be tempted to break rules to create opportunities to demonstrate their skills and knowledge. Of course, this may be achieved by efforts outside of their working environment, such as in their hobbies and sports.

A final fundamental question is, what motivates the motivation seeker? The need for achievement, power and affiliation have been identified. Studies have shown that managers are higher than average in 'achievement motivation' and lower than average in 'affiliation and power motivation'. Their apparent great concern for successful task accomplishment is understandable, as that is the usual criterion against which a successful manager is judged. It is possible that this may become a motive for some senior staff and managers to violate — especially if there are inter-manager conflicts. Organisational factors can become important in developing a climate in which the potential to violate is reduced.

Organisations need to eliminate harmful interpersonal conflict between groups of workers (i.e., that wherein gaining advantage over another person or group is more important than the task itself) as these undermine, rather than aid, the overall task performance.

Ideally, organisations should prevent such conflict at the outset by designing an organisational structure which stimulates collaboration rather than competition. Collaboration which can improve the organisation's effectiveness has been shown to be greater under conditions where individuals freely share information and experiences. This is opposite to working in a competitive situation wherein each group is committed to hiding experiences and information, thereby preventing effective integration of all the organisation's resources. Such conflicts have been effectively reduced through increasing communication within and between workgroups.

The organisation must therefore create conditions which balance its goals with the individual's needs, while minimising any unproductive competition between sets of the workforce or between the workforce and supervision. For example, it may be possible for some form of performance assessment to be made on the basis of the amount of help that a workgroup gives to others. Further, any win-lose situations should be avoided — such as a reward to the best group (either in terms of performance or safety). It is preferable to emphasise the performance of the whole organisation, with any rewards shared across all groups.

Developing any approach along these lines has its difficulties because of the complexity and interactions of the issues considered. The theories strongly point towards a need to develop a system which fully takes into account the different needs and motives of different types of people, both in terms of their position in the organisation and their personalities. The strong influence of organisational factors on some individuals to violate is also clear. These factors must be considered in any attempt to identify and reduce or remove the motives for violations.

11.5   ROUTINE, SITUATIONAL, EXCEPTIONAL AND OPTIMISING VIOLATIONS

A final classification is offered which is based on the mechanism behind the motive rather than the motive itself. This classification was developed following work on the railways [Free 94]. Only a part of it is discussed, as the full classification also covers unintentional acts which subsequently violate procedures.

Within the framework, four classes of intentional violations are described. Although each is unique in terms of the mechanism behind the violation type, many major incidents exhibit a mix of violations from different classes.

11.5.1   Routine Violations

A routine violation is behaviour in opposition to the rule, procedure or instruction that has become the normal way of behaving within the person's peer or workgroup. The violating behaviour is normally automatic and unconscious and is recognised as such by the individual if questioned.

For example, the Clapham Junction rail crash showed that substandard workmanship was not uncommon from electricians employed in signal rewiring. The fact that such violations had become routine is reflected by the public inquiry report [Hidden 89] which stated that 'the concept of absolute safety must be a gospel spread across the whole workforce and paramount in the minds of management. The vital importance of this concept . . . was acknowledged time and again in the evidence which the Court heard . . .', but subsequently it also stated, '. . . the concern for safety was permitted to coexist with working practices which . . . were positively dangerous . . . The best of intentions regarding safe working practices was permitted to go hand in hand with the worst of inaction.'

A similar situation was found in the case of Chernobyl. Gittus et al concluded that 'malpractice by the operator was a contributory cause, and indeed they broke so many rules that one cannot help thinking that this was their regular habit' [Gittus et al 87].

11.5.2   Situational Violations

Situational violations occur as a result of factors dictated by the employees' immediate work space or environment. These include the design and condition of the work area, time pressure, the number of staff, supervision, equipment availability, design, and factors outside the organisation's control, such as weather and the time of day.

An unfortunate mix of situational violations and other errors led to the *Herald of Free Enterprise* disaster [Cullen 90]. Pressure to turn the ship around quickly, personnel not being in post at the bow door, assumptions being made by the captain, and the design of the ship not allowing the captain to know the state of the bow doors all combined to cause this disaster.

11.5.3 Exceptional Violations

Exceptional violations are those which are rare and occur only in particular circumstances, often when something goes wrong. They occur to a large extent at the 'knowledge-based level' (see Chapter 2 for an explanation of this) when an individual is attempting to solve a problem in an unusual situation where existing rules and procedures are considered inapplicable to the specific circumstances [Rasmussen 83] or over-prescriptive [Reason et al 95]. The individual relies entirely on adapting basic knowledge and experiences to deal with the new problem and, in doing so, violates a rule. Such violations are commonly associated with high risk, often because the consequences of the action are not fully understood or because the violation is known to be dangerous.

A combination of single-minded, but non-nuclear, engineers, directing a team of dedicated but overconfident operators, coupled with intense time pressure to complete testing, created a situation at Chernobyl in which a number of exceptional violations were committed. For example, the steam-drum automatic shutdown was overridden, and despite strict rules to the contrary, the reactor was operated with less than the minimum number of control rods in the core. There were undoubtedly pressures on the workforce at the time of the incident. *Nucleonics Week*, 21 August 1986, reported that the Soviets maintained a flat policy of firing any operator who 'scrammed' a reactor. The sequence of violations involved in this incident can be seen as an example of operators believing that the 'end justified the means'. Reason refers to 'unavoidable violations' imposed on those operators by weaknesses in the total system [Reason 87].

11.5.4 Optimising Violations

A final class of violations is created by a motive to optimise a work situation. These violations are usually caused through:

- A need for excitement in jobs which are considered repetitive, unchallenging or boring;
- A desire to explore the boundaries of a system which is thought to be too restrictive;
- Simple inquisitiveness.

An unusual example occurred when a senior manager became bored driving along a long stretch of straight road inside the plant. He developed a game to see how long he could drive with his eyes shut—with predictable eventual results!

11.6 CONTROLLING PROCEDURAL VIOLATIONS

The classification of violations into 'routine', 'situational', 'exceptional' and 'optimising' is especially useful when combined with the direct motivators and behaviour modifiers and the work design theories described earlier. The classifications allow an approach to be developed that can identify those organisational factors present in a system which increase the potential for violations. The approach can also identify practical management strategies to limit or remove the factors from within an organisation.

11.6.1 Attitudes, Behaviour and Solutions to Violations

The three classification systems described above give a good insight into the types and causes of violations and the measures which are available to management to eliminate or at least reduce the potential for them in a wide range of industrial situations. It should be apparent that there can be no golden rules which are applicable to all problems where human error, and in particular violations, occur. This is because the exact blend of motives behind any deliberate unsafe behaviour is likely to be specific to the attitudes of the individual and complex in nature.

In principle, various classification systems, such as those described above, could be used to attempt to 'measure' the attitudes and motives of each individual. In practice this would prove prohibitively expensive in time and effort, in addition to which the ability of such basic measures accurately to represent true attitudes must be questioned. It is also becoming increasingly accepted that it is difficult and unpredictable to attempt to change attitudes directly.

Fortunately, we need not be overly concerned with trying to measure individual attitudes if our objective is simply to identify and reduce the likelihood of potential violations in a work setting. Practical management strategies for reducing violations can be developed by identifying the organisational factors which are likely to influence attitudes and hence the behaviour of the workforce. Such organisational factors include:

- Training;
- Management and supervision;
- Job design;
- Equipment design.

All of these can be controlled by management. It is more effective to change the factors which influence individual attitudes than to attempt to change the attitudes themselves.

Management therefore need a simple procedure that 'measures' the organisational factors which can increase the likelihood of violations. They also need to be able to convert this into practical management actions which can eliminate or reduce the strength of these factors.

Managers' understanding of the problems must be based on the actual views of the workforce. For example, it is generally accepted (as discussed above) that the attitudes and behaviour of an individual are strongly influenced by the degree to which senior management are committed to safety. While senior management may genuinely place safety as their first priority, this commitment may not have been effectively communicated to the workforce. Some individuals may have the perception that they are committed solely to production. When looking at attitudes and behaviour, it is not what the manager actually says which is important but what he is perceived as saying and how he acts in line with this.

The classifications described in Sections 11.2 to 11.6 provide a basis on which to build up a methodology. They have been used to develop a comprehensive question set which can be employed to identify the presence in the organisation of any factors likely to lead to poor attitudes and unsafe behaviour.

11.6.2 Reactive or Proactive Safety Management

Some general-purpose methodologies have recently been developed which help analysts investigate incidents involving human errors — including violations. For example, MORT (Management Oversight and Risk Tree) was developed by the US Atomic Energy Commission [e.g., Johnson 73]. MORT and techniques like it have been used to determine the causes and contributing factors of major incidents by providing decision points in an accident analysis which help detect omission, oversights and defects. Indeed, MORT has also been employed for the analysis of lesser incidents, as described in detail in Chapter 10.

STEP (Sequentially Timed Events Plotting Procedure) [Hendrick and Benner 87] is another technique which identifies the principal people, equipment, and substances relative to the critical actions and events in an accident sequence.

In the main, however, these techniques are only likely to be effective when used by human factors specialists and they only aim to identify the critical factors in the chain of events which have led to an incident. A problem with this approach is that any response is primarily aimed at reducing the 'carbon copy' incident. (An exception to this is the method and tool developed for the use of non-experts in the analysis of incidents at work, as described in Chapter 10.)

An exception to these methodologies is one which was developed in the UK Mining Industry [Simpson 93] to identify factors likely to cause human error and which links the findings to various management strategies for the control of such risks. The 'BeSafe' methodology systematically addresses all forms of human error — including violations — and although initially reliant on specialists for its industrial application, it can now be used by non-specialists.

As suggested above, management need not wait until there is an incident involving a violation before they act to reduce the likelihood of a repeat incident. A better approach is to conduct an audit of potential violations in a work environment. The outcome is a profile of the organisational factors which increase the potential for violations. Management intervention can then be proactive and can address the wider hazards identified to exist in the organisation, rather than only those specific to an incident being investigated. As these organisational factors may be influential in a wider range of potential violations, the resulting remedial actions would be more effective and wide ranging than any identified in a reactive manner from investigations of a small number of incidents.

11.7 THE HFRG VIOLATION APPROACH

The HFRG violation approach was developed by the Human Factors in Reliability Group (HFRG) and published jointly with the HSE [HSE 95]. It is designed to identify not only the main factors present within an organisation which would be likely to influence a person's willingness to commit a violation, but also management strategies for their elimination or reduction by addressing the motives behind them. The approach was designed to be usable by the non-specialist and applicable to a wide range of industries.

The HFRG methodology can be applied at plant commissioning, as part of an accident investigation, or within a routine human error audit. Violations are one type of human error, so it is recommended that this exercise should be complemented by an assessment of the potential for inadvertent human errors which could also result in a heightened safety risk — for example, by using the BeSafe system [Simpson 93]. A full description of the methodology is given in the HSE publication *Improving Compliance with Safety Procedures — Reducing Industrial Violations* [HSE 95]. The methodology involves questionnaires and interviews. It sets out to identify those sets of rules within the organisation which are considered to have the greatest potential impact on safety (or production, if that is relevant) if they are not followed. Each rule set is then assessed using a checklist — shown in the first column of Figure 11.1.

A questionnaire based on the checklist should be completed for each set of rules affecting safety, such as those concerned with the operation of safety-critical systems, the isolation of equipment, slinging loads, working on scaffolding, the use of permit-to-work systems, and entry into confined spaces. A single checklist cannot be applied meaningfully to all the safety rules, as answers vary depending on the rule, so the wording of checklists needs to be tailored to the application. Conversely, completing the checklist

HFRG Question Set:	Score	Solutions												
		A	B	C	D	E	F	G	H	I	J	K	L	M
1 The rules do not always describe the best way of working	24	24		24	24					24		24		
2 Supervision recognises that deviations from rules are unavoidable	11	11					11		11	11		11		
3 Schedules seldom allow enough time to do the job according to the rules	4	4					4			4			4	
4 There are some rules which would make the job less safe/efficient	10	10				10				10	10			
5 I sometimes can't get the equipment needed to work to the rules	21		21				21			21		21		
6 Some rules are impossible or extremely difficult to apply	13		13							13	13	13		
7 It is necessary to bend some rules to achieve a target	13			13	13	13	13							
8 The rules are not written in a simple language	0		0	0										
9 Some rules are very difficult to understand	3	3	3	3										
10 Rules commonly refer to other rules	10		10										10	
11 Some rules are factually incorrect	7	7												7
12 I have found better ways of doing my job than those given in the rules	21	21			21					21	21			
13 Sometimes the operating limits prescribed in rules are too restrictive	16	16			16						16			
14 I often encounter situations where no prescribed actions are available	21	21											21	21
15 There are no general guidelines to use when specific rules do not apply	7	7		7	7				7					7
16 I sometimes don't know why I have to follow rules	3			3	3				3					
17 Some rules do not need to be followed to get the job done safely	13	13			13									
18 Some rules are only for inexperienced workers	12				12	12								
19 Some rules are so complex that I loose track	3		3	3										
20 Some rules are only of value to protect management's back	3	3					3							3
21 Sometimes conditions at the workplace stop me working to the rules	16						16	16		16		16		
22 No systems check people understand procedures before they are used	9			9			9	9						9
23 Infringements of rules occur all the time	10				10	10	10	10	10					
24 There are incentives to ignore some rules	4						4	4			4			4
25 I can get the job done quicker by ignoring some rules	24	24			24	24	24			24		24		
26 Deviations from rules are not always corrected by a superior	18						18	18	18					
27 Short cuts are acceptable when they involve little or no risk	12				12	12	12							
28 There are circumstances where managers will support rules being broken	3	3					3							
29 Management sometimes pressure people to break rules	4				4		4			4				4
30 The workforce sometimes pressure people to break rules	4				4	4	4							4
31 Staff shortages sometimes result in rules being broken to get the job done	24						24	24				24		
32 There are some rules where your natural reaction would be to break them	13	13								13				

Figure 11.1 *Continued on the next page*

33 Contractors are allowed different safety standards	10	10				10							10
34 There is no efficient procedure to monitor that rules are kept to	14					14	14						14
35 Supervisors seldom discipline workers who break rules	10					10	10	10					
36 It is unlikely that somebody would be detected if they broke the rules	10				10	10	10		10	10			
37 There are no personal benefits from strictly following rules and procedures	21			21			21		21		21		
38 There are financial rewards to be gained from breaking the rules	4					4					4		
39 I am sometimes tempted to do work that is not my responsibility	25			25		25		25					
40 I am not given regular break periods when I do repetitive and boring jobs	3									3	3		
41 Working to the rules removes skills	6	6								6			
42 Deviating from some rules demonstrates knowledge of the job	4			4	4	4		4					
43 I sometimes have difficulty getting hold of written rules and procedures	3					3	3				3		
44 I sometimes come across a rule I did not know about	6		6							6			
45 I have rules for tasks I will never have to do	4	4	4	4									
46 I have not been trained in the rules to be used in unusual circumstances	9		9	9									
47 I often come across situations with which I am unfamiliar	6		6							6	6		
48 I sometimes fail to fully understand which rules to apply	4	4	4	4									
TOTAL SCORES		204	67	91	91	99	170	205	83	171	129	101	95	108

Figure 11.1 HFRG Analysis Chart for one of the four scoring methods (Total Scores)

for every safety rule would be prohibitively expensive. Using the checklists for generic rule sets provides a satisfactory working compromise.

A selection of the workforce is asked to rate the 'degree of agreement' with 48 statements. Dependent on their responses, each statement receives a score between 0 and 6 (0 for disagree, 1 for agree slightly, 3 for agree and 6 for strongly agree). The average scores are then determined across the sample of the workforce.

Three scores are obtained:

(i)   The total scores against each of the 48 statements;
(ii)   The number of 'agree' entries (i.e., those with slight agreement, agreement, or strong agreement entries);
(iii) The number of strongly agree (6) scores.

Each of the three scores is entered into a matrix. Figure 11.1 shows an example with the 'total scores' entered. The matrix links each potential factor (rows) with possible generic solutions (referred to as 'solution avenues') which are considered to be appropriate management strategies for minimising the violation potential in the organisation. The generic solutions are adjustments which can be initiated by senior management and are therefore defined in

Table 11.4 HFRG matrix scoring procedure

General Avenues for Solutions	A Total Score	B Nos entries	C Nos '6'	D Mean Score	Selection
A Rules & Procedures — Aims & Objectives	204	66	(8)	55	PRIORITY
B Rules & Procedures — Application	67	24	3	23	
C Training — Rules & Procedures	91	31	4	34	
D Training — Hazards & Risks	91	64	11	49	PRIORITY
E Safety Commitment — Workforce	99	35	2	24	
F Safety Commitment — Management	(170)	(62)	6	48	secondary
G Supervision — Monitoring & Detection	205	63	14	(47)	PRIORITY
H Supervision — Style	83	27	4	24	
I Plant & Equip Design & Modification	171	54	9	37	secondary
J Job Design	129	40	7	33	
K Work Conditions	101	33	4	21	
L Logistic Support	95	28	7	26	
M Organisation	108	38	6	32	

broad terms in the rows of Table 11.4. They are then listed as 'A' to 'M' in the rows of Figure 11.1.

For each scoring method, the scores in each of the columns A to M in Figure 11.1 are simply added and transferred onto the summary chart shown in Table 11.4. A final score is obtained by dividing the total score by the number of entries under each solution avenue. For example, the total score for solution A (Rules and Procedures — Aims and Objectives) is divided by 19 — which is the number of questions linked with this solution in the matrix of Figure 11.1. The result is an average and offsets any bias caused by the fact that some solutions may not be applicable in some cases, resulting in different numbers of entries being possible for the various generic solutions.

The following analysis method is employed, as shown in Table 11.4:

- Generic solutions (A to M) with the three highest scores are identified using each of four scoring methods;
- Generic solution avenues are selected for priority management consideration where at least three of the four scoring methods show

a 'top three' score:

* Solutions for 'secondary' consideration are shown where only one or two scoring methods gives a top three score.

The generic solution avenues which have been selected for either priority or secondary consideration are then expanded [see the appendices to HSE 95] to offer a range of recommendations, hints or suggestions to management. Those considered most relevant in the particular organisation at the time may then be developed into a specific management action plan.

The methodology can be applied to a single rule set giving concern to management, to any number of rule sets which management choose to address, or to a number of rule sets considered to be indicative of the potential for violations within an organisation.

In the first application, a management strategy would be generated specifically for promoting compliance with the specific rule set which was addressed. By applying the checklist to a representative sample of generic rule sets, management can address the deeper-seated organisational factors influencing the whole organisation. For example, the checklist could be applied, in turn, to all the generic rule sets involved with the safety of a category of the workforce, such as electricians, or with the safety of a particular type of operation, such as within a control room. The results could then be combined across those rule sets to identify the underlying organisational factors which apply to the given situation. This wider approach would potentially offer the most powerful solution avenues for the organisation.

The exact response to any management actions can never, however, totally be predicted, as the response to their influences are shaped by personal criteria. Any fixed solutions will therefore, by definition, have variable effects, and any desired results may or may not be achieved in practice. Even if success is proven, the effect may not be as great as was hoped for or envisaged. It is therefore important that management intervention is undertaken along with a means of monitoring the results of any changes. Without regular monitoring information, management can never be sure that their actions are achieving the desired results.

11.8 CONCLUSIONS

The potential of violations of working practices as a significant contributory factor to accidents, ranging from minor plant damage to major disasters, is undeniable. Until recently it has been neglected or dealt with in a highly simplistic manner (such as by the allocation of individual blame).

Recent thinking has now provided a detailed framework for the understanding of the nature of violations and the factors which predispose their occurrence. In addition, this framework is beginning to provide not only the basis for a better theoretical understanding but also a range of practical tools which can identify the potential for violations and suggest proactive initiatives to reduce their likelihood.

The challenge now emerges as to whether these developments will be fully exploited in industry because, when it comes to violations, we need to address the irony that it is management who both need to be the prime mover in reducing the potential for violations, and are often also the primary creators of the potential to violate.

ACKNOWLEDGEMENTS

The author would like to acknowledge the work of the HFRG (Human Factors in Reliability Group) Sub-Group on Violations (which is chaired by the author). Acknowledgement is also given to the European Coal and Steel Community (ECSC) Ergonomics Action Programme and to British Coal for the joint funding of various projects from which much of the material in this chapter was derived.

REFERENCES

[Ahearne 87] Ahearne J F: *Nuclear Power after Chernobyl*. Science, 673-679, 8 May 1987

[Argyris 60]   Argyris C: *Understanding Organisational Behaviour*. Dorsey Press, Homewood, Ill., 1960

[ACSNI 93] Advisory Committee on Safety of Nuclear Installations, Study Group on Human Factors: *Third report: Organising for safety*. Health and Safety Commission, HMSO, London, 1993

[Cullen 90] Cullen The Hon. Lord: *The Public Enquiry into the Piper Alpha Disaster*. HMSO, London, 1990

[Free 94] Free B: *The Role of Procedural Violations in Railway Accidents*. PhD thesis, Department of Psychology, University of Manchester, 1994

[Gittus et al 87] Gittus J H: *The Chernobyl Accident and its Consequences*. UKAEA Report NOR 4200, HMSO, March 1987

[Harper 91] Harper J G: *Traffic Violation Detection and Deterrence: Implications for Automatic Policing*. Applied Ergonomics, 22 (3), 189-197, 1991

[Hendrick and Benner 87] Hendrick K and Benner L Jr: *Investigating Accidents with STEP*. Dekker, New York, 1987

[Hidden 89]   Hidden A: *Investigation into the Clapham Junction Railway Accident*. HMSO, London, 1989

[Hornick 87] Hornick RJ: *Dreams — Design and Destiny*. Human Factors, 29, 111-122, 1987

[Howieson and Snell 87] Howieson J Q and Snell V G: *Chernobyl — a Canadian*

Technical Perspective. Atomic Energy of Canada Ltd, Report AECL-9334, January 1987

[HSE 85] Health and Safety Executive: *Deadly Maintenance — Plant and Machinery — A Study of Fatal Accidents at Work*. HMSO, UK, 1985

[HSE 95] Health and Safety Executive: *Improving Compliance with Safety Procedures — Reducing Industrial Violations*. HSE Books, UK, 1995

[Johnson 73] Johnson W G: MORT: *The Management Oversight and Risk Tree*. US Atomic Energy Commission SAN 821-2, Government Printing Office, Washington DC, 1973

[Krause et al 84] Krause T R, Hidley J H and Lareau W: *Behavioural Science Applied to Accident Prevention*. Professional Safety, Official publication of the American Society of Safety Engineers, 1984

[Krause et al 90] Krause T R, Hidley J H and Hodson S J: *The Behaviour-based Safety Process — Managing Involvement for an Injury-Free Culture*. Van Nostrand Reinhold, New York, 1090

[Krause and Hidley 92] Krause T R and Hidley J H: *On Their Best Behaviour*. Accident Prevention, 11-14, June 1992

[Levine et al 76] Levine J B, Lee J D, Ryman D H and Rahe R H: *Attitudes and Accidents Aboard an Aircraft Carrier*. Report No. 75-68, Sponsored by Naval and Medical Research and Development Command Department of the Navy, Aviation and Space Environmental Medicine 47(1), 82-85, 1976

[Maslow 54] Maslow A H: *Motivation and Personality*. Harper and Row, New York, 1954

[Mason 92] Mason S: *Practical Guidelines for Improving Safety through the Reduction of Human Error*. The Safety & Health Practitioner, 24-30, May 1992

[Mason and Rushworth 92] Mason S and Rushworth A M: *Human Aspects of Maintenance*. Maintenance 7(3), September 1992

[Mason and Simpson 92] Mason S and Simpson G C: *Ergonomics Aspects in the Design of Integrated Face Control and Monitoring Systems*. British Coal Corporation, Final Report on CEC Contract 7249/11/055, 1992

[Mason et al 94] Mason S, King K, Basford K, Fitzakerly H, Peach G and Simpson G: *Improving the Human Reliability of Electrical Isolation on the Surface and Underground*. Final Report of CEC Contract 7250/13/036, Midlands Group, British Coal Corporation, 1994

[Meister 92] Meister D: *Some Comments on the Future of Ergonomics*. International Journal of Industrial Ergonomics, 257-260, 1992

[Munipov 92] Munipov VM: *Chernobyl Operators: Criminals or Victims?* Applied Ergonomics, 23(5), 337-342, 1992

[Rasmussen 83] Rasmussen J: *Skills, Rules, Knowledge, Signals, Signs and Symbols and other Distinctions in Human Performance Models*. IEEE

Transactions on Systems, Man and Cybernetics, SMC-13, No. 3, 1983

[Reason 87] Reason J T: *The Chernobyl Errors*. Bulletin of the British Psychological Society, 40, 201-206, 1987

[Reason et al 95] Reason J T, Parker D, Lawton R and Pollock C:*Organisational Controls and the Varieties of Rule-based Behaviour*. Risk in Organisational Settings Conference, ESRC Risk and Human Behaviour Programme, London, 1995

[Rimington 88] Rimington J: Health and Safety Executive Annual Report. HMSO, UK, 1988

[Rushworth et al 86] Rushworth A M, Best C F, Coleman G J, Graveling R A, Mason S and Simpson G C: *Study of the Ergonomic Principles in Accident Prevention for Bunkers*. Institute of Occupational Medicine Report, TM 86/05 (Final Report on CEC contract no. 7249/12/049), 1986

[Simpson 93] Simpson G C: *Promoting Safety Improvements via Potential Human Error Audits*. In: Proceedings of the 25th Conference of Safety in Mines Research Institutes, Pretoria, SA: SA Chamber of Mines, 1993

[Steel 87] Steel D: *Formal Investigation into the MV Herald of Free Enterprise Ferry Disaster*. HMSO, London, 1987

[USSR 86] USSR: *The Accident at the Chernobyl Nuclear Power Plant and its Consequences*. Information complied for the IAEA Experts' Meeting, Vienna, 25-29 August 1986

12

The treatment
of human factors
in safety cases

12.1 REGULATION AND SAFETY CASES

Previous chapters have reviewed the range of application of human factors to the development of safety-critical systems. This chapter proposes a logical framework for the reporting of this work to a regulator within safety case submissions, and discusses some of the 'project management' human factors issues involved in the preparation of safety case submissions and subsequent negotiations with the regulator.

It is important to start by understanding why regulatory safety cases are needed and what they aim to achieve. People rightly expect the law to protect them and to ensure that they are not being exposed to unreasonable levels of risk. They rely on the law because they do not have the means to assess every risk themselves, and as individuals they often have little control over activities which pose too high a risk. Employees are entitled to be confident that they are not running a significant risk in providing their labour or services. Similarly, those who happen to live or work in the vicinity of an activity which has the potential to harm are also entitled to protection and, generally, the level of risk to them considered reasonable would be lower than that to employees because it is being run involuntarily and the benefits are at best indirect. The regulatory framework has therefore developed beyond the protection of employees and the public to take account of these broader risks. Industries considered to have the potential to subject their employees, customers or the public to unreasonable levels of risk are now increasingly to be regulated in respect of all risks associated with their activities, including indirect risks and damage to the environment.

Until relatively recently, the approach taken has been to build up a body of codes and prescriptive regulation, primarily based on experience. For example, regulations might govern the number and type of lifeboats that a ship must carry, where they are situated, how often they should be tested, etc. Over the years, a set of regulations to allow a ship to be evacuated has been developed, and refined as shortcomings have been revealed by accidents such as the sinking of the *Titanic*. Some regulations are concerned with limiting the hazard potential, for example speed limits, and others with moderating risk by limiting the consequences of an accident, for example the wearing of seat belts. In both cases, the approach is essentially reactive, in that regulations are developed in response to accidents or near misses, and compliance is established primarily by inspection.

In most situations, compliance with prescriptive regulations is an indicator of good practice, but it cannot be expected to result in a consistent (low) level of risk from the various hazards covered, especially in complex or novel systems. Neither can it ensure that resources are applied where they will have the most effect in terms of risk reduction. There may be other equally valid, more cost-effective, and equally safe, ways of evacuating a ship than those which would be defined by prescriptive regulations.

These shortcomings are gradually leading to the adoption of a more sophisticated approach to the regulation of potentially risky industries, for example the off-shore industry [Cullen 90]. Rather than specifying solutions to design problems, the regulator may chose to specify goals which correspond to acceptable levels of risk, and leave it up to the designer to decide how best to satisfy those goals. The equivalent to the lifeboat regulations might include a requirement to be able to evacuate the ship in a given time under specified circumstances. The goal is specified, but not the means by which it is achieved. Compliance is more difficult to establish, though. Every ship could, in principle, have a different solution to the problem. Inspection alone would not be enough to ensure safety, and the operators would not want to wait for the first inspection before finding out that there was a flaw in their proposal. When a goal-setting regulatory regime is established, the operators have to set out their proposed means of meeting each goal in advance of commissioning, demonstrate that it does indeed satisfy the goal, and, because they would not be able to rely to the same extent on external inspection, explain how they propose to ensure that the intent will be met in practice. The regulatory authority then has a technical assessment function, and grants permission to proceed with commissioning on the basis of their assessment of the operator's submissions. The arguments set out in these submissions form the 'safety case'.

A safety case does not provide a justification for an activity. It argues that an installation will be acceptably safe, but it does not demonstrate any national need. A process plant might be safe, but it would probably not be acceptable to build it in the middle of a national park! These issues would be reviewed during the local planning processes to establish whether or not the proposed development was appropriate. The safety of the installation will often be discussed at a planning inquiry and an outline safety case may be required, but it is most unlikely that the case would be fully developed at this stage. The operator would still have to prepare a full safety case to satisfy the regulator before the commencement of operation, even if planning consent had been granted.

A goal-setting regime allows the operator more flexibility to allocate resources, but it does not of itself ensure that different hazards are treated in a balanced way, or that expenditure is matched to risk. This is where the concept of ALARP (As Low As Reasonably Practicable) becomes applicable (see Chapter 1 for a discussion of this). The principle of ALARP is that money should be spent to reduce risk until the cost of further reduction is disproportionate to the actual reduction achieved — with the proviso that there is an upper limit to the level of risk which can be tolerated, no matter what the expense of reducing it. The ALARP principle also governs the amount of resources invested in risk assessment. The greater the risk, the more resources should be directed to its assessment and to the identification of opportunities for reduction. A greater degree of confidence in the results of risk assessment is also required when the limits of tolerability are approached, which implies additional investment in assessment at the margin. ALARP is enshrined in UK legislation; safety cases must therefore convince the regulator that this principle has been satisfied where applicable, as well as demonstrating that the specific goals have been satisfied.

In the context of this chapter, we are concerned with regulatory safety cases, i.e., those required by legislation. Although a full set of safety case submissions would probably only be drawn up in a situation where it was a legal requirement, there are other reasons for carrying out a comprehensive risk assessment. Even setting aside risks to the public and workforce, the financial losses to a company which experiences a major accident, or has to withdraw a product, can be crippling. Understanding the risks is a commercial necessity, and techniques which match expenditure to risk serve to ensure that reliability is achieved in the most cost-effective manner. Finally, even when a safety case is not a legal requirement, it does provide a convenient framework for negotiations with the regulator and can more than pay for itself if it allows the company to demonstrate that proposed changes to prescriptive regulations would not be cost-effective.

Where an activity is judged to be 'high-hazard', the goals set by the regulator may well include numerical targets for the total risk or for the risk posed by individual hazards (e.g., in terms of statistical fatalities per annum). Alternatively, the regulator might ask the operator to propose his own targets. The safety case submissions would then need to contain calculations to demonstrate that these numerical targets were met. For example, if the regulator were to take a probabilistic approach to safety at sea, the operator might be called on to estimate the total risk of death (from drowning, fire, etc.)

per passenger mile and show that it was below the target figure. A numerical probabilistic or semi-probabilistic approach is flexible and provides a sensible way to proceed, particularly as the basis of international standards, for it would prove difficult to reconcile differing national prescriptive regimes in high-hazard and heavily regulated industries. The numerical estimation of risk can be a very imprecise and expensive process, but the benefits mean that it is now routine in high-hazard industries, and the results would probably be included in safety case submissions even if numerical targets were not specified by legislation. Numerical demonstration of compliance with the ALARP principle might also be requested in specific instances.

The nuclear power industry has generally been at the forefront of numerical risk assessment because of the onus on it to demonstrate safe operation, and the arguments regarding risk targets and ALARP are better developed in this context [HSE 92a]. The US NRC's WASH 1400 project was probably the turning point [NRC 75], and probabilistic safety analysis (PSA, see Chapter 3) has been a feature of nuclear industry safety cases ever since. Work on nuclear plant PSAs has been in progress for decades and large sums of money have been invested in data collection and the development of appropriate methods, including human reliability assessment (HRA) techniques. Application of the techniques is now more extensive (human factors in PSA is discussed in Chapter 2), although the relative scarcity of data means that the approach taken outside the nuclear industry is more generic.

12.2   THE CONTENT OF A SAFETY CASE

12.2.1 Introduction

The purpose of a safety case is to convince the regulator that the activity described meets the regulatory requirements for risk levels. In the UK, safety cases have for some time been explicitly required for the licensing of nuclear power plants and off-shore installations. EC Directive 82/501 on the Major Accident Hazards of Certain Industrial Activities, implemented in the UK by the CIMAH regulations [HSE 84], requires the submission of safety cases for a much wider tranche of installations considered potentially hazardous. In each case, the assessing body in the UK is the HSE (Health and Safety Executive). More recently, the safety case concept has been extended to rail transport and some marine applications.

The format, content and level of detail of a safety case depends on the precise requirements under each set of regulations. The safety case for an off-shore heavy lift operation will look different in detail to one for a nuclear power station. The level of detail of the human factors coverage will also vary, though the basic principles are similar. This chapter concerns itself primarily with the basic principles as they apply to safety cases for process plant, taking the nuclear installation as the reference case since it is likely to be the most comprehensive. Components may need to be added, deleted or reordered for other applications but all safety cases have the same fundamental purpose, and similar principles apply when it comes to determining their structure and content, whatever the application or jurisdiction.

Before looking in detail at the treatment of human factors, it is worth looking at the structure of a typical safety case. Hawkesley [Hawkesley 89] put forward the view that the purpose of the safety case is to answer three questions: What could go wrong? Why won't it? and But what if it did? Precede this with the information required by the regulator to put the answers to these questions into context, and sufficient information to convince the regulator of the validity of the work carried out to generate these answers, and we have the outline of a safety case submission.

Note that safety cases are not simply descriptions of the safety systems with a chapter on human factors added at the end. A safety case has more in common with a tender document than with a technical specification; it is not simply a set of facts, it has to tell a complete and logical story, combining fact and argument to lead the regulator towards the conclusion that the installation meets the requirements of the legislation. The safety case needs to be persuasive — a theme we will return to in Section 12.3.

The safety case and its structure need to be designed to meet the circumstances, but the basic four-part structure outlined above usually provides a sound basis on which to proceed. Although some topics will require special treatment (e.g., training), human factors should not generally be consigned to a completely separate section of the safety case submission. People and the actions they take, or fail to take, are a feature of all parts of a safety case. An integrated 'systems engineering' approach is required; otherwise, the reader will draw the logical conclusion that the human factors effort was not integrated into the design and assessment processes.

The components of the safety case are discussed below, with the emphasis on those with the greatest human factors content. As the topics discussed are, in some cases, subdivisions of the structure proposed above, their relationship to that structure is shown in Table 12.1.

Table 12.1 The contents of a safety case

1 The context	Process and installation overview Location and local environment
2 What could go wrong?	Process description Major hazards
3 Why won't it?	Design description Control of operations Maintenance and testing Design, construction and commissioning Risk assessment
4 But what if it did?	Emergency arrangements

12.2.2 Structural Overview

The first part of the submission explains the nature and structure of the documentation and should provide an overview and 'route map'. Safety case submissions are complex documents which may run to many volumes and have to be easy to use and to understand if regulatory consent for the system in question is to be obtained. The same qualities will ensure that design and operations staff can understand the case. These issues are covered in Section 12.3.3 below.

12.2.3 The Context

The first section of the safety case proper — the context — provides the regulator with the information required to understand the nature of the installation and the intended operations, and a description of the location and local environment. This is an important part of the submissions, but human factors would not normally feature to any great extent.

12.2.4 Process Description and Major Hazards

The contextual information is normally followed by an outline description of the installation and the processes carried out by or within it. This is not limited solely to safety-related systems and activities. The regulator needs an appreciation of the complete picture, if only to assure himself that all the activities within the scope of the safety case have been included, that interactions between hazardous and non-hazardous activities have been recognised, and that (for instance) all the calls on the time of the control room operator have been taken into account in any workload analysis.

In describing the process, potential hazards to the workforce and the public need to be identified. Hazards are agents with the potential for doing harm. Chlorine is a hazard. Before a hazard materialises into an incident which causes harm, it has to be triggered in some way. Thus the hazard might be triggered if somebody drove a truck into a chlorine storage tank. The driver is not the hazard, the chlorine is. This distinction is important. Analysts who label human beings as hazards find great difficulty in according proper credit to the strengths of humans and corrective actions taken by them. Also, focusing on the driver predisposes the analyst to solutions which prevent driver error instead of solutions which address the hazard, i.e., the chlorine. Risk is the combination of the probability and the consequence of a specified hazardous event (see Chapter 1).

In describing the process, the safety case submissions should identify those features which have been included to reduce risk. Credit should be sought for introducing low-inventory processes or those which substitute less hazardous materials. Subsequent sections of the safety case must go on to explain why those hazards which do remain will not be realised, or at least not to an extent that contravenes the ALARP principle.

12.2.5 Description of the Installation

The process description is usually followed by a more detailed description of the installation which should include its main systems, processes and intended activities. In particular, one expects to find descriptions of any systems and design principles which have a bearing on the control of risks. This is where the coverage of human factors issues starts in earnest.

Many of the systems described will have had considerable human factors input. The central control room and its various alarm and display systems, for instance, are likely to be key features in a process plant safety case and will usually deserve extended treatment. Many of the systems described will rely on human action, and the role of the operator must be described. Safeguards and interlocks which are part of these arrangements ought to be covered. Operating procedures are an integral part of the system design, being closely analogous to the operating software for computer-controlled functions, and they therefore count as part of the design.

It must be stressed that this part of the safety case is concerned only with a description of the major systems, not with the human factors engineering that has gone into the design, nor with the reliability analysis, although there should be cross-references to the sections which address these issues. Mixing up descriptions of the design and the design process merely serves to confuse. The design process is of interest in the safety case submissions only in as much as it gives confidence in the validity of the end product, and it is secondary to the design description. A typical framework for the presentation of the overall design is shown in Table 12.2.

Table 12.2 Coverage of the design description

Design principles	
Site layout	
Process units	
Service systems and passive features	Utilities Fire protection
Control and instrumentation	C&I design principles Protection systems Control rooms Local interfaces Interlocks
Operating procedures	Structure Content

(a)   Design principles

The key design principles should be described separately so that they serve as a checklist for the design as a whole. It is much more efficient to write the submissions this way, because then only exceptions have to be documented in later sections. The alternative, to list the design principles for each process unit, is tedious, and since it is likely that different authors will write different sections, it makes for inconsistency of treatment. Inconsistency invites comment from the regulator. The problem may be further compounded if different people assess different sections of the safety case submissions. Inconsistency leads to unnecessary debate and delay while matters are resolved. It therefore makes sense to separate the design principles and submit them for comment as early in the design process as practicable. This applies to the design description as a whole as well as to the treatment of individual topics. Thus, policy on the allocation of functions between man and machine would conventionally be set out at the start of the C&I (Control and Instrumentation) section, rather than being described separately for each system.

(b)   Site layout

The layout of a process plant can be critical to safe operation, and particularly to limiting the effects of any incident that might occur. Typical issues to be considered include the proximity of flammable materials to sources of ignition, physical protection for the plant (for example, the protection of chlorine tanks from errant truck drivers) and operators, and access for on- and off-site emergency services. Human factors issues are likely to feature to some extent in this section.

(c)   Process units

The different functions that make up the overall process are normally integrated to form the process descriptions. The submissions show how the design fulfils the functional safety requirements and reviews those design features most relevant to safety. The general principles and common services of the C&I systems will probably be covered separately, but control, instrumentation and protection functions should feature in the description of process unit operation. The role of the control room and local plant operators also need to be explained. The process unit description must include those features which guard against human error or performance shortfall and those which facilitate routine or emergency operations (Chapter 8 provides more specific discussion of abnormal, including emergency, operation). Features relevant to the control of error during maintenance might also be worthy of mention.

(d)   Service systems and passive safety features

The integrity of systems supplying power and cooling water to the process units is vital, as is the operation of emergency systems such as fire detection equipment, water sprays, and gas detection systems. The treatment is analogous to that of the process units, again including those features which guard against human error or performance shortfall, and those which facilitate routine or emergency operations. Some passive features such as fire barriers could be compromised by human action, typical examples being the leaving open of fire doors or the 'temporary' removal of flood protection. Comments on the measures taken to maintain integrity may briefly be discussed.

The design of lighting systems have human factors implications. Lighting levels affect task performance and personal safety and, while safety cases usually address these issues for the key man-machine interfaces, they sometimes fail to address the requirements for maintenance or local operations elsewhere on the installation.

(e)   Control and instrumentation

The control and instrumentation (C&I) systems are service systems, but they are the main focus of interaction between the plant and its operators. Human factors issues therefore usually merit extended treatment in this section.

Once again, generic features and design approaches should be set out first. Human factors issues which would normally be treated generically include:

- Interlock philosophy;
- Approach to diversity within the interface;
- Data control philosophy (data amendment, QA);
- Coding conventions;
- Display data integrity;
- Testing and maintenance.

Specific features of the C&I system should then be described. Examples of features of interest to the human factors engineer are listed in Table 12.3. In describing their arrangements and functionality, sufficient detail should be given, in either the main text or supporting documents, to allow the regulator to assess their functional adequacy and ergonomics against best practice. The regulator will look at key areas, perhaps on a sampled basis, in considerable detail and will want to examine design specifications. He will also review the design process to ensure that appropriate methods have been used and that due weight has been given to the usability of the system under all its modes of operation. Coverage of the design process in the submissions is discussed below in 12.2.8 (and see Chapter 6 for a discussion of man-machine interface design).

Communications are also commonly dealt with under the C&I heading.

Table 12.3 *Examples of human factors topics within the scope of C&I*

Main control room	Functionality: • Role and Functions performed • Staffing arrangements • Allocation of functions between man and machine • Allocation of functions between operators Environment: • Layout and structural provisions • Environment (lighting, etc.) Telecommunications & support systems Interfaces: • Physical interface • Control systems functionality, equipment and interface • Data display systems functionality, operator support systems, equipment and interface Alarm systems functionality, equipment and interface Alternative control locations
Central protection system	Protection systems functionality, equipment and interface Maintenance and testing
Distributed data networks	Network functionality, equipment and interface
Site security systems	

They are important to safety in most applications, and would normally feature significantly in the safety case. A considerable amount of human factors effort will probably have been put into the design of the communications systems and their integration with the man-machine interfaces. The design principles adopted should be mentioned in this part of the safety case submissions, and the human factors work programmes described as part of the design process description (see Section 12.2.8). There will be human factors issues associated with the ergonomics and functionality of the equipment and its local environment, but the measures put in place to ensure that the correct message is sent and that it is correctly interpreted also need to be included.

(f)   Operating procedures

Operating procedures are part of the installation design. The combination of human operator and procedure is functionally equivalent to an automatic control system. They work together to control the process, so it is unhelpful to describe the operation of a process unit and its automatic control in one section and the role of the operator in another. However, because the procedures form a uniform set and the principles that apply across procedures for all process units are generic, they usually deserve a section of their own within the safety case submissions. Coverage should not be restricted to control room procedures. Maintenance procedures and other operational procedures are important, although they are rarely given the attention they deserve in the safety case submissions (see Chapter 7). The submissions must convince the regulator that in practice they are given due weight, and that they really will be used on site and kept up to date (see also Section 12.2.6 below). Examples of topics likely to be discussed include:

- Format (flow chart, text, etc.);
- Status (mandatory or advisory);
- Coverage and scope;
- Ergonomics (features for ease of use and error reduction).

Procedure design and the drafting process would normally be covered in the Design Process section (see 12.2.8 below).

12.2.6 The Control of Operations

The safety case submissions have not only to demonstrate that the installation is inherently safe and that it has been designed so that it can be operated safely, but also to convince the regulator that it will in fact be operated safely — and in accordance with the safety case. Recognising the difference in reliability between the operations of 'cowboys' and those of quality-minded professionals, the regulator needs to be convinced that high-hazard plant will not fall into the hands of the cowboys — no matter how good the design.

The safety case submissions must therefore include details of the arrangements made on site to manage operations and develop an appropriate quality culture. They must also consider corporate policy. Enough information on the wider corporate organisation and culture has to be supplied to give confidence that the site arrangements will be maintained and that formal reporting and supervision arrangements are in place to enable any deviation to be detected and corrected. The extent of the available technical off-site back-up will be an issue, including the availability of human factors expertise for accident investigation. To this end, there needs to be a significant operational content in both the safety case and the control procedures. One possible framework for reporting the arrangements for the control of operations is given in Table 12.4.

The regulator rightly places considerable importance on the safety management systems to be employed. The submissions should include a comprehensive description of the approach to be taken and the tools and methodologies be be used. Safety management systems that rely solely on reactive methodologies are unlikely to find favour, and evidence will be required to show that standards are continually monitored and compared with past performance and best practice in the industry as it develops. The

Table 12.4 *Examples of human factors issues related to the control of operations*

Control procedures	Documentation structure, quality control, compliance monitoring, etc.
Monitoring and maintaining quality and safety culture	Safety management systems Audit against best practice Performance indicators
Monitoring compliance (statutory and operational)	Discharges (quantity and environmental impact) Occupational health monitoring Record keeping
Selection and training	Selection processes for technical and managerial staff Team building and development of leadership capability Personal development Training facilities (simulators, etc.)
Staffing and management structure	Structures Manning levels (normal & minimum) Shift patterns
Operational feedback	On-site reporting and analysis of events Inter-site systems] Statutory reporting
Safety case maintenance	Responsibility for maintenance and mechanisms for flagging the need for updates
Corporate arrangements	Senior management responsibilities Arrangements for supervision of operations Technical and training support Central safety functions
Reliability monitoring	Arrangements for data collection and analysis Mechanisms for updating the safety case if necessary
Control of access	Security arrangements Screening and supervision of staff

submissions need to convey a commitment to a policy of continuous improvement in all aspects of operation.

12.2.7 Maintenance and Testing

Although the principal focus of the safety case is likely to be the operation of the installation, maintenance is vital to safety. It must be carried out to maintain plant availability, yet it carries with it the risk of introducing errors or latent failures into the system (see Chapter 2 for a consideration of latent failures). Many aspects of the safety case will be relevant to hardware and software maintenance, including layout, procedures, training, interfaces used for reinstatement, facilities for diagnostics and routine testing and calibration. It may be worth summarising them in the submissions under a single heading. This approach can be useful for any subject which the regulator is likely to consider as a separate topic, although it can cause problems with cross-referencing and updates. Care should be taken to avoid turning the summary into a detailed discussion as this would invalidate the structure of the submissions.

12.2.8 The Design, Construction and Commissioning Processes

(a) Introduction

The purpose of this part of the safety case is to substantiate the claim that the installation has been designed and specified in a manner which ensures safe operation, and that the installation will be constructed and commissioned according to that specification. As such, it provides support for the Design Description, and in practice much of the material might be provided as supporting documentation rather than as part of the main submission. There may also need to be a section covering decommissioning, including waste disposal and the return of the site to a pure or 'green-field' state.

The safety of an installation depends on the quality of the design work as well as on the quality of operation. Systems are therefore required for the control of the work of all design teams. Similar considerations apply to construction and commissioning. The safety case submissions must therefore address the human factors aspects of these processes.

(b) Human factors issues in the design process

Human error within a system's design, construction and commissioning has been the cause of many disasters, and the parts of the safety case which cover these life cycle stages are likely to receive significant scrutiny. However, the safety case submissions would usually provide only an outline of project management systems, referring instead to the development procedures and project quality assurance manuals (which may be provided as supporting documents). There is considerable overlap between commercial and safety considerations here, and most of the measures important to safety have also been shown by experience to be necessary to limit commercial risk. Relevant human factors issues to be covered include:

- An outline of the project risk analysis, including a discussion of the preventive measures implemented;
- The quality control and assurance actions taken during design, construction and commissioning;
- Communications issues throughout the project (including feedback during commissioning);
- Documentation and the use of standards;
- Selection and training of design, construction and commissioning

staff;
- Project management;
- Corporate quality and safety culture.

In some fields, particularly the development of computer software, more detailed treatment may be required. Complex software is a particular problem because of the difficulty of testing it. Considerable emphasis therefore has to be placed on the structured methods used in the software specification, design and coding to minimise the risk of human error.

(c)   Reporting the human factors input to the design

Most human factors work during the design process is an integral part of system design, and the overall description of the installation provided for the regulator will reflect this by covering many of the areas where human factors engineers have contributed.

While the regulator may be able to form a judgement as to whether the human factors aspects of the design match best practice and meet the functional requirements, most systems are too complex for there to be any chance of spotting anything but relatively obvious shortcomings. Therefore, the regulator is most likely to examine the human factors work programmes and the way in which the work was integrated with the systems design process. The work programmes require explicit treatment within the safety case. The regulator also looks for evidence that the general importance of human factors in the achievement of safety has been fully recognised, that the right work has been done, and that it has been carried out systematically to a high standard. The safety case submissions should therefore allow the regulator to satisfy himself that, for example:

- The human factors programme has been fully integrated into the design process;
- There are adequate resources for the human factors programme;
- Quality plans exist and were adhered to;
- The human factors personnel involved were suitably qualified and experienced;
- Appropriate user-centred human factors methodologies and techniques were used;
- Operations under abnormal and emergency conditions were fully considered;
- Operations with degraded control and implementation systems, environments or staffing were fully considered.

There must also have been suitable input to all external specifications by appropriate human factors experts, and QA arrangements should exist for human factors aspects of work outside the scope of dedicated human factors programmes.

The regulator expects to see high standards of design right across the installation, and concentration on safety aspects to the exclusion of other issues will not be acceptable. The workload on the operator from non-safety functions and the usability of the non-safety aspects of the man-machine interface have an impact on the performance of safety functions. It is not unusual to find installations where safety-related equipment has been audited against human factors guidelines and checked for consistency of coding, but where non-safety-related equipment has been relatively poorly designed (for guidance on consistency of coding see Chapter 6). Compliance with human factors standards is commonly included in specifications for safety-related equipment, but sometimes it is either omitted from the specifications for the balance of plant or non-compliance is accepted too readily. This has the potential to diminish the safety of the installation.

Human factors work programmes usually begin at the conceptual design stage and continue through to early operation when they will be concerned with final validation of the design and the inevitable fine tuning of such things as display format layouts and systems integration. The elements of these programmes are inevitably iterative and interlinked, and vary from installation to installation. Nevertheless, some framework has to be set out to enable the regulator to appreciate what has been done or is planned for the future, even if it oversimplifies the situation. A description of a typical human factors work programme should include:

- Specification of the human factors aspects of the design. The basis of the specifications can be important. Work carried out to establish any shortcomings in earlier installations should be mentioned if relevant.
- Justification of the guidelines and standards used. In many cases, existing documentation will be used, but project-specific guidelines may require additional human factors input. Common examples include the design of procedures, the layout of interfaces, and display conventions.
- Any fundamental research or development work carried out in support of the design work, including human factors reviews to establish the adequacy of proprietary equipment.
- The design process itself. A description of the involvement of human factors personnel in the detailed design of the features which will involve and influence human factors during operation, such as interfaces, procedures, and management systems.
- Verification of the design to confirm that it conforms to specifications and project guidelines and standards. Verification may involve audits by independent experts.
- Functional validation to confirm that the design is correct. Verification alone is inadequate because the specification itself may have been inadequate. For complex systems, this may only be completed as part of the commissioning process, so the possibility of late changes and 'fine tuning' of the design need to be recognised. Simulations can help.
- A review of systems integration, which ensures that the various

interfaces are consistent, that alarm systems are properly integrated without being compromised by 'add-ons', and that safety-related and other interfaces are consistent.

- An outline of anticipated commissioning phase activities. These would usually encompass formal and informal validation. The informal validation is vital, and commissioning engineers should accept, for example, that the telephone is in a noisy spot. The safety case submissions will be in advance of this phase, so the processes and procedures which will apply have to be spelt out, and the regulator will have to be advised of the results separately.

12.2.9 Risk Assessment

(a)   The objectives of analysis

Risk assessment, which incorporates analysis and evaluation, usually proceeds in parallel with the design work in an iterative fashion. The purpose of the analysis is to identify the range of potential hazards and the likelihood of these hazards being realised. Assessment is carried out to examine the potential impact on plant operation and the consequences for site staff and the general public. The conclusions are then evaluated to decide whether the risks (the product of consequence and likelihood) posed by the installation are acceptable without modification. If they are not, countermeasures must be built into the design and the exercise repeated. The safety case submissions describe the work programme, the methods used and the results obtained. They then justify the residual levels of risk.

The objective is to ensure that:

- The installation meets safety targets;
- The installation is ALARP, i.e., that further improvements on safety can only be obtained at disproportionate expense.

Risk reduction beyond this point is not justified on safety grounds, although it may well be in terms of corporate risk. Best practice should be employed in all areas, but risk reduction effort should be concentrated on the areas which really need it.

(b)   Coverage in the safety case

There are some general principles to follow in presenting the human factors aspects of the risk assessment and the human factors content of the work programme in the safety case submissions. They are similar to those discussed above for the design work. Again, the regulator cannot check all the detail and so needs to be given confidence in the processes. Thus the safety case submissions require discussion of resources, quality plans, and integration of the various aspects of the assessment.

Risk assessment used to be thought of as something required for the safety case but peripheral to the design process. The installation was designed and then calculations carried out to show it was acceptably safe. However, this proved to be an extremely expensive approach for several reasons, including:

- It did not identify over-design; it was too late to incorporate simplifications where reliability targets were exceeded. Simplicity brings safety in practice, and complexity reduces reliability, so the simplest design to meet reliability targets is often to be preferred.
- Work was unnecessarily repeated, e.g., scenario analysis for procedure design and error probability assessment.
- The insights gained from the risk analysis programme could not readily be transferred. Some error reduction opportunities identified during the risk analysis may not be important to safety, but they could have economic significance.

Risk assessment must therefore be properly integrated with the design process, where it does the most good. After all, it is — in principle at least — being carried out in order to get the design right, not just because the regulator requires it. It is a commercial as well as a safety matter. This approach forces the designer to define what he requires from the risk assessment. This in turn strengthens the organisation's position when it comes to negotiations with the regulator. Organisations who ask the regulator to set the agenda usually end up spending a lot of money.

The regulator looks for proper treatment of human factors within the assessment. Human error assessment should not be grafted on after the systems analysis has been completed. The safety case has to be comprehensive from the start (at a level of detail appropriate to the application). Being forced to insert additional considerations later tends to be expensive.

One of the project management problems commonly faced during PSA work (see Chapter 3) is the extent to which generalists can sensibly carry out analysis involving human factors issues. When should human factors specialists be called in? As software packages become more complex, less knowledge is required to use them, but users may not understand what the software is actually doing or recognise results which ought to be investigated in more detail. A little knowledge is a dangerous thing! In practice, non-specialists can probably do much of the work, but specialists are needed to screen the results and to ensure that all aspects are considered in an integrated way. It is important to convince the regulator that the analysis has been competently carried out and that the appropriate skills have been employed.

The vast majority of any human factors risk analysis is qualitative in nature, with explicit probabilistic estimates made on the basis of the qualitative analysis as required. The qualitative studies would usually be reported separately, with the numerical human factors work being integrated with the reporting of the overall probabilistic analysis. However, one point applies to both types of analysis. The long-term validity of a safety case has to be demonstrated by monitoring the operational performance of the installation against the assumptions made in it. This does not mean that vast amounts of human error data have to be collected — it would probably be impossible or unrealistically expensive in any case — merely that enough has to be done to give confidence that overall reliability assessments are valid. Exactly what this means in practice is application-specific and a matter for judgement, but the safety case will only be accepted if such arrangements are part of the management procedures.

(c)   Qualitative studies

Even if it is not explicitly probabilistic, all risk analysis involves judgements of relative likelihood of occurrence and gravity of consequence. Nevertheless, the bulk of the human factors work in support of a risk analysis is predominantly qualitative. The distinction between work carried out in support of the design and in support of the risk analysis is often blurred, but in general, that reported in the safety case submissions under the heading of risk analysis should be concerned with events of major safety significance. The scope, methods and resources deployed in task and scenario analyses should also be described, with the results being summarised in the safety case submissions and the detail being made available in supporting documents.

Safety is influenced by the potential for human error in many ways outside the narrow operational context. For instance, the regulator needs to be convinced that maintenance will keep the installation in working order and that maintenance or modification will not introduce latent defects which would invalidate the safety case. Software-based systems pose particular challenges in this respect because of the difficulties of re-proving them after maintenance. Similarly, errors may occur during the off-line calculation of set points or derived plant measurements, or during the update of control or protection data sets. Some explicit analysis of these wider risks may be required.

(d)   Probabilistic analysis

The information required within the safety case submissions for the probabilistic human reliability analysis is comparable with that required for the hardware side of the analysis. The sources of data, techniques used, and assumptions made have to be documented. The way in which qualitative and probabilistic studies relate must be explained. All branches of probabilistic risk analysis rely heavily on expert judgement, but this is particularly so in the case of HRA (see Chapter 3). For instance, task classification, dependency assessment, and application of screening criteria are all subjective. The quality of the analysis therefore depends to a significant extent on the ability of the person carrying it out, how well they understand the activities being analysed, and how closely they work with the designers and with those carrying out qualitative risk analyses. The safety case must address these issues if it is to convince the regulator that the assessment is valid.

12.2.10 Emergency Arrangements

(a)   Treatment within the safety case

It is generally accepted that the safety case submissions must contain an explicit description of the on- and off-site arrangements to be activated in the event of an emergency. The plans referred to may address issues beyond the system design, and should answer the third of Hawkesley's questions: 'But what if the worst did occur, despite the precautions taken?' (See Chapter 8 for coverage of issues related to various types of abnormal situation.)

The writer of the safety case and those drawing up the plans are presented with two questions to resolve: what constitutes an emergency? and what part of the emergency response should be covered by a separate plan? To some extent emergency response is treated separately within the submissions because it has traditionally been done that way: before the advent of modern risk assessment methods, operations tended to be categorised as either normal or emergency, without taking into account the wide variety of off-normal situations which the operators have to deal with in practice.

Getting the right balance in the submissions and making sure that the relevant human factors issues are dealt with requires an appreciation of the reasons why planning for the worst is so important. Emergency plans are required because there is always a chance that the risk is in fact greater than estimated, because of some unforeseen mechanism, or because things do not go as anticipated. The *Titanic* was unsinkable, so it was unnecessary to provide sufficient lifeboats for a full disembarkation! We should not place total reliance on our risk analysis. A major hazard may be tolerable, provided that the probability of it being realised is very low; but 'very low' is not zero, and low-frequency accidents do occur. The public naturally tends to tolerate industrial risk more readily when they can see that emergency plans are in place to contain the consequences.

Finally, the fact that plans, equipment and manpower exist for coping with the worst that could occur gives confidence that less serious events could be dealt with in an effective manner and prevented from escalating. Preparing for emergencies is preparing for the unexpected, and well-trained emergency teams are a necessity because so many of the incidents that occur in real life fall into that category. The effectiveness with which the operators implement the Emergency Plan, even under exercise conditions, is therefore a good indicator of the professionalism of the work force and management, and it is common for the regulator to want to witness regular emergency exercises. Emergency exercises can, however, put people under considerable stress, both as individuals and team members, and any lack of attention to the complex human factors involved in team building are likely to be revealed.

(b)   Off-site emergencies

'Emergency' may be defined as 'a serious situation needing prompt action', but this definition is not sufficient in the safety case context because it would encompass many off-normal conditions which require urgent attention but which do not warrant initiation of emergency plans (see Chapter 8). In practice, off-site emergencies are abnormal events which put the public at risk. They therefore tend to require more attention at process plants and nuclear power stations than off-shore installations. The public and the authorities would normally have no part in the operation of the installation, so special procedures are required, typically allowing for a graduated response depending on the severity of the incident.

If the worst has suddenly happened, there will not be any doubt as to whether emergency plans should be activated, but in a slowly developing situation or where there is still hope of recovering the situation, the decision may not be clear cut, and the installation manager will be subject to conflicting pressures. The criteria for declaring a full or partial emergency therefore have to be absolutely clear.

Evacuating the local population and establishing communications with the authorities and the media are examples of off-site emergency actions requiring attention to human factors. The plan needs to be created and justified, but the submissions must also convince the regulator that the plan is practical and will work under difficult or adverse conditions. An appreciation of the realistic response of the public to an emergency ought to be included. For example, people do not wait around conveniently to be evacuated. Communications with the public and the measures taken to tell local residents what would be expected of them need to be covered.

Where outside authorities or services are involved, reference needs to be made to their procedures and to any specialist training or exercises which they might undertake. The same applies to technical or support services provided from other company locations or by external contractors. The safety case has to be kept up to date, and this is not always straightforward where references to external documentation or practices are involved. This suggests two things: keep such references to the minimum required to make the case, and make sure that there is a mechanism which flags the need for reassessment and revision.

(c)   On-site arrangements

This section of the safety case considers the human factors issues related to the actions of on-site staff during emergencies. If the emergency poses an off-site threat, the off-site arrangements discussed above will also need to be activated. On-site emergency procedures typically address issues such as fire fighting, damage repair, team operations, and control and command arrangements. Accident management plans or facilities which may be used in practice but which are not formally claimed as part of the safety case may also be referred to.

In an emergency, people may be operating near the limit of their capability. They may have to make the transition from well-informed routine to a situation full of danger and uncertainty, and they may have to do it almost instantly. Major accidents have more in common with military action than they have with routine plant operation, and it may be completely beyond their previous experience. Improvisation, knowledge, initiative and courage will be required under hostile physical conditions. Achieving effective emergency responses under these circumstances requires considerable human factors input, and the regulator will be looking for evidence in the submissions that the appropriate analysis has been carried out. Selection, training, and routine personnel appraisal procedures must take account of emergency team duties.

The essence of a good Emergency Plan is its appropriateness to a wide range of unforeseen situations, so the plans themselves must be simple, and will usually be formulated to achieve generally applicable objectives rather than being written against specific predefined scenarios.

12.3   WRITING SAFETY CASES AND OBTAINING REGULATORY APPROVAL

12.3.1   Introduction

A detailed analysis of the nature and extent of commercial risk is a prerequisite for any large project, and in the current climate this ought to include an assessment of 'regulatory risk'. The regulatory risk assessment must anticipate the ways in which delays in regulatory approval or in any necessary public consultation can seriously affect project economics. The preparation and submission of the safety case must be organised so that they do not end up on the critical path. The regulatory risk assessment must therefore identify potential areas of difficulty so that they can be resolved as early as possible in the process.

Early attention needs to be paid to the human factors aspects of preparing and submitting safety cases. Many of the problems which crop up during the project have their origins in human factors issues, and expertise has to be applied if they are to be avoided or mitigated. If human factors expertise is available, use it! Particular attention needs to be paid to the following:

*   Safety case submissions project management;
*   The structure and content of the submissions;
*   Dealing with the regulator.

12.3.2   Project Management

Preparing a major safety case and drafting the submissions to the regulator can be a complex and expensive exercise. Editorial costs alone may easily exceed a million pounds sterling, but even these costs may be modest when compared to the potential costs of delays. Preparing the submissions is a major project in itself and should be properly planned, managed, provided for, and professionally executed.

Preparation of submissions is often distributed and can involve a variety of technical and documentation specialists. This makes it difficult to manage, and attention needs to be paid to the human factors problems which 'matrix management' can introduce. As time runs out, it becomes harder to get technical specialists to spend their time on the safety case. Project teams often shed staff as soon as their contribution to the design is complete, so be aware that key technical staff have a tendency to disappear just when the regulator wants 'their' section redrafted. The development of the safety case needs to be integrated with the design process, and the finished product needs to be 'owned' by all those who have a stake in it (technical specialists, editorial staff, operations staff, safety groups — and the regulator) but editorial responsibility would usually be unambiguously vested in a dedicated editorial team. There should be a single focal point for communication with the regulator.

The programme has to recognise the fact that a number of parties will be involved and that obtaining internal and regulatory approval for submissions can be time consuming. Do not underestimate the effort involved in responding to comments and questions — everybody seems to have an opinion on human factors issues! The importance of building a good relationship with the regulator is discussed below, but attention has also to be paid to internal approval procedures. Contractors will almost certainly be involved, e.g., as designers, engineers, risk analysts, or technical authors, and they also have to understand and accept the planned programme and the common approach.

The quality of the submissions must be as high as can practically be achieved, so formal inspection of them is essential [Fagan 76, Redmill 88]. Errors in the submissions hint at the possibility of poor quality assurance in other parts of the system or project.

12.3.3 The Submissions

Safety cases are complex and of broad scope. It is therefore a considerable challenge to make them clear, concise and 'user-friendly'. As well as setting out a logical argument, they must fulfil a reference function. Preparing documentation of this quality against a background of an evolving design and tight timescales is challenging but not impossible, given the power of modern word processing tools.

It is commonly recognised that documentation professionals are required for the production of the user manual for even a consumer item such as a photocopier. Yet it is not unknown for safety case submissions to be planned and written by engineers seconded onto the job without any experience or training in documentation. Producing a 20-volume, million-pound-sterling document is not the same as writing a technical report, and professional expertise is essential to the editorial team.

(a)   Structure

The structure of the submissions varies with the nature of the safety case, but there are some underlying human factors principles. As the submissions have to fulfil so many different functions, they are usually laid out in a hierarchical manner with overviews of the structure and 'signposting' in the text. The main document should focus on the description of the plant and its management systems. The number of supporting submissions required will depend on the complexity of the plant and the nature of the hazards. An overview of the risk analysis and design processes would normally be included in the top level document, with the detail reserved for the supporting submissions. Documents at lower levels in the hierarchy normally count as part of the safety case and should follow the same development and approval route. As general principles, keep the volume of material within the formal submissions to the minimum required, and keep the structure as simple as possible.

Material which is outside the scope of the formal submissions but which substantiates the claims made, for example by providing further detail, should be identified on a separate document reference list. This provides flexibility and reduces the amount of material which has to be specially written, cross-referenced and submitted for formal comment, while giving the regulator confidence that the claims made will be reflected in the operational system. A library ensures that the references remain available.

Submission documents should include a detailed index. An index can nowadays be generated automatically, and its absence increases the difficulty of achieving complete and consistent amendments, when they have to be made. When deciding on the structure and the approach to cross-referencing, account should be taken of the need for frequent amendment during the development phase and occasional updates after that.

(b)   Editorial issues

The safety case is judged primarily through the submissions, and although it is wrong to judge a book by its cover, the editorial quality of the submissions almost certainly has a large impact. Invest in good quality software for word processing, text editing and reproduction. Having the submissions laid out professionally makes a huge difference to the impression made on the 'customer'. Effective editing improves clarity and reduces documentation errors. Ambiguity in the text invites alternative interpretations and should be avoided. Errors frustrate the reader and will inevitably make the regulator wonder about the care that has been taken in the analysis being reported on.

Diagrams should be simple and unnecessary drawings should not be included: they are likely to need to be updated in the future, and this is costly.

Make drawing reference lists. Ensure that the style is simple and that the text is written in clear concise language. Make sure guidelines have been prepared for authors and that they all understand what has to be achieved and who is going to do what.

Quality assurance needs to be effective but 'fit for purpose', i.e., that which is necessary but no more. It has to be demonstrably reliable, so that by following an audit trail the regulator can derive confidence that the submissions genuinely reflect reality.

12.3.4 Dealing with the Regulator

Even if there was no regulator, there are may good reasons for preparing a safety case and its associated documentation. Realistically, however, the main driving force for most companies is the need to win regulatory approval for their installation. Too many fail to take this to heart and lose sight of the fact that their product must be 'sold' to their customer. The safety case is the product and the regulator is very much the customer. It may sound inappropriate to treat the safety case as something to be marketed, but this leads to good discipline.

(a)   Selling the safety case

Treating regulators as customers puts proper emphasis on establishing what their requirements really are, and on forming good relationships with them. In the UK, the regulator is unlikely to specify in any detail what is required, or to commit themselves to a position in advance of receiving the submissions. Nevertheless, they do not want to waste anybody's time and money unnecessarily and so will usually do their best within established protocols to ensure that submissions are not hopelessly inadequate. Early provision of information, and communication and cooperation throughout the process, help the regulator to do the job, improve feedback from the regulator, and have the added advantage of fostering a sense of involvement in the creative process that makes the eventual 'sale' more straightforward.

For this reason, presentations and reviews with the regulator of the developing safety case and the treatment of key issues within it should be a feature of the programme. Those assessing specialist submissions — e.g., human error assessment — will not necessarily be completely familiar with the hazards, and engineering assessors may not be familiar with the subtleties of human factors risk analysis methods. Background technical presentations can therefore be a considerable help in cutting down on the number of comments — comments which would otherwise take time to deal with. Questions can be resolved and misconceptions cleared up much more easily. Personal contact helps build confidence as well.

No customer likes to be bullied or hurried by a salesman, so the programme for the preparation and submission of draft and final issue safety case material should be agreed in advance. Goodwill is important. It might not get a poor product accepted, but it can smooth the process and achieve faster delivery of a good one. Goodwill is more likely to be forthcoming if the regulator's staff feel that their contribution is valued and appreciated. An atmosphere of trust allows both sides to reveal the underlying reasons for their views or the basis of their concerns, and this is a valuable prize.

Don't forget that the regulator and staff are unlikely to be sitting idly waiting for submissions. They need to plan resource utilisation, particularly in specialist areas such as human factors, and if submission target dates are not met they may not be able to meet target dates for comments, no matter how much goodwill has been cultivated. At the same time, the regulator does have a responsibility to the agreed programme and, barring exceptional circumstances, should adhere to it.

(b) Changes to the design

Pressure to alter the design or provide additional equipment does sometimes result from the regulator's examination of the safety case. It is inevitable given the degree of judgement present in any risk assessment. Regulators may be experienced in the industry, and they naturally want to contribute. Some may be mildly frustrated by their inability to do any of the design or analysis themselves. This also needs to be acknowledged and defused in a constructive manner. A view independent of commercial pressures can be valuable, and suggestions should not be rejected out of hand, but do not lose sight of commercial reality. The regulator's comments must be reviewed by technical people as well as by those handling the administration and editorial side of the safety case process; there should be a clear distinction between suggestions for alternative ways of doing things and true shortcomings.

The regulator almost inevitably has different priorities from those of the owner of the installation. Both want a satisfactory outcome to the safety case submission process, but the regulator is likely to have a broader perspective. The regulator's performance is to a large extent judged by the public and their representatives on the basis of the number of accidents, and the regulator will have a longer-term aim of improving safety standards. The cost-effectiveness or innovation of the design is of consequence, but the regulator receives little credit for it. Thus, where there is doubt, there is almost always a tendency to respond conservatively to technological change and to err on the side of safety. 'Ratchetting' is the process by which requirements are pushed up across an industry by tightening standards on one site, perhaps where there is no real cost penalty, and then working to bring the others into line. It can result in considerable expenditure, so a central overview of the safety case process within a company may be necessary to ensure that the implications of recommendations for other sites can be identified and a considered response made.

Human factors expertise is essential to the preparation of a safe design and to the subsequent management of safety on site, and equally so to the achievement of regulatory approval.

DISCLAIMER

The views expressed in this chapter are those of the author alone and do not necessarily represent those of any current or past employer.

REFERENCES

[Cullen 90] Cullen The Hon. Lord: *The Public Enquiry into the Piper Alpha Disaster.* HMSO, London, 1990

[Fagan 76]   Fagan M E: *Design and Code Inspections to Reduce Errors in Program Development.* IBM Systems Journal, 15(3), 1976

[HSE 84]   Health and Safety Executive: *Control of Major Accident Hazards (CIM H) Regulations.* HMSO, London, 1984

[HSE 92a]   Health and Safety Executive: *The Tolerability of Risk from Nuclear Power.* HMSO, London, 1992

[HSE 92b]   Health and Safety Executive:*Safety Management and Safety Case (Off shore Installations) Regulations.* HMSO, London, 1992

[Hawkesley 89]   Hawkesley J L: *A View from ICI.* Part (d) of Chapter 7 of Lees F P and Ang M L (eds): *Safety Cases within the Control of Industrial Major Accident Hazards (CIMAH) Regulations 1984.* Butterworth, 1989

[NRC 75]   NRC: *Reactor Safety Study — An Assessment of Accident Risks in US Commercial Nuclear Power Plants.* NUREG-75-014. NRC, Washington DC, 1975

[Redmill 88]   Redmill F J, Johnson E A and Runge B: *Document Quality — Inspection.* British Telecommunications Engineering, Vol. 6, January 1988

Authors' biographies

Jonathan Berman
Jonathan Berman Associates, 23 Picton Way, Caversham, Reading, Berks RG4 8NJ, UK.
Jonathan Berman has over 17 years experience in applying psychological and human factors knowledge to solve problems in industries ranging from aerospace and railways to nuclear power and off-shore oil exploration. In 1994 he formed Jonathan Berman Associates, to offer tailored human factors and safety management consulting services to maximise the benefits of the human contribution to system performance while minimising the cost of doing so. Prior to that he had a successful career in the aerospace industry, working first for British Aerospace and subsequently at the RAF Institute of Aviation Medicine, before joining the UK electricity supply industry in 1986. He left Nuclear Electric in 1991 to move into risk management consultancy. He continues to undertake human factors research for the UK nuclear industry, in addition to work for many other industries. He is a Chartered Psychologist, a Fellow of the Ergonomics Society, and an Associate Fellow of the British Psychological Society.

Carlo Cacciabue
Commissione Delle Comunita Europee, Centro Comune di Ricerca, 1-21020 Ispra (Va), Italy.
Pietro Carlo Cacciabue was born in Nizza Monferrato (Asti), Italy, on 4th November 1949. He graduated in Nuclear Engineering at the Politecnico di Torino and received the PhD in Nuclear Engineering from the Politecnico di Milano. Since 1975 he has been a researcher of the Commission of the European Union. In 1976 he entered the Joint Research Centre (JRC) at Ispra and was engaged in theoretical studies and in reliability and probabilistic safety assessment analysis of water reactors. Since 1982 his major interest has been the study of man-machine systems, with particular emphasis on the development of models of human behaviour to be included in safety evaluations as well as in decision support tools. Since 1991 he has been responsible, within the Institute of Systems Engineering and Informatics at JRC, for the research activity on working environments. This has focused on

the domain of aviation safety, with particular reference to air traffic control and pilots' behaviour analysis. He is the author of several publications in journals and in a book, on the subjects of safety engineering, human-machine interaction, and nuclear and aviation safety.

Jeremy Clare
Cambridge Consultants Ltd, Science Park, Milton Road, Cambridge CB4 4DW, UK.
Jeremy Clare is a principal consultant at Cambridge Consultants, the European Laboratory of Arthur D. Little. He has spent 25 years doing research and consultancy in Human Factors. His principal concern has been with the role of humans as part of total systems, i.e., complex configurations of people and machines working together. Inevitably, this has led to the consideration of the safety of such systems and in this area he has worked closely with Arthur D. Little colleagues who specialise in safety assessment and management. He holds a BSc in Psychology and Physics. Earlier appointments included a 2 year research fellowship at the Physiological Laboratory, Cambridge University, carrying out an investigation into the mechanism of visual recognition.

David Collier
Greenstreet Consultancy, 2 Bramley Cottages, Claverton Down Road, Bath BA2 7AP, UK.
David Collier is a chemical engineer and a member of the Ergonomics Society. He is 38 and lives in Bath, England. He spent a varied 15 years in the UK nuclear industry, working in reactor physics, plant commissioning, ergonomics, health and safety, and strategic planning, and was one of the editorial team for the Sizewell B NPP safety case. He moved on in 1994 to found Green Street Consultancy Limited. His current portfolio of work includes human factors, safety and environmental management, and public consultation programmes.

Andrew Hale
Delft University of Technology, Faculty of Philosophy and Technical Social Sciences, Kanaalweg 2B, 2628 EB Delft, The Netherlands.
Andrew Hale is professor of Safety Science at the Delft University of Technology in the Netherlands. His background is as an occupational psychologist working on human error, accidents and the perception of danger. He is director of the post-graduate course Management of Safety, Health and Environment (MoSHE). More recently his work has concentrated on safety management and organisation and regulation and certification in safety. He is a member of numerous professional and scientific associations in safety and occupational hygiene and is a member of or has chaired various committees of standards bodies and research and advisory bodies on machinery safety, safety management, professional certification and accreditation. He is co-author of the book 'Individual Behaviour in the

Control of Danger' and co-editor of the book 'Near Miss Reporting as a Safety Tool'.

Florus Koornneef
Delft University of Technology, Faculty of Philosophy and Technical Social Sciences, Kanaalweg 2B, 2628 EB Delft, The Netherlands.
Florus Koornneef is a staff member of the Safety Science Group at the Delft University of Technology in the Netherlands. His backgrounds are electrical engineering and educational technology. He co-designed the professional post-graduate course 'Management of Safety, Health and Environment' (MoSHE). More recently, his main work has concentrated on real-time risk management in public health care, e.g. in anaesthesia, and knowledge-based analysis of small-scale incidents in industry and in general hospitals. He is a member of the System Safety Society and of the European Workshop on Industrial Computer Systems, Technical Committee 7 on safety, reliability and security (EWICS TC7). He is editor of the book 'Veiligheid in het Ziekenhuis' (Safety in the Hospital) and is author of various other publications in the above mentioned areas, including chapters in books and a series of educational video productions for professional in health care.

Deborah Lucas
Health and Safety Executive, Magdalen House, Stanley Precinct, Bootle, Merseyside L20 3QZ, UK.
Deborah Lucas is a psychologist with the Health and Safety Executive (UK). For ten years prior to joining the HSE she was a consultant in human factors. She has consulted extensively across a range of industries, including railways, petrochemicals, nuclear, and service sectors. She has also been a university lecturer in psychology. Her areas of experience include modelling of human error in industrial safety, incident and near miss reporting, human reliability assessment, ergonomic design of the work place, and issues relating to fitness for work. Dr Lucas holds degrees from the Universities of Bristol and Manchester and is a Chartered Psychologist.

Steve Mason
Health, Safety and Engineering Consultants Limited, Bretby Business Park, Stanhope Bretby, Burton on Trent DE15 0YZ, UK.
Steve Mason is a Principal Consultant with Health, Safety and Engineering Consultants Limited. Previously he worked for the ergonomics unit of British Coal. He has a degree in Engineering from the University of Leicester and a Masters degree in Work Design and Ergonomics from the University of Birmingham. He is a Fellow of the Ergonomics Society and a Member of the Institution of Occupational Safety and Health. He is also a member of the Human Factors in Reliability Group (HFRG). He has published and researched extensively in the areas of rule violations, maintainability and design aids of designers.

Felix Redmill
Redmill Consultancy, 22 Onslow Gardens, London N10 3JU, UK.
Felix Redmill is an independent consultant in project management, software engineering and quality improvement. Prior to entering consulting, he spent over 20 years as engineer and manager in the telecommunications and IT industries. He has been the Co-ordinator of the UK's Safety-Critical Systems Club since its inauguration in 1991 and is the author and editor of a number of books.

Andrew Shepherd
Dept of Human Sciences, University of Technology, Loughborough, Leis. LE11 3TU, UK.
Andrew Shepherd has been a researcher and practitioner in applied ergonomics and industrial psychology for over 25 years, working in a wide range of industries. He has a particular interest in task analysis as a basis for pursuing his other interests in interface design, operator support and training. He is currently a senior lecturer in Human Sciences at Loughborough University.

Graham Storrs
Logica Informatik AG, Reppischtalstrasse, 8914 Aengstertal, CH-8050 Zurich, Switzerland.
Graham Storrs has 15 years experience in the IT industry. He had done research in HCI and artificial intelligence as well as working as a freelance programmer, an auditor and a human factors consultant before joining Logica Cambridge Ltd., the research and development centre for Logica world-wide. He works within the User Interfaces Division and, over the years, has participated with the company in a wide range of consultancy and development projects. In addition, he has worked on a number of research projects in the areas of intelligent training and help, argumentation support in safety-critical software development, and methods for analysing group work-support systems.

Ron Westrum
AEROCONCEPT, 19017 Saxon Drive, Beverly Hills, Michigan 48025, USA.
Ron Westrum is Professor of Sociology and Interdisciplinary Technology at Eastern Michigan University. He is also the President of Aeroconcept, a management consultancy. A sociologist of science and technology, he has published numerous papers as well as two books: _Complex Organisations: Growth, Struggle and Change_ (with Khalil Samaha) and _Technologies and Society: The Shaping of People and Things_. He is currently working on a history of the Sidewinder missile. He is A.B. Harvard (Social Relations) 1966 cum laude and his degrees include a Ph.D in Sociology from the University of Chicago in 1972.

Index